SOUTH
REA

Inner Mongolia

Hohhot

Baotou

Ordos Desert

Huang He

Ningxia
Hui

Shanxi

Hebei

Beijing

Tianjin

Qingdao

Shandong

Yellow Sea

Huang He

Jiangsu

Xi'an

Henan

Shanghai

Shaanxi

Anhui

Ningbo

Hangzhou

Chengkou

Hubei

Jiujiang

Zhejiang

Yangtze

Yichang

Sichuan

Chengdu

Chongqing

Hunan

Jiangxi

Fujian

xing

Shan

Guizhou

TAIWAN

Guiyang

Tropic of Cancer

Guangdong

Guangzhou

Kunming

Guangxi Zhuang

Hong Kong

Macau

Yunnan

VIETNAM

Leizhou
Peninsular

South China Sea

N

LAOS

Hainan

kong

100°

110°

FATHERS OF BOTANY

The discovery of Chinese plants by European missionaries

JANE KILPATRICK

Kew Publishing
Royal Botanic Gardens, Kew

The University of Chicago Press
Chicago and London

First published in 2014 by
Royal Botanic Gardens, Kew
Richmond, Surrey, TW9 3AB, UK
www.kew.org

and

The University of Chicago Press, Chicago 60637, USA

23 22 21 20 19 18 17 16 15 14   1 2 3 4 5

Kew Publishing ISBN 978 1 84246 514 1   eISBN 978 1 84246 590 5
The University of Chicago Press ISBN-13: 978 0 226 20670 7 (cloth)

British Library Cataloguing in Publication Data
A catalogue record for this book is available from the British Library.

Library of Congress Cataloging-in-Publication Data

Kilpatrick, Jane, author.
    Fathers of Botany : the discovery of Chinese plants by European missionaries / Jane Kilpatrick.
       pages cm
    Includes bibliographical references.
    ISBN 978-0-226-20670-7 (cloth : alkaline paper) 1. Botanists—France—Biography. 2. Botany—China—
       History—19th century. 3. Plants—China. I. Title.
    QK26.K55 2014
    580.92—dc23
                                                                                        2014013979

Cover design: Christine Beard
Map design: John Stone
Design, typesetting and page layout: Nick Otway
Project editor: Michelle Payne
Copy editor: Sharon Whitehead
Production Manager: Georgina Smith

Printed in Italy by Printer Trento

For information or to purchase all Kew titles please visit
shop.kew.org/kewbooksonline or email publishing@kew.org

Kew's mission is to inspire and deliver science-based plant conservation worldwide, enhancing the quality of life.

Kew receives half of its running costs from Government through the Department for Environment, Food and Rural Affairs (Defra). All other funding needed to support Kew's vital work comes from members, foundations, donors and commercial activities including book sales.

DEDICATION

For Drew, Ginny and Sally

Contents

Acknowledgements

I AM VERY GRATEFUL TO ROWENA BARTLETT who came with me to Yunnan and to Edward He of Edward Adventures in Dali who looked after us so well on our journey to Dapingzi and Cigu. I must also thank Father Richard Leonard SJ who discussed with me the theological beliefs that lay behind the nineteenth century missionary impetus and for his help with various passages. Roy Lancaster has provided unfailing encouragement and support, and has let me use many of his own photographs.

I am most grateful to David Boufford who painstakingly read through the manuscript and made many helpful suggestions. I am also grateful to Seamus O'Brien for his help with Chapter 11. Their comments have saved me from many errors. Paolo Cuccuini of the Botany Department, Museo di Storia Naturale in Florence kindly supplied information about the Italian missionary-botanists. David and Stella Rankin of Kevock Garden Plants patiently answered my questions and generously provided several photographs. Peter Cox of Glendoick Gardens identified several rhododendrons for me and showed me *Rhododendron davidii* in flower; and Ken Cox has kindly supplied photographs. Chris Reynolds and Daniel Luscombe of Bedgebury Arboretum showed me several conifer species discovered by the missionary-botanists and Chris has let me use several of his photographs. Martyn Rix was very helpful and provided photographs of Baoxing. I am grateful to Tony Marden of Shady Plants who talked to me about his collection of *Arisaema*; to Hugh McAllister who answered my *Sorbus* questions; and to Marc Colombel who provided information about Adrien Franchet. Thank you to Mikinori Ogisu for his help and to his colleagues Dr Liu Bo and Ye Jianfei in Beijing. Cédric Basset, Raymond Evison, Jeanette Fryer, Gail Harland, Harry Jans, Nick Macer, Seamus O'Brien, Keith Rushforth, Julian Sutton and Toshio Yoshida have kindly provided photographs.

I must also thank the librarians without whose help the research for this book would have been so much harder: particularly Elizabeth Gilbert and Elizabeth Koper in the RHS Lindley Library in London; Andrea Hart and Armando Mendez in the Botany Library, and Lisa di Tommaso in the General Library, at the Natural History Museum, London; and Julia Buckley in the Library, Art and Archives at Kew.

I am very grateful to Jennifer Harmer, and to Hilary Lenton, for their help with photographs. Thank you also to Simonne Frissen; and a special thank you to Jane Crawley who gave me the title.

Botanists and Explorers

MISSIONARY-BOTANISTS

Bodinier, Père Émile (1842-1901) collected around Beijing and in Hong Kong before collecting in the Guiyang area after his return to Guizhou in 1897

Cavalerie, Père Pierre (1869-1927) collected in southern Guizhou

David, Père Armand (1826-1900) collected around Beijing and made three extended journeys between 1866 and 1874 to Inner Mongolia, to Baoxing on the borders of western Sichuan, and to the Qinling and eastern Jiangxi

Delavay, Père Jean Marie (1834-1895) collected extensively in north-west Yunnan, particularly in the Cang Shan and Heishanmen ranges, and around Lijiang

Ducloux, Père François (1864-1945) collected around Kunming, Yunnan after 1897

Esquirol, Père Joseph (1870-1934) collected in Guizhou

Faber, Pastor Ernst (1839-1899) German Protestant missionary who collected in Guangdong and was the first to collect plants on Emei Shan in 1887

Farges, Père Paul Guillaume (1844-1912) collected in the Daba Shan in north-east Sichuan

Incarville, Père Nicholas le Cheron d', SJ (1706-1757) Jesuit missionary at Beijing 1741-57 and the first missionary to collect plants in China

Genestier, Père Annet (1858-1937) Père Soulié's travelling companion to the Tibetan borderlands

Giraldi, Padre Giuseppe (1848-1901) Italian Franciscan who collected in northern Shaanxi

Maire, Père Édouard Ernest (1848-1932) collected plants in north-east Yunnan after Père Delavay's visit to Longki in 1894

Martin, Père Léon (1866-1919) botanized with Père Bodinier in Guizhou after 1897

Monbeig, Père Théodore (1875-1914) collected at Cigu and Cizhong in the Tibetan borderlands – both now in Yunnan

Perny, Père Paul (1818-1907) the first missionary to collect in Guizhou

Scallan, Father Hugh (1851-1927) Irish Franciscan who collected in northern Shaanxi

Silvestri, Padre Cipriano (1872-1955) Italian Franciscan who collected in north-west Hubei

Soulié, Père Jean André (1858-1905) made extensive collections around Kangding, Tongolo, Yaregong (now in Sichuan) and Cigu (now in Yunnan) then in the Tibetan borderlands

BOTANISTS AND PLANTSMEN

FRANCE

Bureau, Professeur Édouard (1830-1914), botanist and Director of the Herbarium at the Muséum d'Histoire Naturelle, Paris

Decaisne, Joseph (1807-1882) botanist and Director of the Jardin des Plantes

Franchet, Adrien René (1834-1900) botanist at the Muséum d'Histoire Naturelle who classified the missionaries' plant collections

Henry, Louis plant-breeder at Jardin des Plantes, Paris

Lemoine, Victor (1823-1911) nurseryman at Nancy and gifted plant-breeder

Milne-Edwards, Professeur Henri (1800-1885) zoologist and Director of the Muséum d'Histoire Naturelle, Paris

Vilmorin, Maurice Lévèque de (1849-1918) nurseryman with an Arboretum at Les Barres in the Loire valley who received seeds from the missionaries and introduced many of their discoveries

RUSSIA

Maximowicz, Carl (1827-1891) botanist and Keeper of the Herbarium at the Imperial Botanic Gardens in St Petersburg, who classified the collections Potanin made in China in 1884-1886

BRITAIN

Bulley, Arthur K. (1861-1942) wealthy alpine enthusiast who founded Bees Nursery in Cheshire in 1903 and sponsored plant-hunters including George Forrest

Hemsley, William Botting (1843-1924), botanist and Keeper of Herbarium at Kew and author of *Index Florae Sinensis*

Hooker, Sir Joseph Dalton (1817-1911) botanist and Director of Royal Botanic Gardens at Kew who collected rhododendrons in Sikkim in 1848-1851

Thistleton-Dyer, Sir William (1843-1928) succeeded Hooker as Director of Royal Botanic Gardens at Kew 1885-1905

Veitch, Sir Harry (1840-1924) nurseryman, and head of the firm of James Veitch and Son of Chelsea and Coombe Wood near London, who sponsored E.H. Wilson's first two expeditions

ITALY

Biondi, Antonio (1848-1929) wealthy botanist who worked with the Botanical Institute in Florence and inspired Padre Giraldi and Padre Silvestri to collect plants

Sprenger, Carl (1846-1917) German plantsman with a large nursery in Naples who received seed from the Italian missionaries

AMERICA

Fairchild, David (1869-1954) botanist and explorer responsible for foreign plant introductions for the US Department of Agriculture

Gray, Asa (1810-1888) botanist and Professor of Natural History at Harvard University who formulated modern theories of plant distribution

Rehder, Alfred (1863-1949) German-born botanist who worked on Wilson's specimens at the Arnold Arboretum

Sargent, Charles Sprague (1841-1917) Director of the Arnold Arboretum in Boston who employed E.H. Wilson from 1906 onwards

EXPLORERS AND PLANT COLLECTORS IN CHINA

Berezhovsky, Mikhail M. (1848-1912) Russian naturalist who accompanied G.N. Potanin and also collected independently

Bretschneider, Dr Emil (1833-1901) physician and botanist at the Russian Legation in Beijing

Bunge, Alexander von (1803-1890) German-Estonian botanist who investigated the flora of the Beijing area

Cooper, Thomas (1839-1878) British merchant and consular official who travelled in the Tibetan borders in 1868

Davies, Major H.R. British cartographer who travelled through Sichuan and Yunnan in 1894-1900 and made the first detailed map of the region

Farrer, Reginald (1880-1920) British plant-hunter in Gansu and northern Burma

Forbes, Francis member of the American firm of Russell & Co. who collected plants in China

Forrest, George (1873-1932) Scottish plant-hunter in Yunnan and the Tibetan borders who collected initially for Bees Nursery and introduced many of the plants discovered by Père Delavay

Fortune, Robert (1812-1880) first professional British plant-hunter in China

Gill, Captain William (1843-1882) British traveller who visited the Tibetan borders in 1877

Hance, Henry (1827-1886) British customs official in Guangzhou and expert botanist

Handel-Mazzetti, Freiherr Heinrich (1882-1940) Austrian botanist who collected plants in western China 1914-1919

Henry, Augustine (1857-1930) Irish plant-collector who worked for the Customs Service in China and made an extensive collection of the flora of Hubei for Kew

Kingdon-Ward, Frank (1884-1958) collected in the Tibetan borders for Bees Nursery in 1911-1913

Kirilov, Porphyri V. (1801-1864) physician and botanist at the Russian Legation in Beijing

Maries, Charles (1851-1902) collected in Hubei for Veitch's nursery in 1877-1879

Mesny, William (1842-1919) originally from Jersey and a Major-General in the Chinese army who travelled with William Gill and collected plants for Henry Hance

Meyer, Frank (1875-1918) Dutch plantsman who made four expeditions between 1905-1918 for the US Department of Agriculture and the Arnold Arboretum

Orléans, Prince Henri d' (1867-1901) French explorer and plant-collector who met Père Soulié at Kangding 1890 and Cigu in 1895

Pratt, Antwerp E. (1852-1924) British naturalist who met Augustine Henry in 1887 and Père Soulié in 1889 and 1890

Przewalski, Nicolai (1839-1888) Russian explorer who made three expeditions to northern China 1870-1885

Potanin, Grigory (1835-1920) Russian explorer and plant-collector who travelled through Gansu and northern Sichuan in 1884-1886; and through Sichuan to Kangding, where he met Père Soulié in 1893

Purdom, William (1880-1821) collected for Veitch and the Arnold Arboretum, and then travelled with Reginald Farrar

Rock, Joseph (1884-1962) Austrian-born American botanist and ethnographer who lived near Lijiang from 1922-1949

Rockhill, William Woodville (1854-1914) American traveller who met Père Soulié in 1889 and 1892

Wilson, Ernest Henry (1876-1930) British plant-hunter who collected plants for Veitch and then the Arnold Arboretum, principally in Hubei, Sichuan and the Tibetan borders, and introduced many plants discovered by Père David and Père Farges

Gazeteer

Note: 'Q' is pronounced – approximately – 'ch' hence Qinling = *Ch*inling, etc.

CITIES, TOWNS AND VILLAGES

Pinyin	Old Style	Pinyin	Old Style
Badong	Patung	Kangding	Tatsien-lu/Tachien-lu
Baihanluo (Beixia luo)	Balhang/Lou-tse-kiang	Kunming	Yunnan-fu
Baotou	Paotow	Lijiang	Likiang
Baoxing	Moupine/Moupin	Malu, Guangxi	La-fou, Kwangtung
Beijing	Peking	Mengzi	Mongtse/Mengtse
Binchunglu	Champutong	Menhuoying	Mo-so-yn
Caka'lho (Yanjing)	Yerkalo	Moxi	Mosimien
Chengde	Jehol	Pianjiao	Pin kio
Chengdu	Chengtu	Qutong	Chu-tung
Chongqing	Chungking	Shiyan	Siang-yang
Cigu	Tsekou/Tsiku	Tengchong	Tengyueh
Cizhong	Tsedjerong/Tsed-rong	Tianjin	Tientsin
Dali	Tali	Tumd Zuoqi	Saratsi
Dapingzi	Tapintze	Weixi	Wei-his
Deqin	Atuntse/Atuntzu	Xiao-Weixi	Hsiao-weihsi
Eryuan	Lankong	Xinjiang	Sinkiang
Gansu	Kansu	Xuanhua	Suanhwa
Gualapo	Koua la po	Yantai	Chefoo
Guangzhou	Canton	Yibin	Sui fu
Hankou, now part of Wuhan	Hankow	Yichang	Ichang
		Zhongdian	Chung-tien
Heqing	Hoking/Hoching	Zunyi	Tsen-y/Tsuni
Hohhot	Kweisui		
Hongshanding	Hongchantin/Hung-shan-ting	Tongking (now north Vietnam)	Tonkin
Huanjiapin	Houang kia pin		

PROVINCES

Pinyin	Old Style
Anhui	Anhwei
Guangdong	Kwangtung/Kuangtun
Guangxi	Kwangsi
Guizhou	Kweichou/Kueichou
Henan	Honan
Hubei	Hupeh
Jiangxi	Kiangsi
Jiujiang	Kiukiang
Qinghai	Tsinghai
Shaanxi	Shensi
Shandong	Shantung
Shanxi	Shansi
Sichuan	Szechwan/Szechuan
Zhejiang	Chekiang

MOUNTAINS, RIVERS AND PASSES

Pinyin	Old Style
Dadu River	Tung Ho
Emei Shan	Omei Shan
Heishanmen	Heechanmen
Huang He	Hwang Ho/Yellow River
Lake Qinghai	Koko Nor
Lancang Jiang	Upper Mekong River
Luoping Shan	Lopinchan
Maogushan	Mao Kou Shan
Nu Jiang	Upper Salween River
Qinling	Tsinling
Xi La Pass	Sela

INTRODUCTION

—

It was most pleasing to find that many of the Catholic missionaries busied themselves in collecting material for the advancement of science. In so doing they were certainly not motivated by any desire to supplement their meagre stipends, but by genuine enthusiasm for their subject, of which they often had considerable knowledge.

HEINRICH HANDEL-MAZZETTI[1]

—

So MANY OF OUR FINEST ORNAMENTAL plants come from China that it would be difficult to find a park or garden anywhere in the temperate world without at least one Chinese plant on display. The first plants to arrive in the West were those grown in Chinese gardens such as wisteria, magnolias and roses. This only changed after Britain and France forced China to grant access to the Chinese interior at the end of the Opium Wars in 1860. Westerners were then able to investigate the plants of previously inaccessible areas of western and central China, and they quickly realised that this vast country was home to one of the richest and most varied floras in the world. When dried plant specimens began to reach Europe and America, they proved a revelation to botanists and horticulturalists alike. Western plantsmen immediately wanted to grow the new discoveries and professional plant-hunters were swiftly despatched to western and central China to collect living material. When we discuss the discovery of Chinese plants today, we usually do so in connection with the plant-hunters and nurserymen who first introduced them to our gardens.

In many cases, however, professional plant-hunters were not the first Westerners to explore the richest botanical areas and discover eye-catching new plants. The botanical pioneers were a number of Catholic missionary-priests sent to live in China's remote heartlands in the decades after 1860. Several of them were so fascinated by the diverse flora they saw around them

that they spent their limited free time exploring their districts, collecting and drying thousands of plants which they sent back to European botanists for study. It was quickly apparent to these botanists just how important the missionaries' investigations were, and they marked their appreciation by naming several plants after their discoverers. Today, many of these plants are well-known to gardeners: *Davidia involucrata, Buddleja davidii, Paeonia delavayi, Ilex pernyi, Ilex fargesii, Rosa soulieana, Rhododendron souliei, Callicarpa bodinieri* var. *giraldii, Rosa hugonis, Acer fabri, Deutzia monbeigii*… But who now remembers the men behind the names?

Perhaps the chief reason why the missionary-botanists commemorated in these names have been forgotten is that little is known about the individual priests. The missionaries were self-effacing men who lived isolated lives deep in the Chinese countryside. They were devoted to the wellbeing of their scattered communities and uninterested in the botanical fame that was rightfully theirs. They did not write about their experiences, or describe the plants they found and, as their discoveries were reported in specialist botanical periodicals usually only read by experts, few outside a narrow circle of professional plantsmen heard of them.

This is in sharp contrast to the professional plant-hunters who followed in the missionaries' footsteps. Their discoveries and adventures were widely reported: they helped classify the

Rosa soulieana in midsummer.

Some of these later plant-hunters – particularly E. H. Wilson, George Forrest and Frank Kingdon-Ward – are frequently mentioned in this account as their activities so often overlapped with those of the missionary-botanists. This is hardly surprising as they arrived in China hard on the missionary-botanists' heels, explored many of the same areas and introduced to cultivation several noteworthy plants first discovered by the missionaries. Essential biographical information and brief details about the plant-hunters' various expeditions have been provided in order to clarify the narrative; readers who are interested in learning more are referred to the plant-hunters' own writings, various biographies, and histories of plant-hunting in China.

The missionary-botanists' story is more than just an interesting historical curiosity as their discoveries continue to be important in a number of different disciplines, particularly conservation, medical and scientific research, and horticulture. Specific examples have been provided throughout the narrative but some general points can be made here.

Conservation

Without the priests' determined collecting in all weathers and in all seasons and the long hours they spent patiently preparing their specimens, we would know far less about the extraordinarily rich and diverse native flora of western China and – because habitat destruction proceeded as fast as it did – many of the plants they found between 1869 and 1914 would now have vanished without trace. It is thanks to these largely forgotten priests that so much of the native flora was preserved, classified and described. The thousands of herbarium specimens they collected provide an invaluable record of the native Chinese flora that is used by botanists today as a benchmark against which they can compare their own collections.

So many of the plants collected by the missionary-botanists proved to be new to science that contemporary theories of plant classification had to be considerably amended to accommodate the discoveries. The range of adaptations revealed was so extensive that botanists were able to pinpoint the origin of several important genera such as *Rhododendron*, *Primula*, *Gentiana* and *Pedicularis*. Further detailed study of the missionaries' collections led to a better appreciation of the evolutionary links between species and genera, and this led, in turn, to a greater understanding of global plant distribution patterns and to further developments in the new science of plant geography. Distribution patterns based on the missionaries' initial discoveries

plants they collected, gave lectures, and often wrote books and articles about their travels. The wealth of information about the adventures and discoveries of plant-hunters like E. H. Wilson and George Forrest has proved invaluable for biographers and historians and, over the years, these men and others like them have come to dominate accounts of plant discovery in China. Their achievements were certainly outstanding, but glossing over the missionary-botanists' contribution distorts the true picture. This narrative aims to redress the balance, as only when the extent of the missionary-botanists' achievements is recognised, can we properly appreciate the debt we owe them.

The account focuses particularly on the lives of four great French missionary-botanists – **Père Armand David**, **Père Jean Marie Delavay**, **Père Paul Guillaume Farges** and **Père Jean André Soulié** – but also includes other French priests who collected plants, particularly **Père Paul Perny**, **Père Édouard Maire**, **Père François Ducloux**, **Père Émile Bodinier**, **Père Pierre Cavalerie** and **Père Théodore Monbeig**. Franciscan missionaries were also involved – Italian **Padre Guiseppe Giraldi** and **Padre Cipriano Silvestri**, and Irish **Father Hugh Scallan** – as was German **Pastor Ernst Faber**, the only Protestant missionary to make significant plant collections.

It is true that Père Armand David is still relatively well-known: but then *Davidia involucrata*, the dove or handkerchief tree, is especially celebrated, and he is also famous as the discoverer of the giant panda and Père David's deer. His remarkable talents as a naturalist were recognised by his contemporaries and his superiors permitted him to devote himself to investigat-

are still being refined today, as new locations are fitted into the existing scheme.

The descriptions in botanical journals of the plants found by the missionary-botanists helped to inspire nurserymen and wealthy gardeners – who wanted to grow the new plants themselves – to fund plant-collecting expeditions by professional plant-hunters, which led to the introduction of hundreds of new plants to Western parks and gardens, and to the discovery of even more hitherto unknown plants. Without the missionaries' initial explorations, plant-hunters like George Forrest would not have arrived in China as soon as they did and many plants subsequently introduced to cultivation might already have disappeared from the wild. The nurserymen and private individuals who received the new plants were quick to supply them to botanic gardens on both sides of the Atlantic, where curators further distributed the plants to colleagues around the world. This rapid circulation of newly-introduced plants among scientific establishments, where they could be cared for by highly-skilled professionals, has contributed markedly to their preservation: today many wild populations are under threat due to the degradation and destruction of their native habitat.

Research

The importance of conservation is emphasised by the fact that some of the missionaries' plant discoveries, which were not of particular significance at the time, have now become the focus of medical and scientific research into human and plant diseases.

Horticulture

Nurserymen found that many of the newly-discovered Chinese plants had considerable ornamental value and they ensured that, within a remarkably short time, they were commercially available. Gardeners were so anxious to acquire the latest introductions that nurserymen expended every effort to meet the demand and saw to it that the new plants, which had at first been expensive rarities, were soon generally obtainable. They are now so widely disseminated among Western parks and gardens that their survival is secure, even if they should disappear from the wild; and for those of us who cannot visit the areas explored by the missionary-botanists, growing the garden-worthy plants they discovered is the next best thing.

Many of the missionaries' discoveries proved excellent garden plants in their own right, but Western nurserymen were also quick to recognise their breeding potential and used them to develop new ornamental plants such as *Deutzia* x *elegantissima* 'Rosealind' and *D.* x *hybrida* 'Magicien', Asiatic Hybrid Lilies and North Lilies, and varieties of *Astilbe* x *arendsii*. Gardeners today continue to benefit from plants bred from those first introduced over a century ago.

Corydalis flexuosa 'Père David' in midsummer in Marion Pollard's Somerset garden.

Some plants discovered by the missionary-botanists are still being introduced to our gardens: *Corydalis flexuosa* discovered by Père David in 1869 and introduced to the West in the 1990s has since proved a fine perennial in cultivation, as has Père Delavay's *Podophyllum delavayi* and *Epimedium* species such as *E. fargesii* and *E. leptorrhizum* discovered by Père Farges and Père Bodinier and introduced in the past twenty years. Trees like *Nothaphoebe cavaleriei* discovered by Père Cavalerie, *Acer fabri* found by Pastor Faber, *Mahonia duclouxiana* found by Père Ducloux and *Euonymus phellomanus* discovered by Padre Giraldi proved such fine ornamental plants when recently introduced to cultivation that horticulturalists can anticipate them becoming more widely available.

CHAPTER 1

A Dangerous Vocation

—

It is impossible not to admire the zeal and unselfishness which induces the

Catholic missionary to come to China, there to spend the rest of his days in the

cause of his religion, without hope of ever seeing his native country again…

He makes up his mind from the beginning to live and die in China.

MAJOR H. R. DAVIES[1]

—

Treaties opened China to missionaries

In the seventeenth century, when the Emperor allowed Jesuits to live in Beijing, a few priests had also lived secretly in some Chinese provinces; but waves of anti-Christian persecution meant their situation was always precarious. The opportunity to evangelise openly in China only came with the Treaty of Nanking at the end of the Opium War in 1842, and the concessions that China was forced to make then were further extended by various clauses in the Treaty of Tientsin made between France and China in 1860. These and subsequent agreements, which allowed missionaries to live and work openly throughout China, revolutionised the status of European missionaries and Chinese Christians, and made possible a large expansion of the Church.

The 1860 Treaty made the French minister in Beijing the official intermediary between Catholic missionaries and the Chinese government, thus establishing France as the protector of all Roman Catholic missions throughout China. France adopted the role with alacrity as a way of enhancing her prestige and influence and became a fierce advocate of the right of European missionaries to live and work in the midst of Chinese society without restrictions. In effect, this established a foreign protectorate within the Chinese state as Catholic missionaries were not subject to Chinese laws and were able to appeal directly to the French minister in Beijing for help and support whenever they came into conflict with the authorities. These arrange-

ments were understandably detested by the Chinese and gave rise to increasingly antagonistic relations with foreigners.

The strong anti-foreign feelings fuelled by the Treaties led to frequent attacks on the missionaries and their communities and Rome came to rely on France to force reluctant Chinese officials to protect the priests living in their midst. The missionaries were encouraged to appeal to the French minister in Beijing at the first sign of danger. The minister would then impress upon the Imperial authorities the need for Treaty provisions to be enforced at local level, and demand redress for any damage caused to Church property. This access to an external power was bitterly resented by the Chinese and resulted in a growing antipathy to the missionaries who came to be seen, not primarily as religious figures, but as foreign agents.[2]

Attacks on missionaries

On Christmas Eve 1868, **Père Jean Marie Delavay** was saying Mass in his chapel at Leizhou on the coast of southern China, when the evening quiet was shattered by a fusillade of bricks and stones that smashed against the building. He found the exit barred and the chapel surrounded by a hostile crowd. The General in charge of the soldiers stationed in the town refused to help and when the congregation tried to escape many were wounded or killed. Local officials managed to rescue Père Delavay from the mêlée, but he had already been severely hurt.

We do not know precisely what sparked this incident but it did not take much to incite violence against the missionaries, who were a focus for anti-foreign feeling. Tales of priests killing children and other lurid stories were widely current and would have been enough to enrage the inhabitants. Attacks such as this were not uncommon and Père Delavay was lucky to escape with his life: only a month later one of his confrères was murdered in Sichuan province.

Père Delavay had only arrived in China in September 1867, a little over a year earlier, and the mission station at Leizhou had been his first assignment. The chapel there had just been restored after an earlier attack and, although there were rumours in the town of fresh ill feeling against the missionaries, Père Delavay thought the situation was calm enough and he trusted the officials to protect the Christian community if there was any trouble. His inexperience made him too sanguine – the General to whom he appealed for help was an inveterate enemy of the missionaries and had encouraged the rioters. This hatred of missionaries was shared by many of the Chinese elite who never accepted the Treaty provisions and remained extremely hostile to the foreign missionaries imposed upon them.

It took until April for Père Delavay to recover enough to evade capture and make his way to the safety of the bishop's residence at Canton (Guangzhou). He did not convalesce there for long and had returned to work by the beginning of July.[3]

The missionaries
Père Delavay's commitment to his missionary duties was typical of the growing number of young French men who felt called to missionary work in the Far East. The end of the Napoleonic era in France saw a resurgence of religious fervour, and the imagination of Christians was haunted by the contemporary belief that those who had no knowledge of Christianity were condemned to eternal damnation. In the light of this, devout young men became obsessed with the plight of the vast population of China, which they believed was destined for Hell unless something was done to save all those millions of souls from their certain fate. This led to a rise in French missionary zeal that was orchestrated by the Association de la Propagation de la Foi, founded in 1822, which worked closely with the church hierarchy to publicise the need for missionaries. So effective was their collaboration that, by the middle of the century, the message was reaching every French parish. The response was remarkable. Accounts of attacks such as those suffered by Père Delavay were quickly disseminated and each assault was interpreted as persecution of a devoted Christian who had sacrificed everything for the sake of his faith. Dozens of young men, fired with a longing for similar self-sacrifice and 'hungry for an imagined martyrdom' joined missionary societies dedicated to international evangelism.[4] So great was the enthusiasm with which the call for missionaries was embraced that by the 1890s one society alone – the Missions Étrangères de Paris – was sending forty to fifty priests a year out to China and the rest of the Far East.

Missions Étrangères de Paris
Père Delavay and all the French missionary-botanists discussed in this account, apart from Père David, belonged to the Missions Étrangères de Paris, one of the largest of the French missionary societies. When Rome divided the Chinese provinces up among the various missionary organisations, the Missions Étrangères was given responsibility for converting a vast territory in southern and western China that included the provinces of Sichuan, Yunnan, and Guizhou, as well as the Tibetan borderlands: a region later found to contain one of the richest and most varied floras in the world.

The Missions Étrangères was not a religious order as such but rather a society of priests bound by their shared devotion to the service of missions. One had to be under thirty-five to join and many members came straight from school; but Père Delavay was thirty-two and had been working as a parish priest in his home diocese of Haute Savoie for six years before he entered the Society's seminary in Paris in November 1866. His family was devout: his elder brother Joseph was also a priest and a sister became a nun, but Père Delavay's decision to join the Missions Étrangères some years after having entered the priesthood showed an extraordinary commitment to the Christian ideal. Those who chose to become missionaries knew that their choice involved permanent exile from France. Missionaries were only ever permitted to come home if their health broke down irretrievably: the vast majority never returned. Today, such self-sacrifice at the beginning of their adult lives seems almost incredible, but the fact that so many young men devoted themselves to a lifetime's exile from their homes, their families, friends and culture testifies to the strength of their faith and to their unshakeable conviction that spreading the message of the Gospels in foreign lands was of paramount importance. They took quite literally Christ's injunction that anyone who could not leave parents and families behind and devote themselves entirely to His service was not worthy of Him.

It was from the ranks of these devout men that the missionary-botanists who are the focus of this narrative were drawn. The driving purpose of their lives was the conversion of the people they had exiled themselves to serve – for them, nothing else mattered. They were well aware of the importance of the botanical discoveries for which we celebrate them today, but botany was always a secondary activity. They remained, first and foremost, missionary priests. It is only when we understand this, and realise that the time the priests could devote to botany was limited to whatever time they could spare from their missionary and ecclesiastical duties, that we can really appreciate the extent of their botanical achievements.

The Missionary Project

It is important when considering the lives of the missionary-priests to remember that *at the time* what they were doing was considered not only normal but entirely praiseworthy. Today, we find the idea of going to live in a foreign country with the sole purpose of inculcating a new set of spiritual beliefs and undermining its traditional culture wholly unacceptable, but in the nineteenth century the Christian missionary impetus was a strong cultural force and was encouraged, not only by the ecclesiastical authorities, but also by politicians, who saw it as a way of increasing Western influence in countries over which they wished to exercise greater control.

It is hard today to understand why the priests remained in China in the face of such determined and continued hostility on the part of the authorities and much of the populace. However, they interpreted the enmity and aggression they faced, not as an expression of political anger, but as religious persecution, and this led them to believe that they were sharing the sufferings of Christ. Consequently, each assault made them identify ever more closely with Him and strengthened their determination to remain and complete their self-appointed task. They believed so fervently in the rightness of their cause that the greater the hostility they faced, the greater the obstinacy with which they clung to their posts.

Consequences

As the focus of this account is the missionaries' botanical exploration and discoveries, the narrative has not been burdened with comments on the nature of their evangelical work or on the deleterious effect their activities had on Western relations with China: although, as the narrative progresses, it becomes increasingly obvious just how much the missionaries were resented and how their presence destabilised the position of other Westerners living in China.

Photo Credit: Ye Jianfei and Dr Liu Bo

CHAPTER 2

The Open Door

—

In ornamental trees, shrubs, and herbs, suitable for outdoor cultivation in the British Isles, China is the richest country in the world. Our indebtedness to China may in a measure be realised if an imaginary attempt be made to expunge from our gardens all the plants she has given us.

E. H. WILSON[1]

—

THE FIRST AND MOST FAMOUS of the missionaries to investigate the botany of China in the nineteenth century was **Père Armand David**, one of the most accomplished naturalists ever to visit the Far East. As well as collecting a multitude of new plant species during his extensive journeys into unexplored areas of China, he also discovered new birds, mammals, reptiles and insects. He kept comprehensive journals in which he noted in fascinating detail the minutiae of his journeys. Colourful descriptions of the people he met, the places he stayed and the food he ate were accompanied by precise details of the geological changes in the landscape and soil, together with notes on some of his most remarkable zoological and botanical finds. These included the giant panda and Père David's deer, together with several new monkey and bird species, while gardeners all over the temperate world are familiar with plants such as *Buddleja davidii*, *Corydalis flexuosa* and *Davidia involucrata*, the dove or handkerchief tree.

Armand David was born on 6 September 1826 in Espelette, a small town in the Pyrénées-Atlantiques in the far south-west corner of France. It is an area that prides itself on its Basque heritage and, in later years, Armand David always said that he owed his stamina on his long journeys in China to his upbringing as 'a true Basque'. His father was a doctor and keen naturalist and as a boy Armand delighted in accompanying his

father on visits to patients in the surrounding countryside, looking, listening, asking questions and discussing everything he was told, eager to discover the laws which governed the wonders of nature which he saw around him. At school, he proved an outstanding pupil, devoting himself to languages, especially to Latin – the language of scientific description – and to natural sciences such as botany, ornithology and entomology. He noted later that it was some years before he understood what his fellow pupils found to enjoy in books other than the volumes of natural history with which he filled his own spare time.[2]

Photo Credit: Kew Archive

Photo Credit: Kew Archive

LEFT

Prunus davidiana flowering in its native habitat.

ABOVE

Père Armand David.

ABOVE

Alexander von Bunge.

A promising career as a zoologist or botanist seemed to lie before him but, even before he finished his studies at the Grand Seminaire at Bayonne, Armand David felt called to the priesthood and he later wrote that his greatest ambition had always been to share in the work of the missionaries who had tried for three centuries to win over the teeming populations of the Far East for Christianity. This vocation led him, in 1848, at the age of twenty-two, to enter the Congregation of the Mission, an order of missionary priests based in Paris, whose members were known in France as Lazarists, and more commonly elsewhere as Vincentians after their Patron, St Vincent de Paul. The Lazarists had taken over the Jesuits' role in Beijing and had also been given responsibility for converting north-east China, so the young Armand David hoped that he would be sent out to China without delay. However, his superiors had other plans for him and his proficiency as a naturalist led them to appoint him to one of their colleges in Savona on the coast of northern Italy as a lecturer in natural sciences. He now had a congenial occupation in pleasant surroundings and yet he wrote to his superiors in November 1852 saying that he could not stop dreaming of the Chinese missions and that for a dozen years he had been driven by the wish to die while working for the salvation of non-believers. His pleas had no effect and even after he was ordained to the priesthood in March 1853, he had to remain in his teaching post. It was not until October 1861, when he was thirty-five, that the order for which he longed arrived: he was to be sent out to Beijing with a group of his confrères to found a school where he would teach science.[3]

Père David returned to Paris to prepare for his departure. There, he was taken to meet Stanislas Julien, a Chinese scholar and member of the Institut de France. M. Julien immediately recognised in Père David a man of uncommon abilities and introduced him to several members of the scientific establishment, including the zoologist **Henri Milne-Edwards**, who was head of the Muséum d'Histoire Naturelle, and **Joseph Decaisne**, director of the Jardin des Plantes, the botanic garden connected to the Muséum. They realised immediately that Père David's enthusiasm and knowledge presented them with an unprecedented chance of acquiring zoological and botanical specimens from an area about which they had very little accurate information, and each specialist seized the opportunity to present him with a list of their most pressing requests.

First botanical explorers in China

Before Père David's arrival in Beijing in 1862, what little was known about the flora of north-east China was due to the efforts of Russian enthusiasts. Russia was the only Western power to have had access to northern China prior to 1842, as it was allowed to maintain a permanent Legation in Beijing; and **P. V. Kirilov**, physician at the Legation from 1831–1841 was the first to use the opportunity to investigate the flora of the region. He collected plants in the plains around the city and explored Po Hua Shan, the famous Mountain of a Thousand Flowers in the Western Hills: and his collections included an unusual herbaceous clematis, *Clematis tubulosa*, which has since become an ornamental garden plant.[4]

The Legation personnel was changed every ten years and **Alexander von Bunge**, a talented botanist, took advantage of the change-over in 1830/31 to travel to China. He collected plants in Inner Mongolia and explored the mountains and plain around Beijing, before returning to his home in Dorpat (Tartu) in Estonia where he became Professor of Botany at the University. When he came to examine his collections, he was able to publish descriptions of some 330 new species. This was remarkable, as he had made his collection in the space of a single six-month visit: botanists could only speculate about the number of species he might have discovered had he had more time. Bunge also distributed duplicates of his specimens to the great botanical institutes of Europe so that other botanists could familiarise themselves with the new plants.[5]

At the time, the flora of coastal China was much better known to Western botanists and gardeners, as British enthusiasts and collectors working for the East India Company had been sending plants back to Britain since the early eighteenth century. These efforts were given fresh impetus in 1842 when Britain secured access to several Chinese ports at the end of the first Opium War. The Horticultural Society in London took immediate advantage of the new freedoms by sending **Robert Fortune**, a professional plant-hunter, out to China. The four prolonged collecting trips he made in the next twenty years resulted in the introduction of the majority of traditional Chinese garden plants to Britain.[6]

Fortune also explored inland areas and collected some of the region's hitherto completely unknown wild plants, but it was an amateur botanist called **Henry Hance** who put efforts to

investigate the native flora of the Chinese provinces to which Westerners now had access on a more regular footing. Hance joined the British consular service in China in 1844 and, as an enthusiastic and determined plantsman, took full advantage of the newly-acquired freedoms to conduct botanical explorations in Hong Kong, and the provinces of Guangdong and Fujian. He had a great advantage over Fortune as he actually lived in the area he was exploring and was not limited by the time constraints of a fleeting visit. This was an advantage later shared by the missionary-botanists: they lived in the areas they were investigating and could return to particular sites again and again, building collections up over several years to provide a comprehensive picture of the native flora. In 1861, Henry Hance was promoted to the position of vice-consul at Whampoa (Huangpu), the busy customs post for Guangzhou, and soon found that he had little free time for botanising. Nevertheless, his new position gave him an excellent opportunity to meet merchants and traders, as well as other British officials, and he encouraged all those with an interest in botany to send him plant specimens from the areas they visited, or to which they were posted. In this way, Hance was able to establish connections with amateur collectors throughout southern and eastern

China, who explored their own localities and then sent their carefully-dried specimens back to him to be identified. He published descriptions of any new plants in widely-read botanical journals, which swiftly brought them to the attention of botanists in Europe and America.[7] Hance's connections with collectors were so good that when he died in 1886, his herbarium contained around 22,500 specimens and included almost all the plants then known from China.

Père David's first discoveries and their introduction to cultivation in the West

It was apparent from Fortune's discoveries and from those made by Hance and his colleagues, as well as from the plants collected around Beijing by Bunge and Kirilov, that China had an extraordinarily rich native flora. Botanists began to long for a chance to penetrate even further into the unexplored interior of the vast Chinese Empire, long known as the Flowery Kingdom. Hence the excitement of Joseph Decaisne and his colleagues in Paris at the thought that Père David's sojourn in Beijing might lead to a raft of new botanical discoveries. Père David was just as eager to start exploring, although he soon discovered that the plain around the city was too intensively cultivated to yield

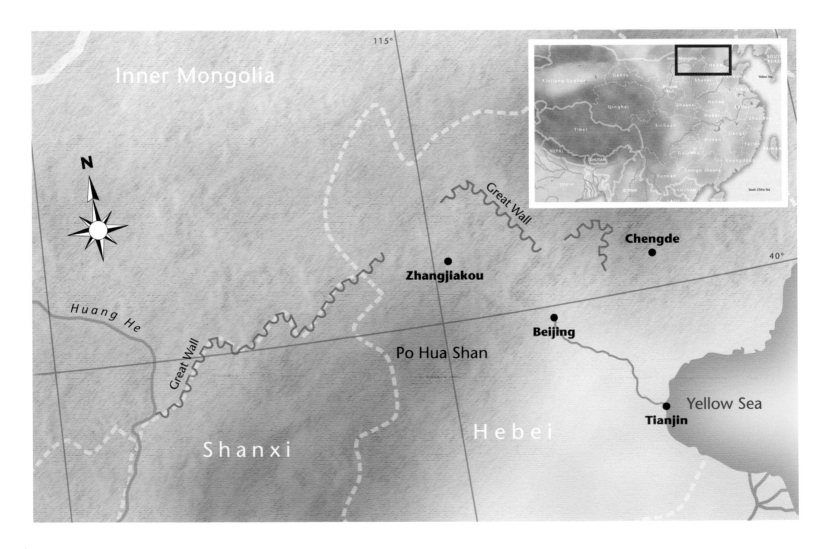

Emil Bretschneider.

Syringa villosa. (*Bot.Mag.* (1929) No.9284)

much of botanical interest, and the neighbouring hillsides were completely deforested. Père David also visited other missions and, in September 1862, travelled north of the Great Wall for the first time when he visited a Lazarist mission at Xiwanzi, about 48 km (30 miles) north-east of Zhangjiakou. The following summer he was able to spend several months exploring the mountains west of Beijing, including Po Hua Shan.[8]

It was during this time that he first collected specimens of *Tilia mongolica*, the Mongolian lime, which was introduced to the West from the Beijing area by Dr Emil Bretschneider, the physician at the Russian Legation.[9] **Emil Bretschneider** was an enthusiastic botanist and from the moment he took up his post in Beijing in 1866, he devoted every spare moment to investigating Chinese plants. As well as studying Chinese botanical texts, he began researching the European discovery of the Chinese flora and in 1898 published the results in a comprehensive two-volume history that remains the definitive work.[10] He also collected seeds of several local plants to send back to various botanical gardens and, in this way, seeds of *Tilia. mongolica*, which he had collected on Po Hua Shan, reached the Jardin des

Plantes in Paris in 1880. In 1882 Bretschneider sent *T. mongolica* seeds to the Arnold Arboretum, the arboretum in Boston connected to Harvard University. Staff at both these gardens were very successful in raising plants from the seeds, and their skill provided botanists interested in the Chinese flora with a chance to study living examples of newly-discovered plants. Botanists can tell a great deal from well-dried plant specimens but having living material to examine is invaluable, especially when trying to understand and identify new species and work out the relationships between unfamiliar plants. The *Tilia mongolica* tree raised in the Jardin des Plantes flowered for the first time in 1896 and in 1904, some of its seeds were sent to the Royal Botanic Gardens at Kew in London. One of the trees derived from this initial sowing still flourishes at Kew and at 14 m (46 ft) is now recognised as a Champion Tree. *Tilia mongolica* is a handsome hardy lime and as it does not suffer from the honeydew secretions that affect many European species it is often planted as a street tree in north-west America. Its leaves are smaller and more attractive than those of many other species of *Tilia* and it has the best autumn colour of any lime as the foliage turns golden yellow.[11]

One of Bretschneider's consignments contained a new lilac, *Syringa villosa*, which first flowered in the Jardin des Plantes in 1886.[12] Although Bretschneider introduced it to cultivation, specimens of this lilac had already been collected by P. V. Kirilov and Père David, and also by an even earlier collector: **Père Nicholas le Cheron d'Incarville**, the first French missionary to collect plants in China.[13] Père d'Incarville was a Jesuit priest who lived in Beijing from 1741 until his death in 1757, and he made several plant collections which he sent back to France. At the time, communications between China and Europe were poor, and some of his consignments never reached France, and even those that did arrive in Paris seem to have been largely ignored by the botanists to whom they were sent. Nevertheless, the specimens had been so well prepared that when Père d'Incarville's forgotten collections were examined by the botanist Adrien Franchet in 1883, some 149 species from Beijing were recognised, including *Syringa villosa*, one of the handful of his specimens that had already been described.[14]

Bretschneider had also sent seeds of *Syringa villosa* to the Arnold Arboretum and in 1889 this newly-introduced lilac was featured in the illustrated American weekly magazine *Garden and Forest*, as 'an ornamental plant of the first class'.[15] **Charles Sprague Sargent,** Director of the Arboretum, had founded *Garden and Forest* in 1888 as a way of ensuring that American gar-

LEFT

Charles Sprague Sargent.

RIGHT

Adrien René Franchet in later life.

deners knew about the latest developments in horticulture and forestry, and it included regular reports on the new introductions from China. In Britain, articles on new plant discoveries and their cultivation were found in magazines like the *Gardener's Chronicle* and *The Garden,* while periodicals such as the *Flore des Serres* and the *Revue Horticole* kept French horticulturalists up-to-date. These reports aroused considerable interest, but although receiving dried specimens of hundreds of new species was very exciting for botanists, plant enthusiasts were more interested in living plants. As it became clear that the flora of China contained a wealth of plants that could be grown in temperate climates, nurserymen and gardeners in the West longed to grow them and find out if they were garden-worthy. To this end, prominent nurserymen in France encouraged those in China to send them seeds from the new plants, and Bretschneider responded enthusiastically. One of his correspondents was **Victor Lemoine**, the French nurseryman, who successfully raised *S. villosa* at his premises in Lyon, and offered it in his 1890 catalogue, albeit under the name *Syringa bretschneideri,* a synonym of *S. villosa.*

Photo Credit: Kew Archive

Adrien Franchet at the Muséum d'Histoire Naturelle

It was not just Père d'Incarville's specimens that had to wait for Adrien Franchet to describe them but also those collected by Père David. Even though French botanists were very excited by Père David's finds, none of the botanists at the Muséum were particularly familiar with the temperate flora of eastern Asia – and Père David's collections demanded the attention of someone with specialist knowledge. Their time came when the remarkable botanist **Adrien René Franchet** joined the Muséum in 1880, and it is fair to say that without his invaluable efforts in working on the collections made by the missionary-botanists who feature in this narrative, their names would today only be known to a few specialists. Franchet was a true enthusiast whose devotion to botany had begun when he was a boy and had begun to learn Latin and collect flowers under the aegis of the local priest. His enthusiasm for plants was so great that when he was twelve his mother had apprenticed him to the local pharmacist, as plants were then the basis for virtually all medicines, and she thought this might be a suitable career for the budding botanist. Young Adrien, though, used his new freedom from regular school hours to spend virtually all his time botanising and rarely appeared at the pharmacy. This happy situation could not last and a month later Adrien found himself back at school, although botany continued to monopolise his spare time. After completing his studies, Franchet became a teacher but when he was twenty-three the Marquis de Vibraye offered him the post of curator at his chateau at Cheverny, with responsibility for the renowned collection of archaeological and geological exhibits housed there. Franchet accepted with alacrity but continued to botanise whenever possible, focusing on collecting French plants; and as he purchased those that he could not collect himself, he eventually created an almost complete collection of the French flora. He began to prepare an account of the flora of his home département, Loir-et-Cher, which was published in 1885.[16]

While at Cheverny, Franchet also worked on a large collection of Japanese plants made by his friend Dr Ludovic Savatier, who had lived for a decade in Japan, and they published a major two-volume work on the Japanese flora in 1875 and 1879. The Marquis' death in 1878 meant that Franchet had to look for a new post. Professeur **Édouard Bureau**, the director of the herbarium at the Muséum d'Histoire Naturelle, had been considerably impressed by Franchet's work on Japanese plants and engaged him, in 1880, to identify Père David's collections. This eventually resulted in Franchet's majestic publication *Plantae Davidianae ex Sinarum Imperio*, published in two volumes, in 1884 and 1888, in which he described all Père David's botanical discoveries. In 1886 Franchet was given an official position as an instructor at the Laboratoire des Hautes-Études attached to the Chair of Botany, and he remained at the herbarium for the

Bot. Mag. 6801.

6801

AB. del.   Clematis tubulosa var. Hookeri.

rest of his life, concentrating on identifying and describing the Muséum's increasing collection of Chinese plants.

Plant classification was Franchet's particular skill as he was a systematic botanist, specialising in identifying and classifying plants in accordance with strict rules based on their form and structure. As Franchet examined the dried plants collected by the missionary-botanists, many of which had never been seen before, he had to work out what these new plants were and how they were related to existing plant groups, and then consider where they best fit within the overall system of plant classification. Once he had determined the genus, he had to describe and name the new species. Some plants that Franchet examined did not fit within the circumscription of known genera and required him to create new ones to accommodate them. Among the new genera that he named were several that honoured the missionary-botanists, including *Delavaya*, *Fargesia* and *Souliea*. The descriptions were published in botanical journals so that other botanists could keep abreast of new discoveries.

Correct identification and classification depends on a wide knowledge of related plants and Franchet's familiarity with the Japanese flora gave him an important advantage when it came to working with the plants from China. His industry was just as important: the sheer volume of work he accomplished in classifying not only the hundreds of specimens collected by Père David but also the thousands sent back from China by later missionary-botanists was extraordinary and his diligence and determination simply cannot be overpraised. Those missionaries with a passion for botany such as Pères Delavay, Farges and Soulié who arrived in China after 1862 were lucky that they were dealing with a botanist of Franchet's calibre, as it was through his labour that so many of the plants they discovered were examined and described. Even the finest and most interesting collections will languish in obscurity if no botanist is available to work on them – as did Père d'Incarville's plants, and those of Père David before Franchet's arrival.

Botanical puzzles

Even though the majority of Père David's specimens had to wait for Franchet's arrival in 1880, the seeds he collected for the Jardin des Plantes were sown straightaway and when he finally returned to France in 1876, he discovered some eighty-odd plants that he had introduced growing in the garden.[17] One of them appeared to be *Clematis tubulosa*, an herbaceous clump-forming clematis with upright blue, hyacinth-like flowers that he had found to the north of Beijing in 1863.[18] As we have

seen, it was originally discovered by P. V. Kirilov in the 1830s, and then introduced to the West where it became popular as a late-flowering border plant. Joseph Decaisne did not believe that Père David's introduction was the same as the *C. tubulosa* plants already growing in the Jardin des Plantes and considered it a completely new species, which he named *C. davidiana* in honour of its discoverer. The argument over whether *C. tubulosa* and *C. davidiana* were the same continued to exercise French botanists but by 1881 Decaisne, who had carefully examined the various Asian herbaceous clematis cultivated in the Jardin des Plantes, felt confident that *C. davidiana* was definitely different from *C. tubulosa*.[19] Franchet disagreed, deciding in 1883 that Père David's plant was merely a variety of *C. tubulosa* that he called var. *davidii*, although he did wonder later if it might be a form of *C. heracleifolia*, another Chinese herbaceous species, and he called this variety var. *davidiana*.[20]

There is still disagreement on the question. In the *Flora of China*, the current account of all Chinese plants, the Beijing species is identified as *C. heracleifolia*, a species also found in Japan and Korea, and *C. tubulosa* is considered a synonym.[21] Horticulturalists, though, still maintain a distinction between the two species.[22] This well illustrates the complexities of naming and classifying completely new and unfamiliar species, even when living plants are available for comparison, and it shows just how difficult it can still be to resolve taxonomic problems when naturally variable species are involved.

Père David, even as he collected new plants, did not forget the list of botanical requests he had originally been given by Decaisne and he spent much time searching for *Cedrela sinensis*, a tree that had first been described in 1830 from one of Père d'Incarville's specimens. Despite his efforts, Père David had to report to Decaisne that he had been unable to find fertile seeds or any young individuals that he could pot up. Unknown to either of them this species had actually been introduced to France from Japan in 1862, but the sapling had been wrongly identified as *Ailanthus flavescens* and it was only when it flowered in 1875 that botanists realised that it was not an *Ailanthus* at all but the *Cedrela sinensis* that Père David had been seeking. This emphasises the importance of cultivating new plant discoveries, as botanists can use living plants to check the accuracy of identifications based on dried specimens. Indeed, minds changed again and *C. sinensis* is now called *Toona sinensis*. Unusually for a member of the tropical mahogany family, *T. sinensis* is quite hardy.[23]

By the middle of 1863, the first of Père David's collections of

Astilbe davidii. (*Bot.Mag.* (1903) No.7880)

animals, birds and plants had arrived at the Muséum d'Histoire Naturelle and the specialists there were delighted by the quantity and quality of Père David's specimens. **Henri Milne-Edwards**, the Director, immediately wrote to him, expressing their gratitude and praising the achievements he had made in so short a time. Milne-Edwards also arranged to send Père David enough money to fund his natural history explorations for another year.[24] Encouraged by the knowledge that his discoveries were proving of real benefit to French science, Père David spent five months the following year at Jehol (Chengde), north-east of Beijing, where the Qing emperors had their summer palace and hunting park. He returned to the same area in 1865 to continue his exploration of the principal mountains and valleys of the region.

Père David's explorations around Beijing

For the naturalist on expedition, evenings are as busy as the days and during Père David's journeys, each evening was fully taken up with tasks connected with the preservation of the day's finds. Freshly-collected plants are preserved by being pressed and dried between sheets of paper that must then be changed frequently to prevent specimens rotting off. This means that, on returning to camp, botanical collectors must first lay out the day's plant finds between sheets of paper, which are then stacked into piles and pressed flat with weights or installed in special presses, before they begin the delicate and time-consuming task of replacing the wet papers from previous collections with dry sheets, without damaging the fragile drying plant material. Labels and field notes also have to be written up.

In this narrative, the focus is on Père David's plant collections but, although he was a fine botanist, his personal preference was for ornithology and zoology and it should not be forgotten that his collections included large numbers of birds, mammals, reptiles and insects. This meant that, at the end of a day's collecting, as well as taking care of any plant finds, Père David also had to see to his ornithological and zoological specimens. As he commented in his diary:

> One of the most unpleasant tasks of collecting natural history specimens consists in the necessity of doing disgusting taxidermic work immediately, when one needs rest and comfort, especially after a fatiguing day. As for me in particular, as a naturalist, I bear a double burden because I am not free from my religious duties.[25]

Preparing the feathered skins and fur pelts of birds and animals so that they did not spoil required considerable specialist knowledge but, while at the college in Savona, Père David had put together a large collection of natural history exhibits and this had taught him a great deal about taxidermy and the preservation of skins. These practical skills stood him in good stead during his extended collecting expeditions in China.

Collecting ornithological and zoological specimens involved killing the creatures, a necessity which Père David regretted, so he made it a rule never to kill an animal unless it was needed for his natural history collections. He delighted too much in the natural world to abuse it, saying that he found it, 'less distressing to feed [himself] with only rice or millet than to kill for the table one of these poor creatures, who revel in life so joyously and do not harm nature, but on the contrary embellish it.'[26]

However, zoological expeditions meant that his guns were always to hand which probably saved his life when, during one of his excursions into the countryside around Chengde, he suddenly found himself facing eight mounted robbers, all armed and some bearing European weapons. Père David stood his ground and the robbers fled at the sight of his guns. He did not make much of his calm response to such incidents in the narratives that he later published, but his accounts attest to the quiet courage with which he faced the dangers of solitary travel in wild isolated country.

For the most part, though, Père David was able to explore the countryside surrounding Chengde without incident and among the plants he collected was one that is now recognised as a fine ornamental garden plant. He came across it in June 1864, while exploring a mountain slope, when he was attracted by clusters of vivid violet-rose plumes. As he got closer he realised that he was looking at the flower panicles of a splendid astilbe growing on the banks of a stream. The plant that Père David collected that day was initially described as *Astilbe chinensis* var. *davidii*, as if it was a naturally occurring form of *A. chinensis*, but it was not until the professional plant-hunter E. H. Wilson sent seed back to Veitch's nursery in south London in 1901 that living plants were first raised in the West.[27] The astilbe grown by Veitch was considered so superior to *A. chinensis* that Augustine Henry, an Irish plantsman very familiar with the Chinese flora, decided it must be a separate

LEFT

Carex siderosticta 'Variegata' in summer at the Sir Harold Hillier Gardens, Hampshire.

species and, when he described it in the *Gardeners' Chronicle* in 1902, he gave it the name *A. davidii* in honour of its discoverer.[28] It is now known that *A. chinensis* is very variable and, as several different forms are found in the wild, botanists generally include them all under *A. chinensis*.[29] Horticulturalists, though, continue to recognise Père David's plant as a distinct variety because it is clump-forming like other astilbes, whereas *A. chinensis* is very vigorous with a running rootstock.

Another of the herbaceous plants Père David found during one of his forays around Chengde was *Carex siderosticta,* a creeping broadleaf sedge, that is widely distributed in northern and central China and also found in Japan. As well as sending plant specimens to Decaisne at the Paris Muséum, Père David also sent duplicates of many of them to Henry Hance in Guangzhou, who published descriptions of some of them, including *C. siderosticta*. One of the wild forms of the species is variegated and several variegated cultivars are now available, many originating in Japan where this handsome sedge has long been a popular ornamental plant.[30]

While exploring the hills, Père David came across a wild cherry that he had first seen in Beijing, where it was valued as a robust ornamental tree that flowers at the end of the long northern Chinese winter. Adrien Franchet believed it to be a new species of cherry that he called *Prunus davidiana*, but not all botanists agree with this and it has also been described as a peach and an almond.[31] Père David sent seeds to the Jardin des Plantes in 1865 from which eight trees were raised. These had flowered and fruited by 1872 but, in spite of this early introduction, *P. davidiana* is very rare in cultivation. Perhaps the problem is that as *P. davidiana* flowers so early in the year – sometimes even in mid-winter – its small white or pink flowers are frequently damaged by frost and although new flowers soon appear, it has never become a favourite.[32]

Père David also found two elms near Chengde. The first of these is *Ulmus macrocarpa*, the large-fruited elm, originally discovered by P. V. Kirilov and introduced to the Arnold Arboretum in 1908 by Frank Meyer, a Dutch plantsman employed as a plant-hunter in China by the US Department of Agriculture. The second, *U. davidiana*, was described by Meyer as, 'a medium-sized tree with a round, spread-out head; …not a common tree at all. Grows in very dry and exposed localities.'[33]

As far as the botany department at the Muséum was concerned, Decaisne could not have been more pleased: not only had he received fascinating dried specimens but also fertile seeds. When he wrote to thank Père David, he explained why he attached such importance to the investigation of the Chinese flora: 'Chinese plants are of particular interest; as they originate in a climate which has much in common with that of France, almost all these plants can be grown here, either commercially, or for ornament, or simply as botanical examples…'[34]

Many of the plants that Père David collected had already been described by Alexander von Bunge, and one that Decaisne particularly wanted to acquire was *Xanthoceras sorbifolium*, an attractive shrub or small tree that had originally been discovered by Père d'Incarville in the eighteenth century. Père David had realised that fine trees and shrubs were often planted around the graves that dotted the countryside, as well as in the larger gardens, and it was among these ornamental plants that he found *X. sorbifolium*. He managed to acquire living plants and early in 1866, he gave them to M. Pinchon, one of the young secretaries at the French Legation, to take back to the Muséum. In June, he learned from Decaisne that they had arrived safely. The young plants were kept in the orangery at the Muséum for the first two years before being planted outside, where they flowered well, producing seeds in 1873. (A specimen derived from cuttings taken from one of those original plants in 1965 can still be seen in the Jardin des Plantes.) Keeping the precious newcomers in the orangery was an understandable but unnecessary precaution as *X. sorbifolium* is robust and will withstand cold winters.

It was not long before *Xanthoceras sorbifolium* was available commercially and gardeners were quick to appreciate the decorative qualities of this new shrub, which flowers in early spring when its white flowers, with yellow centres that darken to reddish-purple, appear at the same time as the young leaves. Reports in French and British horticultural periodicals were glowing and as *X. sorbifolium* is entirely hardy, thrives in hot dry sites, flourishes in poor soils and does not mind lime or chalk, it was no wonder that it was greeted as 'one of the most important introductions made during the last few years'. [35]

Père David's success
As well-chosen and well-prepared specimens continued to arrive at the Muséum, it became clear to Milne-Edwards and his colleagues that in the missionary-priest they had found the ideal natural history collector. Père David had written that he would like to make further exploratory journeys but, as it was obvious that his opportunities to make scientific collections would always be limited by his missionary duties, the Muséum staff realised that a way had to be found to free him from these obligations. Milne-Edwards decided to enlist the Government's help and he asked Victor Duruy, the Minister for Public Instruction, to approach the Superior General of the Lazarists and make an official request for Père David to be allowed to continue with his natural history studies. The French government was only too pleased to help as it was felt that national pride was at stake: although scientists from different countries co-operated admirably in discussing and analysing the latest finds, they were still eager to ensure that their own countries were at the forefront of exploration and discovery, and the French were very

ABOVE

Xanthoceras sorbifolium in early summer.

RIGHT

Ulmus macrocarpa.

Photo Credit: Roy Lancaster

conscious that current scientific honours in China lay with Britain and were keen to amend the imbalance. Père David himself felt this and was always eager to do as much as he could for France and not leave the Far East solely to English explorers.[36] The Superior General agreed to the Minister's request and permission was given for Père David to devote himself to natural history exploration. The decision was momentous and was to result in some of the most important discoveries ever made by a naturalist.

Horticultural and Research Developments

Syringa villosa

It has been said of **Victor Lemoine** that 'he probably gave more to horticulture than any other individual known' and in his case this is not an exaggeration.[37] Lemoine was a brilliant plantsman who seems to have had an intuitive understanding of plant breeding, and he seized the opportunity presented by the introduction of new Chinese species of lilac and then of *Deutzia* and *Philadelphus* to develop a series of fine ornamental hybrids and cultivars. The first plants that he turned his attention to were the lilacs and, from 1876 onwards, he initiated a series of complex crosses that eventually produced many of the large-flowered ornamental lilacs that we know today.[38] He was ably assisted in this work by his wife Marie-Louise and later by his son Émile (1862–1943) and one of the lilac species they worked with was *Syringa villosa*. As the staff of the Arnold Arboretum had recognised, *S. villosa* was already a fine ornamental plant, but the Lemoines were not satisfied and believed it could be improved. They did not cross it with any other species but, by

carefully selecting only the best plants from each generation of seedlings and then using these to raise new plants, they created some excellent cultivars, which are now classified under the name *Syringa* Villosae Group.

Louis Henry of the Jardin des Plantes was impressed by the early results of the Lemoines' breeding programme and he began crossing *Syringa villosa*, which produces pale pink flowers, with the dark-flowered European lilac *S. josikaa*, in an attempt to obtain deeper colours. By 1899, he had succeeded in raising a group of violet and purple hybrids now grouped under *S.* x *henryi*.[39] This and other hybrids developed from *S. villosa* are still being used by breeders today to develop new garden lilacs.[40]

Astilbes

One of those who saw Père David's new astilbe growing at Veitch's Coombe Wood nursery thought it 'certainly the most remarkable hardy plant lately introduced', and a group of these tall astilbes in full flower is very beautiful.[41] At the time, all known astilbes had white flowers and the German plantsman Georg Arends (1863–1952) was so intrigued to learn of a coloured astilbe that he came over to London specially to acquire the new introduction. He then used Père David's find, along with three other Asian species, to develop the ornamental range of *Astilbe* x *arendsii* hybrids that are so familiar to gardeners today. Indeed, we owe almost three-quarters of all astilbe cultivars to Arends, who derived them all from his original *A. chinensis* var. *davidii* plant.[42] As hybrids developed from *Syringa villosa* showed, the introduction of new plants from China provided Western nurserymen with an opportunity to breed new ornamental plants and, in several cases, the new cultivars and hybrids have largely replaced the original introductions in our gardens.

Père David's *Ulmus* discoveries

The current importance of *Ulmus macrocarpa*, and especially of *U. davidiana* and its two varieties– var. *japonica* (long cultivated as *U. japonica*) and the recently-introduced var. *davidiana*– lies in their resistance to Dutch Elm Disease (DED), the fungus infection that has wiped out almost all European and American elms. Of all the Asian species, *U. davidiana* in all its forms seems to be the most promising for the breeding of new DED-resistant cultivars; and this potential is being studied at the Morton Arboretum in Illinois.[43]

RIGHT

Astilbe x *arendsii* cultivar in midsummer at the Sir Harold Hillier Gardens, Hampshire.

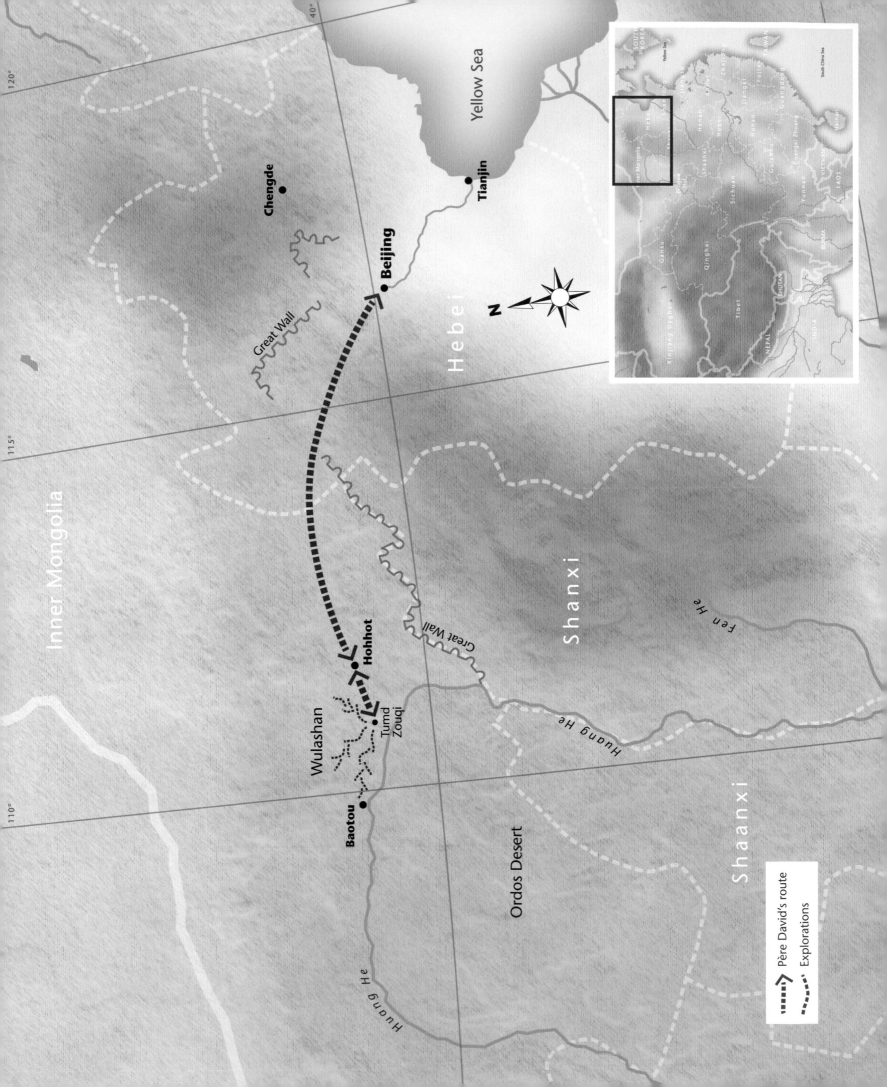

CHAPTER 3

North of the Great Wall

—

I passionately love the beauties of nature; the marvels of the hand of God

transport me with such admiration that in comparison the finest works of man

seem only trivial...

PÈRE ARMAND DAVID, 23 MARCH 1866[1]

—

ONCE INFORMED OF HIS SUPERIOR'S decision, Père David obediently put aside his long-held ambition to work directly towards converting the peoples of the Far East to Christianity and accepted that henceforward his time would be devoted to natural history. Although this change of direction meant that he no longer had a formal evangelical role, he never forgot his religious duties as a priest and, in the midst of all the excitements of new places and new discoveries, he always performed his daily ecclesiastical offices. He viewed his work as a naturalist as a continuing expression of his faith and believed that '...all sciences concerned with the works of creation increase the glory of their Author. They are praiseworthy in themselves and holy in their objective, for to know truth is to know God'.[2] His delight in the beauty and diversity of nature is a constant theme in his diaries and it is apparent that he saw the whole world as 'charged with the grandeur of God'.[3] For him, studying 'the works of creation' was a sure way of drawing nearer their Creator. And although this belief was shared by the other missionary-botanists who feature in this narrative, Père David's official role as a travelling naturalist meant that he was able to devote his time to the study of natural history, which ensured that his experience of life in China was very different from that of his confrères who spent years living on their own in isolated towns and villages, and who had to carve the hours that they devoted to botanical pursuits out of lives spent in the service of their communities.

Père David's deer

Even before Père David embarked on his first journey, he made a remarkable discovery. He had long been intrigued by talk of a strange reindeer-like animal, the *Si bu xiang* or Milou, which lived in the Imperial hunting park outside the city. One day at the beginning of 1865, he persuaded the park guards to let him see over the wall and as soon as he caught sight of the deer-like creatures within, he realised that they were unknown to science. By dint of gifts, bribes and repeated pleading he managed

Photo Credit: La Congrégation de la Mission

LEFT

Père David in travelling dress.

to acquire two individual specimens in January 1866, which he gave to M. Pinchon to take back to the Museum. When Henri Milne-Edwards saw them, he confirmed that the specimens represented a new species, which he called *Elaphurus davidiana* or Père David's deer. Living animals were later sent to Europe, and although all those in the Imperial Park perished during the Boxer rebellion, the species was reintroduced to the central Yangtze in 1985 from the Duke of Bedford's herd at Woburn Abbey. The animals flourished in their new Chinese home and the population has increased to over 2,500.

Père David's first journey of exploration: Inner Mongolia

The scientists at the Muséum d'Histoire Naturelle wanted Père David to explore the Tibetan uplands and the province of Qinghai in north-west China, but a Muslim rebellion in the neighbouring province of Gansu had affected the whole region and made travel in the area impossible. Instead, Père David decided that he would visit the **Ourato region** in Inner Mongolia, which he had been told was covered in forests and full of exciting plants, animals and birds, such as the blue pheasant.

Père David was to be accompanied by **Ouang Thomas**, one of the Chinese Christians at the mission, who would serve him at Mass and help hunt and prepare specimens. By 12 March 1866, they were ready to leave for Inner Mongolia, where Père David intended to find lodgings for the summer in the large village of Saratsi (Tumd Zuoqi) situated on the plain between the southern edge of the Ourato and the great northern bend of the Huang He (Yellow River), which he thought would make a convenient point from which to explore the neighbouring grasslands and low Ourato mountains. His excitement at the prospect of visiting an area that seemed to promise much in the way of natural history discoveries was mixed with sadness at leaving his 'beloved colleagues'. He confided to his diary that, as missionaries had already been irrevocably separated from their closest relatives, their confrères took the place of family, and he found leaving them very hard, especially when he planned to be away for several months in unknown and probably unsafe territory.[4]

The French Government was providing financial support for Père David's scientific expeditions, so he had sufficient funds for a lengthy journey and could afford to buy enough supplies to last for several months. He took little in the way of personal provisions, apart from warm clothing and bedding, as he relied on being able to buy day-to-day requirements from the local inhabitants: 'I believe that with a little good will, one man can live wherever another can. I do not burden myself, therefore, with any food, except a bottle of cognac for emergencies'.[5] He concentrated on acquiring all the equipment he would need to preserve and store his natural history collections, including everything necessary for taxidermy, plus dozens of collecting boxes, glass storage vessels, and reams of paper for plant specimens. The luggage was carried by three pack mules, walking at the steady two-and-a-quarter mile pace of eastern travel, and Père David also rode a mule, although whenever he came to interesting countryside he preferred to walk, so as not to miss anything on the ground.

His first destination was a mission station at Suanhwa (Xuanhua) where he stayed until the beginning of April, exploring the neighbouring countryside and learning as much as he could about the wildlife of the area for which he was bound. On 4 April he crossed the Great Wall and reached Erh-shih-san-hao, a mission station run by the Lazarists, where he met **Frère Louis Chévrier** who was to accompany him. Frère Chévrier was a good shot and would help with the acquisition and preparation of natural history specimens. A few days later, Père David's Chinese guide arrived: an erstwhile lama (Buddhist monk) called **Sambdatchiemda** who had converted to Christianity and accompanied Père Huc and Père Gabet, two Lazarist priests, during their pioneering journey across Tibet some thirty years earlier. Père David and Frère Chévrier were delighted to meet him and to hear first-hand about the famous journey.

On 18 April, with the luggage now loaded into a cart, they set off westwards for the town of Kweisui (Hohhot), their next destination. The first stages were not easy as the road was difficult and ran through country infested with robbers, where they were dependent on finding the scattered dwellings of the nomadic Oirat Mongols for their daily supplies. It was bitterly cold and the countryside was covered in snow and when the cart driver lost his way one night, Père David could hear a leopard calling in the darkness. By 23 April, they had reached Hohhot and the weather had warmed up. They stayed there a week, exploring the vicinity, and met Frère Chévrier's brother, a Lazarist priest who had been visiting the Christian community in the town.

Arrival at Tumd Zuoqi

They left on 1 May, when they saw the Huang He for the first time, and four days later they arrived at Tumd Zuoqi, where they put up at the inn while they looked for suitable lodgings for the next few months. As Père David and Frère Chévrier were the first Europeans to visit the area, they found themselves the centre of much curious attention from the inhabitants and, on 11 May they were visited by the village officials, who were not reassured when they saw the Imperial passport obtained for Père David by the French Minister in Beijing. The officials did not understand his scientific purpose and suspected both missionaries of being Russian spies or financial adventurers; but in the end, they seem to have decided that the strangers were harmless enough and left them to their own devices.

By 7 May, Père David and his companions were able to move

LEFT

A poor family of nomads by their yurt.

into the premises they had hired for the summer: two small isolated cabins covered with earth that faced onto a courtyard, which gave them the space they required. Now they had somewhere to stay, they began to explore the countryside around the village, close to the Huang He. They visited some local coalmines and also a lamasery (Buddhist monastery) situated high up on the steep slopes of the mountain behind the village, from where they could see, stretching away to the south, the wide flood plain of the Huang He and the flat Ordos desert beyond. Up to this point, Père David's collections had been chiefly zoological and included a large number of birds, but the weather was warming up and, as the flora of the region reappeared, he was able to collect several plants, including green-flowered *Aquilegia viridiflora* which had been discovered in Russia in the eighteenth century.

First short expedition

On 28 May, Père David left with Sambdatchiemda to explore the Ourato mountains, which stretched for some 281 km (175 miles) west of Tumd Zuoqi, while Frère Chévrier and Ouang Thomas remained to look after their lodgings and the collections that had been made so far. The flora was now much more interesting and Père David saw iris and fragrant pink *lactiflora* peonies among the acers, birches, small oaks, spiraeas and cotoneasters. The weather was unsettled and one night, as if to prove that the long winter was still not over, there was a snowstorm and the tent itself froze. Another day, there was a tremendous dust storm, which covered everything with gritty sand. There were also wolves around, and so they took the donkey that carried the luggage into the tent to keep it safe for

the night, which caused Père David to note with commendable restraint that 'his presence… is not without inconveniences'.[6]

It was here in the dry foothills of the Ula Shan (Wulashan in the Yin Shan range) that Père David first saw *Xanthoceras sorbifolium* growing in the wild and in June he received a letter from Decaisne informing him that the plants he had sent to Paris at the beginning of the year with Pinchon, the young diplomat, had arrived safely.[7] There were yellow roses everywhere, which Père David recognised as the wild form of a double-flowered yellow rose that he had seen cultivated in Beijing, and he was able to collect ripe hips. This yellow rose was later identified as *Rosa xanthina*: a beautiful robust species with small delicate leaves that produces abundant cup-shaped yellow flowers in early spring.[8] As the name *R. xanthina* had already been applied to the cultivated double form, the single-flowered expression was named forma *normalis*. *R. xanthina* f. *normalis* is widely distributed eastwards from Afghanistan and across northern China, and was originally introduced to Kew in 1880 from seeds collected during a British expedition in Afghanistan. In 1907, the Arnold Arboretum in Boston received seeds collected by Frank Meyer in Shaanxi province.[9]

There were several lamaseries scattered throughout the region and Père David and Sambdatchiemda spent one night in the vicinity of Wu-t'ang-chiao (WuDang), the large Tibetan-style lamasery where Sambdatchiemda had once lived. As they explored the countryside, they met several lamas in their red or yellow robes and with shaven heads, and many Mongol herdsmen, although Père David could see that the nomads were gradually being driven from their traditional grazing lands by Chinese farmers pushing north and settling on the fertile flood

ABOVE

Rosa xanthina. (Willmott, Ellen (1914) *The Genus Rosa* Plate 94)

plain of the Huang He. He was disappointed to learn that the forests of which he had heard so much in Beijing lay even farther west and he realised that, as his provisions were running low after nearly two weeks' travelling, he would have to return to Tumd Zuoqi. A further disappointment was the discovery that the blue pheasant had not been seen in the region for many years.

The pair ran into an armed man one evening whom they took to be one of the robbers known to be in the area, but Père David, whose own appearance with a gun and several days growth of beard was wild enough, treated the stranger just as if he had been another hunter and his boldness frightened off the man and his companions. His courage was further tested as they returned eastwards to Tumd Zuoqi by way of the large town of Paotow (Baotou), as the rebellion in Gansu had now spilled across Inner Mongolia and a detachment of Imperial troops had camped near the town. Père David decided that the best course was to proceed boldly and he was halfway across the town before he was stopped by troops perturbed at the sight of a European appearing from the east where they knew the rebels were based. They insisted that the party accompany them to their General. Père David immediately adopted a confident high-handed tone and after declaring that he was 'a Frenchman and a man of honour', berated the soldiers and sent them off to the General to show his passport, instructing them to return within fifteen minutes. They were obviously impressed by his manner and duly returned within the stipulated time with the General's permission for him to continue. Père David later remarked that, 'often in this country one must have the air of commanding so as not to be crushed.' He goes on to say that such high-handedness was very alien to his nature, but his assumption of a bold confident front was obviously very convincing, and there is no doubt that his refusal to allow himself to be intimidated whatever the circumstances played a large part in the success of his ventures into unexplored territory.[10]

By 10 June Père David was back in Tumd Zuoqi after a fortnight's absence and he felt he could be moderately pleased with what had been achieved so far, even though he had not made the major discoveries for which he had hoped.

Second short expedition

For the rest of June, which was enlivened by continual rumours that the rebels were approaching, Père David explored both banks of the Huang He and prepared for his next trip to the Ourato. This time sufficient supplies for an extended trip of some fifteen to twenty days were packed onto a two-humped Bactrian camel and Frère Chévrier accompanied Père David and Sambdatchiemda. They left on 30 June and travelled west along the sandy plain between the Huang He and the mountains, turning north after a couple of days to explore the wooded valleys of the region. Père David saw conifers, elms, willows, a poplar and an aspen, as well as corylopsis, cornus, cherry and even a hydrangea. Herbaceous plants included *Thalictrum*, yellow *Trollius*, Jacob's ladder and a rose-coloured valerian. Père David does not identify individual species in his journal but he could see that many of the plants were already well known in Europe. Frère Chévrier made the most interesting discovery when he showed Père David some wild silk worms that he had found on a mulberry tree growing amidst almost inaccessible rocks.

The trio continued to climb and eventually found themselves in a series of alpine meadows from which they could see the yellow sands of the Ordos desert stretching away south of the river. Père David was delighted and could almost imagine himself back in the breezy sub-alpine mountains of Europe.[11] The camel, though, found the mountain slopes very difficult to negotiate and the luggage kept slipping so they lost a great deal of time repacking and reloading. Once they turned south to begin their return, they found that much of the forest had already been cut down, and firewood and water became much harder to find. Storms ended the drought but one night the rain was so heavy that the tent flooded and Père David, ever resourceful, had to use his plant trowel to dig a channel to divert the stream away from the collections. He managed to save the plant specimens but his boxes of insects were considerably damaged. This time he was spared any trouble with the soldiers at Baotou, who seemed to have accepted his presence, and on 17 July the little group arrived back at Tumd Zuoqi.

It was now extremely hot, so Père David spent the next few days sorting out his collections and completing his notes. By 24 July he was ready for his next excursion: this time to the area north of the WuDang lamasery. Once again the group found the hills so denuded of trees and shrubs that they had difficulty even finding wood for their fires, and the plains beyond were so assiduously cultivated by Chinese farmers that there was little of interest in the sparse remaining flora.

One of the hardest things he had to cope with on these gruelling collecting trips was the incessant diet of millet, which he found almost indigestible, and so he occasionally broke his own rule and shot game for the pot. He needed to maintain his health and energy for he spent most of the day on foot, often using his butterfly net as a walking stick, as he zigzagged across grassland or clambered among rocks and gullies, searching for plants, lizards and insects. He was also on the watch for new birds, which had to be retrieved from wherever they had fallen once he or Frère Chévrier had shot them. After spending almost a month exploring and collecting, they began the journey back to Tumd Zuoqi. The camel was very heavily laden, with the tent, cooking utensils and provisions, as well as with boxes of zoological specimens, bundles of drying plants and bags of

rock samples, all tied between its humps. Père David recognised that the human members of the party presented an equally odd appearance and he described the scene in his diary:

> Sambdatchiemda, blackened by the summer's sun, leads the docile camel loaded with an enormous amount of baggage… We walk in front at an accelerated pace… our skin has not been spared any more than [Sambdatchiemda's]; it looks no different from that of the desert children; but, with our untidy beards on faces thin from fatigue, and our costumes which are mixtures of Chinese, Mongol, and European, we have a remarkably wild appearance. Frère Chévrier lets me walk at his left, the place of honour in China. He marches along dressed in Chinese fashion, nobly bearing his double bore shotgun and his black game bag of waxed linen. His shoes are old leather ones from the West, in which he has crossed the seventy-two fords on the road from Che-kouen to Saratsi [Tumd Zuoqi] *almost* without getting his feet wet. An old hat of agave pith, souvenir of the Franco-English expedition in China, protects his head fairly well from a sun, which sends the thermometer to 122 °F [50 °C]. My costume, less eccentric, is composed of Chinese and Mongol elements. My hat is a *liang-mao*, cool and shaped like a mushroom, draped with thin cloth but lacking the tuft of red yak-hair. The Chinese hat seems to me healthier and more comfortable than the European one.[12]

Not surprisingly, they attracted some attention when they reached Tumd Zuoqi on 1 August and passed through the village to their lodgings.

They now had a chance to rest while Père David took stock of the collections and wrote up his notes. The herbarium had grown considerably and contained many new species; but Père David lamented the fact that during expeditions, when they were continuously on the move, there was never enough paper to dry the plant specimens properly and they suffered whenever the group was caught out in the rain. On Sunday 5 August, he had time to write letters, which were sent to Hohhot by express the next day but would then take a month to reach Beijing, before being sent on their way to Europe. He made a heartfelt admission in his diary:

> In this distant and somewhat savage country one likes to keep in touch with absent friends; our spirits and hearts are always with them … What a consolation and joy it is to receive letters from relatives, friends and colleagues! They are read and reread … for the moment [we] forget where we are; courage returns to us to continue the life of sacrifice.[13]

9219

L. Snelling del. et lith.

LEFT

Caryopteris mongholica. (Bot.Mag. (1928) No.9219)

autumn.[15] *C. mongholica* is certainly attractive but it prefers warm dry sites and needs protection in cold spells. It was never widely available and is probably no longer cultivated.[14]

Once back in Tumd Zuoqi, Père David considered making a trip into the sandy Ordos desert region but Sambdatchiemda was ill and he could not go without him. Frère Chévrier was also sick and, as the intense heat and drought continued, Père David began to suffer fever, headaches and rheumatic pains. He felt that he had now accomplished all that he could in the region and began preparations for the long trek back to Beijing. The little group left Tumd Zuoqi on 27 August.[15]

Return to Beijing

Père David was now quite ill with a high fever but, although he stopped at Hohhot for three days while transport was arranged, he was anxious about the safety of his collections and so continued straight on to the Lazarist community at Erh-shih-san-hao. He spent twelve days there with his confrères, recovering his health and exploring the neighbourhood. When he reached the Christian community at Xuanhua on 26 September, he decided to stay for a month to take advantage of the flocks of birds migrating south. He arrived back in Beijing on 26 October, after an absence of seven and a half months.

When he came to assess the results of this journey, Père David had to acknowledge that they were not 'brilliant' and that, although he had discovered some new animal and plant species, the high plateaus he had visited were 'desperately poor in every respect'. The results would, he felt, have been very different if he could have fulfilled his original intention and explored Qinghai in north-west China and the Tibetan uplands.[16]

Horticultural Developments

Rosa xanthina

The fine garden rose 'Canary Bird', which produces its highly-scented single yellow flowers very early in the season, was originally believed to be *R. xanthina* f. *normalis* itself, but although its precise origins are unclear, it is now considered to be a hybrid involving either *R. xanthina* f. *normalis* or f. *hugonis*.[17] Naturally enough, given the dry sandy habitat native to wild forms of *R. xanthina*, 'Canary Bird' is an excellent rose for a hot dry site.

Caryopteris mongholica

C. mongholica might no longer be in cultivation but *C.* x *clandonensis,* the *Caryopteris* hybrid with which many gardeners are familiar, owes its very existence to Père David's discovery. Some enthusiasts did grow *C. mongholica* when it was first

The missionaries might have chosen their exile but, even so, loneliness was one of the hardest things the solitary priests in China had to bear, and their memories of all they had left behind were deeply cherished. As one traveller remarked, 'From the different missionaries whom I have met at various times, I have learnt to know the intense yearning which they feel towards their home-ties and associations.'[13]

During one of the short collecting forays Père David made from Tumd Zuoqi in August, he collected seeds and specimens of *Caryopteris mongholica.* This small blue-flowered sub-shrub had first been discovered by Bunge in 1833 and then introduced to cultivation in the West, although it seems to have disappeared fairly quickly. Healthy plants were raised in the Jardin des Plantes from Père David's seed and the nurseryman E. A. Carrière described the new introduction in 1872 as a charming ornamental shrub that flowered from June through to the

introduced from the Jardin des Plantes and it was certainly grown in the early 1930s by Arthur Simmonds, Secretary of the Royal Horticultural Society, in his garden at West Clandon in Surrey. He had received his plant about fourteen years earlier from Charles Musgrave, another prominent member of the Society, who was the only other person Simmonds knew to possess it. In his garden, Simmonds also grew *C. incana*, a Chinese *Caryopteris* that had been introduced many years earlier by Robert Fortune. In the wild, *C. incana* inhabits southern and coastal regions and so would never naturally grow on the same northern hillsides as *C. mongholica*; but when planted as neighbours in Arthur Simmonds' garden, the two species seem to have crossed, producing the spontaneous natural hybrid now known as *C.* x *clandonensis*.[18]

C. x *clandonensis* is extremely attractive and much hardier than either of its parents, with a broader climate tolerance. Several cultivars are now available with flowers in varying shades of blue, and molecular evidence indicates that all are derived from the original plant found in Arthur Simmonds' garden. The cultivars 'Summer Sorbet' and 'Ferndown' are the closest to *C. mongholica* itself.

BELOW

Caryopteris x *clandonensis* 'Ferndown'.

ABOVE

Rosa 'Canary Bird' in the author's garden in spring.

Photo Credit: John Glover

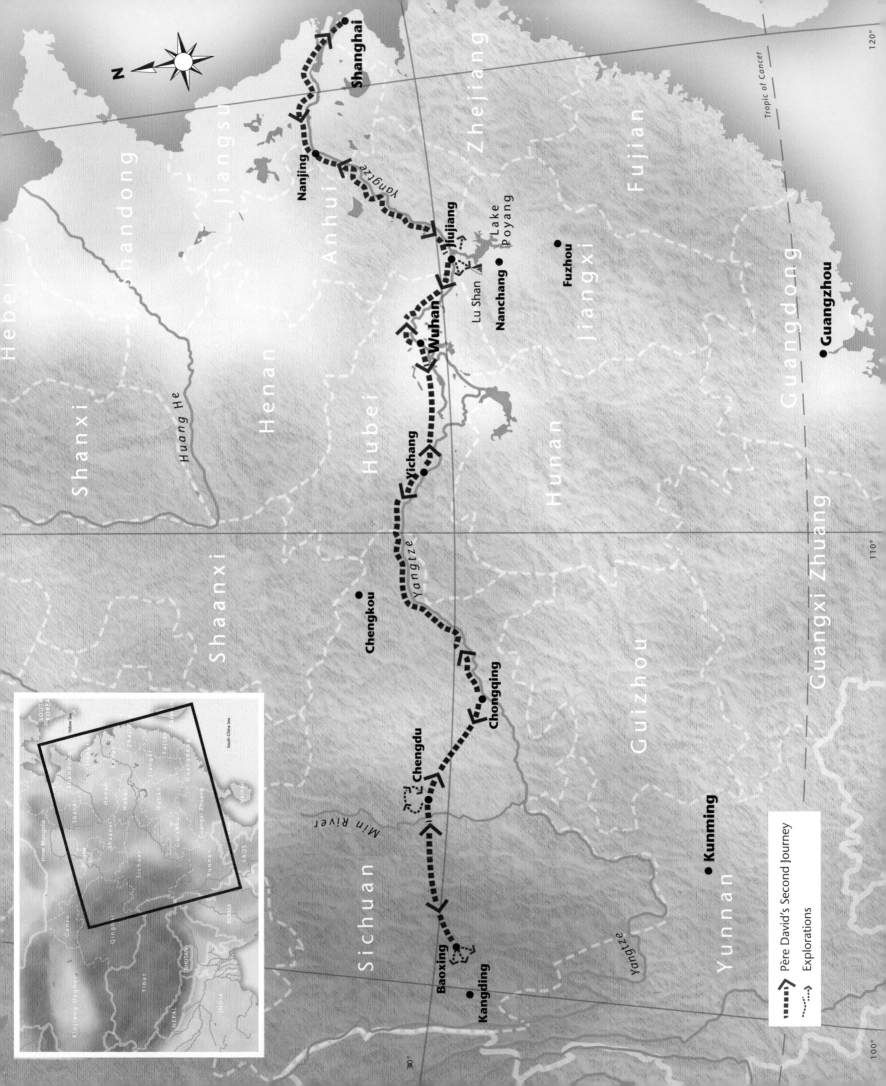

Shanghai

Nanjing

Anhui

Jiangsu

Shandong

Hebei

Shanxi

Shaanxi

Henan

Huang He

Hubei

Yichang

Chengkou

Sichuan

Yangtze

Min River

Chengdu

Chongqing

Baoxing

Kangding

Wuhan

Jiujiang

Lu Shan

Lake
Poyang

Nanchang

Jiangxi

Fuzhou

Zhejiang

Fujian

Hunan

Guizhou

Yunnan

Kunming

Guangxi Zhuang

Guangdong

Guangzhou

Yangtze

Yangtze

Tropic of Cancer

120°

110°

100°

30°

N

Père David's Second Journey

Explorations

CHAPTER 4

Land of the White Bear

—

The range of the mountains is his pasture and he searcheth after

every green thing.

JOB 39 v.8

—

Planning Père David's second journey of exploration

The scientists in Paris were still eager for Père David to explore the Tibetan uplands and the province of Qinghai but when they began to plan his next journey in 1867, access to north-west China from the east was still blocked by the rebellion in the province of Gansu. It was, therefore, decided that Père David should bypass Gansu and the eastern approaches altogether, and try to enter the region from the south. The southern route lay through the mountains west of Sichuan province. To reach it Père David would have to travel up the Yangtze as far as Sichuan and then make his way across the province to **Moupine** (**Baoxing**), one of the semi-independent principalities that lay in the mountains on Sichuan's western border. (These mountains are now part of Sichuan.) Very little was known about the area so no definite plans could be made beyond deciding that he should aim for Baoxing where the Missions Étrangères had a seminary. Père David would have to try to learn more en route and then work out a more precise itinerary. Some of the scientists speculated that as the independent territories lay so close to Tibet and Qinghai, they might themselves prove a rich source of new discoveries, but such hopes were vague and Père David, who had been disappointed by such tales in the past, was not optimistic. As it turned out, the natural history

RIGHT

Camptotheca acuminata fruiting in northern Guizhou.

Photo Credit: Nick Macer

investigations that he carried out in the mountains bordering western Sichuan proved to be some of the most important ever undertaken, and the whole region was revealed to be a far more exciting source of new species, both zoological and botanical, than any of those involved could have imagined: far richer in every respect than Qinghai.

Père David intended the expedition to last about three years and he thought that as his health seemed better since his trip to Mongolia, he would have sufficient stamina to complete the planned itinerary; but the five years he had spent in China had already taken their toll on his health and he no longer had the robust constitution of the boy who had spent days walking in the Pyrenees. He had also contracted a form of pernicious anaemia that affected his memory and sapped his strength – although one would never suspect this from the gruelling schedule recorded in his diary – and at times his weakness and fatigue were such that only his indomitable will enabled him to continue with his explorations.

Père David was again accompanied by Ouang Thomas and the two men left Beijing on the 26 May 1868, heading for the Christian community at Kiukiang (Jiujiang) on the Yangtze in Jiangxi province. At the port of Tianjin, they were offered passage south in a French naval vessel, which took them down the coast to Shanghai at the mouth of the Yangtze. There, Père David met the explorer François Garnier and other members of the French expedition which had recently explored the Mekong river northwards through Burma to Yunnan. Most usefully, he also met a Chinese man who had actually lived in Baoxing and who was able to give him valuable first-hand information about the natural history of the surrounding mountains. On 23 June, he and Ouang Thomas boarded a steamer to take them upriver to Jiujiang, where they arrived three days later. Père David was warmly welcomed by the Bishop and his fellow missionaries, and he expressed his delight at finding himself once again among his confrères and 'back in the family'.[1]

Summer in Jiangxi

As the Yangtze was too dangerous to ascend during the summer, Père David decided to stay until the autumn at Jiujiang. On 8 July he moved into an empty house in open country owned by the seminary, which he thought would be more convenient for collecting expeditions as it was about three miles from the river and closer to the mountains. The house was on the edge of Lake Poyang, the largest of a series of lakes that extend along the southern bank of the Yangtze, making it, in Père David's words, 'a completely aquatic region'. He realised after his very first day of collecting that he would not find many 'novelties' in the region. Nevertheless his time in and around Jiujiang was productive enough and he discovered dozens of new fish and frogs (including a large black one that barked like a dog) and

hundreds of insects, together with a great many birds, small mammals and reptiles. He explored the surrounding territory, and twice climbed Lu Shan, a nearby sacred mountain. As for the flora, he commented that, 'The herbarium I am preparing here contains a goodly number of unusual plants. The ferns are varied, the twining kind abounding everywhere. There are saxifrages, begonias, and primulas. I never found the last in northern China, though there are many species in the south.'[2]

Even though his journey had only just begun, his preliminary observations already provided new insights into various plant distribution patterns within China.

He was not lucky with the weather for the summer proved exceptionally wet and storms and downpours constantly disrupted his plans, while the excessive humidity produced by the heavy rainfall made it very difficult to dry plants or prepare zoological specimens. On dry days, when he was able to explore the countryside, he found the heat exceptionally trying, and he ascribed his lack of strength to the restricted diet of badly cooked gourds and cucumbers on which he was forced to subsist. Nevertheless, in spite of all the difficulties, he was eventually able to send back to Paris some 194 species of plants from northern Jiangxi, including a plant that he had found during his ascent of Lu Shan, which proved to be not only a new species but also a member of a completely new genus. This was *Camptotheca acuminata*, a handsome fast-growing tree with large glossy leaves and creamy-white flowers arranged in a globose head, which are followed by unusual yellow fruit clusters resembling round bunches of tiny bananas. *C. acuminata* is attractive in flower and, since its introduction in 1934 to the US Department of Agriculture Station at Chico, California has been recognised as a fine ornamental shade tree for warm regions. Its name in Chinese is *Xi shu*, which means Happy Tree, but in the West it is also called the Tree of Life or, more prosaically, the False Tupelo.[3]

In September, Père David went back to Shanghai to meet the Procurator-General of his Order who was visiting the Lazarist missions in China. On his return towards the end of the month, he found that heavy rain and flooding had effectively put a stop to any further exploring and he moved back to the mission house in Jiujiang, which was itself surrounded by the rising floodwaters, forcing the missionaries to travel everywhere by boat. Père David was now eager to finish his work on his specimens, but drying anything at all was very difficult in the high humidity, and the situation was not improved when the floodwaters invaded the kitchen and the refectory and he found 'not only frogs and toads but even fish… disporting themselves there'.[4] Nevertheless, by 13 October, he had managed to pack up his collections for despatch to Paris and was ready to join the four young missionaries with whom he was to travel upriver to Sichuan.

Westwards to Sichuan

A steamer took them as far as Hankow (now part of the city of Wuhan), the main city in Hubei province, which had been made a Treaty Port in 1860 and had a small population of European traders and officials. It lies some 996 km (600 miles) from the mouth of the Yangtze and, at the time, was the farthest point upstream navigable by steamers. Père David stayed at the Italian mission, where he met an Italian priest who was also a keen zoologist and able to give him more information about the wildlife of the central region. His thirst for any details at all about the natural history of China helps us appreciate the complete lack of information available at the time; and also the consequent achievement of his three pioneering expeditions, in which he amassed not only comprehensive collections of zoological and botanical specimens but also accurate information about the geography, geology and climate of the regions he visited.

Père David and his companions joined a Chinese boat to continue their journey west to Yichang, using the network of lakes and canals that provided a more direct route, rather than following the looping course of the main river. One night, Père David fell ill with such severe stomach pains that he passed out. After recovering from the attack, he wondered if the episode had been caused by poisoning, as he knew that missionaries had been killed in this way in the past. Such fears illustrate his courage in venturing alone into the Chinese interior, knowing that he would be living among strangers and that, on occasion, he would have to entrust his life to those who were not just indifferent to the missionaries but were deeply hostile, bitterly resenting their presence in the country and seizing any chance to do them a mischief. Missionaries in China knew they were only there on sufferance and that their safety was precarious. Indeed, just a few months later Père Delavay was attacked in Guangdong and another missionary murdered in southern Sichuan.

As they travelled west, the flora began to change, with orange trees and palms appearing. By the time they reached Yichang, Père David thought the landscape had a semi-tropical appearance, although the weather became colder again as their boat was hauled laboriously up through the gorges. He explored along the banks whenever the slow passage upriver to Chongqing gave him time to leave the boat and, during one of these excursions, he collected a remarkable small willow, *Salix variegata*,[5] which, unusually for the genus, flowers late in the season.[5] They put-up as often as they could at the missions run by the Missions Étrangères de Paris, and Père David quizzed all those he met about the best place for natural history exploration on the western borders of Sichuan. Everything he learned indicated that he was right to make for Baoxing, where there was a seminary founded by the Missions Étrangères earlier in the century when the local prince had given them refuge during a period of severe persecution in Sichuan. He learned more from a Chinese priest who had studied at the seminary and, when he reached Chongqing, he stayed with Monsignor Deflèches, the bishop of eastern Sichuan, who confirmed that Baoxing would make a good centre for his natural history investigations and suggested that he base himself at the seminary.

Arrival at Chengdu

To effect this plan, he would need permission from Monsignor Pinchon, the bishop of northern Sichuan, who had his episcopal residence at Chengdu, the provincial capital; so Père David left his companions to continue their journey by river and took the shorter overland route to Chengdu, arriving on 7 January 1869.

BELOW

A porter resting his load on the upright support attached to the traditional porter's back-frame.

Photo Credit: Courtellemont, G. (1904) *Voyage au Yunnan*

Bishop Pinchon knew Baoxing well as he had lived there for several years and was able to give Père David accurate details about the whole area. There had been trouble in the small territory some years earlier when the long-running Muslim rebellion in Yunnan had affected Sichuan, leading to attacks on the seminary and the students. Relative calm had now been restored, allowing the missionaries and the Christian community to resume their former quiet existence, and Bishop Pinchon was happy to allow Père David to stay at the seminary for as long as he wanted.

Père David could do little until his baggage, which was coming by river, arrived; so he took the opportunity to make an excursion beyond Chengdu. He had been warned about thefts and murders committed by bands of brigands, but he commented that, 'if I am to be held back by fears of this kind I can do no exploring, since the wild places, reputed to the be the haunts of thieves and malefactors, are precisely the ones that offer the most in the way of natural history in China'.[6] The short journey passed off without mishap and by 6 February he was back at the episcopal residence and making preparations in earnest for what he hoped would be a year's sojourn at Baoxing.[7] It was at this juncture that he learned that a French missionary and many Christians had been murdered the month before in eastern Sichuan.

Journey westward

Such bad news must have made it doubly hard to leave his confrères and travel alone to a remote location but, nevertheless, Père David left Chengdu as planned on 22 February. Four days later he reached the foothills of the vast mountain ranges that rise steadily westward towards the Qinghai-Tibetan plateau and, as the way wound swiftly upwards, he soon found himself in a jagged land of steep ridges and deep, narrow, thickly-forested river valleys, punctuated by occasional flatter terraces. His baggage was carried by heavily-laden mountain porters who travelled over the high passes at an efficient but necessarily steady pace and, early on 28 February, Père David, who was now keen to reach his final destination, decided to go on ahead of the main group. The track led through dense woods and he noticed many tall tree-like rhododendrons with great thick trunks, as well as several conifers, including one called *thié-sha* or iron fir because of its very hard wood. This was later identified as a hemlock, *Tsuga chinensis* (see chapter 11).

Arrival at Baoxing

Père David reached the seminary early that afternoon and Père Anatole Dugrité, the young priest in charge, installed him in 'a comfortable little room'. He also provided a large workroom where Père David was able to set up his laboratory and store his collecting boxes and other equipment.

The seminary, which was located in a side valley about 8 km (5 miles) north-east of the small village of Baoxing, was situated on Dengchigou Mountain at 2,129 m (6,985 ft) and surrounded by much higher mountains that reached 5,200 m (17,000 ft). Their slopes were wooded and although the forests were foggy and damp, the rich meadows above 3,100 m (10,170 ft) were generally dry and clear. The Buddhist beliefs of the local people and their hostility towards any Chinese settlers had helped preserve the native plants and animals of the area, and during the next eight months Père David took full advantage of the natural history opportunities presented by the region. However, as he had seen in Inner Mongolia, the situation was beginning to change as more and more Chinese farmers moved in and began to cultivate the land, destroying the old-growth forests and denuding the accessible slopes of their vegetation. The valley and slopes around the village, and the hillside around the seminary, were already deforested and increasing numbers of settlers in the narrow valleys would only increase the pressure

BELOW
Primula moupinensis collected by Martyn Rix.

RIGHT
Rhododendron moupinense in early spring.

Photo Credit: Julian Sutton

on the surrounding forests.[8] Fortunately the process was only just beginning, and it did not take Père David long to realise just how rich the region still was in both plant and animal life.

First discoveries

Père David knew that if his zoological collections were to be as comprehensive as possible he would need the help of the local people, who knew where to find the birds and animals of the area. The seminarians were adept at hunting small birds and were able to supply him with several specimens; he also met the local hunters who ranged further afield, and engaged them to bring him whatever birds and mammals they could find, especially the famous white bear of which he had heard a good deal as he journeyed west. Père David immediately set about exploring and discovered that, even though thick snow still covered much of the area, there were botanical finds to be made, such as *Salix moupinensis*, a large-leaved willow with brilliant scarlet leaf buds borne during winter.[9] The damp sheltered woods were home to various primroses and some were already in flower. His finds included *Primula moupinensis*, a small lilac-flowered species that spreads by runners: hence its common name of strawberry primula. Another discovery, *P. ovalifolia*, was described by Reginald Farrer, a plant collector and alpine specialist, as 'a magnificent species suggesting a great violet-blue Polyanthus in general effect'. *P. ovalifolia* had already died out in cultivation when Farrer wrote in 1919; although it has recently been reintroduced from Baoxing, along with *P. davidii*, another of Père David's finds that did not persist in cultivation for very long after its original introduction. Later on he discovered *P. obconica*, a most attractive primrose that was once popular as a decorative plant for cool rooms and glasshouses.[10]

On 11 March, Père David saw his first flowering rhododendron, 'a little shrub two feet high growing on decayed tree trunks and on rocks near streams. The leaves, which are persistent, rather resemble those of tea [i.e. *Camellia*] and the large white flowers are elegantly spotted with rose...' This gem was *Rhododendron moupinense*, usually seen in its pure white form.[11]

Three days later, he discovered *Helleborus thibetanus* flowering profusely on some of the north-facing slopes. This beautiful species is found in only three other Chinese provinces and is exceptionally isolated, with its nearest relatives over 5,000 km (3,000 miles) to the west. *H. thibetanus* was collected again in 1885 in Gansu by the Russian explorer M. M. Berezhovsky, but the Russian botanist Carl Maximowicz, who examined this specimen, did not realise it was the same as Père David's *H. thibetanus* and called it *H. chinensis*. William Purdom, who collected in Gansu in 1909–1911, might have sent seed to Veitch's nursery near London, but there is no record of any successful introduction and *H. thibetanus* remained a botanical curiosity for decades, known only from herbarium specimens.[12]

Narrow escape

Père David was eager to investigate nearby Hongshanding, the peak whose sharp crests dominated the vicinity. Early on 17 March he set out with Ouang Thomas to climb the mountain and, as they had not experienced any difficulties when exploring the hillsides around the seminary, they did not take a guide with them. They soon discovered their error, as the ascent of Hongshanding proved far from straightforward. Even reaching the lower slopes was difficult. The approach valley was blocked by a succession of icy fast-flowing streams and waterfalls, so they needed to find a way forward by climbing the steep valley sides. Père David described the ensuing ordeal in his diary:

> For four whole hours we pull ourselves up from rock to rock as high as we can go by clinging to trees and roots. All that is not vertical is covered with frozen snow. Twenty times our courage fails us continuing this horribly uncomfortable ascent. These immense steep walls are capable of frightening the boldest. Fortunately the trees and shrubs prevent us from seeing too clearly the depths over which we are suspended, sometimes holding only by our hands... after we have reached a certain height it becomes impossible to descend without slipping and falling on the ice... we are badly scratched and our clothes and equipment are soaked. Our strength is exhausted as never before... we see that it is a matter of life and death. The danger is extreme. Sometimes we are plunged into half-melted snow, or the trees which we clutch, break and we roll to another tree or near-by rock.[13]

BELOW

Rhododendron orbiculare var. *orbiculare* in late spring.

Rhododendron lutescens in early spring.

Thick fog fell during the afternoon and they were soon lost. They had had nothing to eat all day but a crust of bread moistened with freezing water and their guns and ammunition were sodden. They were hungry and afraid, but it was too cold for them to risk spending a night out in the open so they were forced to clamber and slide down the precipitous rocky wall they had climbed with such difficulty – later Père David estimated its height at 915 m (3,000 ft) – only to find themselves on the banks of a racing torrent and still some 3 km (2 miles) from the seminary. Night had fallen and they struggled along the 'abominable path' in the darkness for two more hours. When it began to rain, the rising river waters forced them to grope their way into a rocky cavity for shelter. It was at this parlous juncture, when they were 'at risk of death from cold and hunger', that they heard voices in the darkness – almost miraculously, it seemed. They called out and a local man, who heard their cries, took them to his house nearby. He gave them food and let them spend the rest of the night drying out by his fire. The following day, the stricken explorers returned to the college without further mishap. It had been a terrifying episode and Père David knew that he and his companion had come close to losing their lives.[14]

Nevertheless, even in the midst of the day's horrors, Père David did not forget his duties as a naturalist. He collected a few zoological specimens and observed the flora, in particular, 'a remarkable rhododendron with round leaves'. This was *Rhododendron orbiculare*, which does indeed have almost round leaves, rather like those of some eucalyptus, and bell-shaped pink flowers. *R. orbiculare* was introduced in 1904 by the plant-hunter E. H. Wilson and has proved a most ornamental rhododendron, especially as its fine foliage makes it an asset even when not in flower.[15] A few days later, Père David collected yellow-flowered *R. lutescens*, a very early-flowering evergreen species common in the woods around Baoxing.[16] He also found purple *R. polylepis* in March, although this has not proved to be a species with much ornamental value.[17]

The White Bear

On 23 March his hunters brought Père David, 'a young white bear, which they took alive but unfortunately killed so it could be carried more easily... [it] is all white except the legs, ears, and around the eyes, which are deep black.'[18] By the middle of April he had acquired specimens of an adult male and female, and his examination of them convinced him that they represented a new species. He was quite right and today the white bear is called the giant panda (*Ailuropoda melanoleuca*). He made haste to prepare the specimens for the Muséum and said in the accompanying note that the panda was elusive and very rarely seen as it never descended from the inaccessible areas in which it lived.[19] As is now well known, the giant panda is an endangered species and several special panda reserves have been established in the mountains of western Sichuan to protect the species and its habitat. The giant panda was certainly the most

Deutzia glomeruliflora in early summer at Kew Gardens.

Chrysosplenium davidianum in spring at Glendoick Gardens, Perthshire.

spectacular new animal Père David discovered while at Baoxing, but he also found the lesser or red panda (*Ailurus fulgens*), new monkeys including the golden snub-nosed monkey (*Rhinopithecus roxellana*), wild cats and many smaller mammals, as well as several new birds and numerous insects.

The prince of Baoxing was a severe ruler and had issued various edicts against hunting and, as he was opposed to the missionaries' presence in his domain (unlike his parents, who had originally given them refuge), Père Dugrité was afraid that he would hear about the hunting that was being carried out for Père David and use it as an excuse to punish the Christians. The prince's antipathy towards the missionaries was encouraged by the Imperial authorities, and Père David remarked bitterly that they were constantly pursued by the ill will of the Chinese.[20] Père David, aware that his time in Baoxing was too limited to allow even a temporary halt in his collecting operations, pressed on, watching the dormant plants come to life around him as he did so. Their abundance and variety astonished him and by 19 April he had realised 'more and more how rich the

flora of this region is and I foresee that my preoccupation with zoology will not give me time to make a complete herbarium.'[21] His zoological activities might have prevented him from collecting as many plants as he would have liked but, even so, the plant collections he did manage to make revealed a rich and complex flora, which included an extraordinarily high number of new species. His discoveries fascinated botanists and still inspire plant-collecting expeditions; and because so many of the plants he found were extremely ornamental, they are now cultivated in parks and gardens across the temperate world.

Plant discoveries

One of the first of his finds was *Acer davidii*, a medium-sized deciduous tree that has proved a particularly decorative member of the snake bark group of maples. Its striking bark is strongly marked with narrow vertical white and green stripes and this vivid colouring is particularly eye-catching in young trees, especially if they are grown as multi-stemmed specimens.[22]

It was also in April that Père David discovered the evergreen *Viburnum davidii*, a compact shrub with glossy foliage and white flowers followed by beautiful blue berries on red stalks (when male and female plants are grown together). *V. davidii* has proved an extremely useful and adaptable ornamental, as it thrives in sun or shade and can be grown as a medium-sized specimen, or kept short and used as under-planting in shrubberies. It was introduced by E. H. Wilson in 1908 while collecting for the Arnold Arboretum in Boston.[23]

By now, more and more plants were coming into flower. Some of the shrubs, especially the deutzias, were very fine. *Deutzia glomeruliflora*, with clusters of white star-like flowers, has proved exceptionally garden-worthy, as has *D. longifolia*, another deutzia that Père David collected in June, with long pointed leaves and pink-flushed flowers. An evergreen clematis with thin narrow leaves and pure white vanilla-scented flowers was also prominent in sunny spots and Franchet named it *Clematis armandii* in Père David's honour. When it was introduced by E. H. Wilson in 1900, it was quickly recognised as a first-class climbing plant; but *C. armandii's* native range includes the sub-tropical provinces of southern China and it is not fully hardy, requiring warmth and shelter to flourish, although when these are provided, it is outstanding.[24] Père David later found another white-flowered clematis, which has been variously associated with both *C. chrysocoma* and *C. montana*, but is now considered by many botanists to be a separate species called *C. spooneri*.[25]

One of the first woodland plants that Père David discovered was *Chrysosplenium davidianum*, a spreading evergreen saxifrage relative that provides effective ground cover in damp shady places, where its heads of small yellow flowers 'make a diffused effect of sunlight in the darkness.' It was not introduced until 1981, when the Sino–British expedition found it again in western Yunnan, over a hundred years after its original discovery.[26] April also brought *Corydalis flexuosa*, a most attractive spring-flowering perennial with delicate foliage and striking blue flower clusters; but *C. flexuosa*, like *Chrysosplenium davidianum*, also had to wait over a century before being introduced to cultivation.

Later in April, Père David discovered *Pleione bulbocodioides*, a species of ground orchid that pushes up through the leaf litter to produce its showy flowers, which vary in colour from pale pink to deep rosy purple, before the single leaf appears. *P. bulbocodioides* is a very variable species that is widely distributed throughout the uplands of western China, and was subsequently collected by several of the missionary-botanists. Consequently, it was described by several different botanists who each thought it a new species and gave it a new name. *P. bulbocodioides* was actually introduced to cultivation in 1906 by Suttons of Reading under the name *P. yunnanensis*.[27]

As the temperature rose and the surrounding flora came to life, Père David's botanising resulted in growing piles of plants that had to be prepared and dried: but the high rainfall created such damp humid conditions that many plants rotted off before

ABOVE
Rhododendron calophytum in early spring at Westonbirt National Arboretum, Gloucestershire.

BELOW
Rhododendron davidii in late spring at Glendoick Gardens, Perthshire.

they could be properly dried out. It was even harder to prepare his zoological specimens, although he kept a fire lit all day in the workroom, and all too many of his skins were destroyed by the damp or eaten by beetles. His Chinese assistants, in spite of all his training, did not always treat the skins sufficiently and their carelessness also led to losses. There were no cupboards in which he could store specimens and it proved impossible to have chests made as he could not find a carpenter or even the necessary timber. Considering the difficulties, it is remarkable that Père David managed to put together any collections at all: and yet he had already acquired twenty-three new species of mammal, as well as numerous birds and insects, and had made a good start on his herbarium. He confided his frustrations to his diary; but he was much cheered by the arrival of letters from Europe on 15 May, reporting the safe arrival of some of the previous year's despatches, and he resolved to stay to the end of the year to complete the work at Baoxing.

Rhododendrons

At the beginning of May, Père David noted that he had found at least twelve new rhododendrons: this was a surprise as the majority of rhododendron species had hitherto been found in the Himalaya, and it provided one of the first indications of the strong connection between the flora of western China and the Himalayan massif.[28] Several of Père David rhododendron discoveries were highly ornamental, especially *Rhododendron calophytum*, an imposing large-leaved species that has proved extremely hardy in cultivation, although it needs space to flourish. White-flowered *R. decorum*, a vigorous species that is widespread throughout south-west China, is shorter and has established itself as a decorative mainstay of many garden displays; *R. oreodoxa* with white flowers flushed with pink is another useful free-flowering species. Some, such as white-flowered *R. argyrophyllum* and the rare *R. strigillosum* with rich red flow-

ers are much less well known. *R. davidii*, the rhododendron that bears Père David's name, although a very fine medium-sized species, has always been extremely rare in cultivation. It was reintroduced in the 1990s by the late Edward Needham and is now represented in specialist collections, where its ornamental qualities can be properly appreciated. It is to be hoped that it will soon be more widely available. *R. pachytrichum*, another find, is variable and only garden-worthy in its best forms.[29] Most of Père David's rhododendron discoveries were first introduced by E. H. Wilson during expeditions to western Sichuan between 1901 and 1910.[30]

Another of May's discoveries was *Epimedium davidii*, a fine herbaceous perennial with copper-tinted young foliage and long-spurred yellow flowers held on thin stems. Like *Corydalis flexuosa* and *Chrysosplenium davidianum*, *E. davidii* had a long wait before reaching cultivation, as it was not collected again until 1985 when the British botanist Martyn Rix visited Baoxing. As *E. davidii* is very attractive and relatively easy to grow, it is now becoming common in cultivation.[31] By the end of the month, the snow was melting rapidly and Père David was able to reach the higher slopes where he found fritillaries, including one with yellow flowers 'spotted in chequerboard fashion' that he had first collected in April and which Franchet named *Fritillaria davidii*.[32]

Ill-health

Père David's health had been generally good for the last few months, and he no doubt benefited from the varied diet enjoyed at the seminary, which included a daily serving of pork, as well as eggs and occasionally chicken. However, towards the end of May, a worsening intestinal illness made him feel so unwell that he decided he really could not accompany his hunters on an expedition to territory some five days' journey to the north-east. By 11 June, he was too ill to leave his room and he lay listening to the sound of landslides caused by a recent severe storm, and fretting at the loss of time.[33] On 16 June, fine weather tempted him outside and he made a day-long collecting foray to a distant valley. This was a mistake. He was so ill when he returned that he passed out for several hours and woke with a high fever. He tried everything he could think of to alleviate his symptoms, including warm baths in an old animal trough, but nothing seemed to work and by 19 June he was desperate. Then he remembered that he had seen the annual weed *Chenopodium album* or white goosefoot (fat hen in Britain) growing in the garden. He had some picked and boiled in water, which he dressed with a little oil and vinegar and ate as an invalid food.

He felt stronger, perhaps 'because of the improvised remedy or from a fortunate and natural turn for the better'.[34] Whatever the cause, he continued to improve, eating only the goosefoot, and he returned to work on 22 June. Two days later he considered himself cured. Letters from Beijing arrived, which contributed to his recovery.

In spite of feeling so ill for most of the month, Père David had continued to botanise whenever he could and had made a number of discoveries. One of the most striking was *Cypripedium flavum*, a splendid slipper orchid with large ribbed leaves that form a natural 'vase' out of which an erect stem rises bearing a single flower with a large well-formed pouch, which is a bright clear yellow in the best forms, although paler shades are known. It is common in the mountains of west and southwest China, where it is found in cool shady woodland glades, unlike the habitats preferred by its close relative, the North American *C. reginae*, which requires bog-like conditions to thrive. *C. flavum* was collected by Wilson for the Arnold Arboretum where it first flowered in May 1911, and Reginald Farrer

also found it in Gansu. Farrer admired the species enormously, calling it 'Proud Margaret' and describing it – rather improbably – as having an 'especial look of well-fed intelligence'.[35] In spite of these and subsequent introductions, *C. flavum* has not become established in cultivation – replicating the conditions of its native slopes has proved extremely difficult.

On 1 July, even though he did not feel particularly strong after his recent illness, Père David went out collecting for the day and was rewarded by finding 'several unknown plants' and, a week later, he botanised in the midst of 'a very interesting flora'.[36] His perseverance was rewarded with some fine discoveries, including a striking lily with several spotted orange turkscap flowers, which grew on the steep slopes. Pierre Duchartre, Professor of Botany at the Sorbonne, named it *Lilium davidii* in 1877. *L. davidii* resembles a delicate tiger lily and is very attractive. Despite this, it has never been extensively cultivated.

Père David had actually found his first lily a month earlier in June: a turkscap species with white recurved petals strongly edged and spotted with maroon that Franchet named *L. duchartrei*. It was first introduced in 1903 by E. H. Wilson who found it growing on all the higher mountain ranges of western Sichuan, and again in 1920 by Reginald Farrer, who called it the 'Marbled Martagon'.

LEFT

Lilium davidii in June at Barkham, Sichuan, 2,750 m (9,022 ft).

BELOW

Davidia involucrata in early summer.

Discovery of the dove tree

The most famous botanical discovery Père David made at Baoxing was the tree that bears his name: *Davidia involucrata*, the dove or handkerchief tree. It is a deciduous species with leaves like a *Tilia,* but it is related to the tupelos and dogwoods, although distinct from both. The genus, which had to be specially created to accommodate Père David's discovery, contains only this one species. *D. involucrata* comes into its own when it flowers as the small and relatively inconspicuous round flowerheads are framed by two white bracts of unequal size, the largest of which can reach up 15 cm (6 inches), and these create a unique and visually arresting inflorescence. The species made a great impression on E. H. Wilson when he first saw trees in flower in Hubei some thirty years later: 'The flowers and their attendant bracts are pendulous on fairly long stalks, and when stirred by the slightest breeze they resemble huge butterflies or small doves hovering among the trees.'[37]

The fact that the bracts hang down only became apparent much later and, as the illustration accompanying the first description of the dove tree was prepared from dried herbarium specimens, it showed the bracts pointing upwards.[38] *D. involucrata* was only introduced at the end of the nineteenth century, but it quickly became popular and is now one of the most familiar hardy exotic trees (see chapter 10 p.141). Specimens always collect groups of admirers when they flower in early summer in the sheltered parks and gardens with the deep rich soil they require to thrive.

The roses were now in flower, including *Rosa davidii*, a new species with pink flowers and delicate foliage similar to that of *R. xanthina*, and clusters of red flagon-shaped hips. *R. davidii* was introduced by E. H. Wilson who collected it at Baoxing during his 1908 expedition for the Arnold Arboretum, when he also came across a form with elongated fruits that is now

BELOW

Cotoneaster salicifolius at Westonbirt National Arboretum, Gloucestershire.

known as var. *elongata*.[39] Other woody species included *Photinia davidiana*, a large shrub that Père David found in June, which becomes increasingly handsome through the seasons, as its white flowers are followed by striking clusters of scarlet berries and the foliage turns a rich red in autumn.[40] *Cotoneaster salicifolius*, the willow leaf cotoneaster, is another good berrying plant that flowers in June, and, like *R. davidii*, it was also introduced by Wilson from Baoxing in 1908. *C. horizontalis*, easily recognised by the distinctive fishbone pattern of its branches, is another of Père David's cotoneasters that is now very widely cultivated. It was described by Decaisne in 1877 – not from dried specimens but from a living plant in the Jardin des Plantes, which had been raised from seeds collected by Père David in Sichuan, possibly in Baoxing. *C. horizontalis* was introduced to commerce in 1885 by M. Morel, a nurseryman in Lyons, and its mass of pink and white flowers followed by persistent red berries and vivid autumn foliage led to its rapid recognition as a valuable new ornamental plant. It is a vigorous species, thriving in poor dry soils, and has become naturalised in parts of Britain.[41]

In July, Père David found a handsome and unusual evergreen shrub with trifoliate leaves that is now known as *Metapanax davidii*, which has recently been rediscovered, notably by Dan Hinkley, the American plantsman, who collected seeds near Baoxing in 1996. *M. davidii* appears entirely hardy and eminently garden-worthy.[42] *Rubus thibetanus*, a new bramble that Père David had already noticed, was now covered in small pink flowers. It remains attractive all year round as the arching purple-brown stems are covered with a pale 'bloom', which gives them a ghostly appearance indicated by the species' common name of ghost bramble. Plants grown from seeds collected by E. H. Wilson flowered at Veitch's nursery in 1908, and were sold for many years as *R. veitchii* before anyone realised that they were the same as Père David's *R. thibetanus*.[43] A dwarf shrub that Père David had seen growing on some of the tree trunks proved to be a new evergreen species of *Vaccinium*, *V. moupinense*; its bright red flower stalks and purple berries have since proved effective in rock gardens.[44]

A vigorous climbing shrub closely resembling a hydrangea was now coming into flower, and Franchet initially identified it as a form of the Japanese climber *Schizophragma hydrangeoides*: but Daniel Oliver of the Kew Herbarium later recognised that it was an entirely separate species that he called *S. integrifolium*. *S. integrifolium* is very variable but remarkably handsome in its best forms, reaching up to 12 m (39 ft), with deep green ovate leaves that turn yellow before falling. The foliage provides an effective foil for the wide creamy inflorescences, which appear in midsummer and are formed by numerous tear-shaped sepals surrounding lacy flower heads. A mature well-grown individual in full flower is magnificent. Many con-

sider *S. integrifolium* to be the loveliest of temperate climbers, particularly as the fine display continues for several weeks over the summer.[45] Wilson, who first introduced *S. integrifolium* in 1901 for Veitch's nursery, always had an eye for a good plant and when he collected the species for the Arnold Arboretum in 1908, he remarked presciently that *S. integrifolium* had 'an extraordinary potential as an ornamental plant'.[46]

In various places under the trees, Père David saw that thick vertical stalks had emerged and at the top of each one, a single large leaf had unfurled, which sheltered a hooded spathe-tube (a sheath protecting the flower spike). These belonged to a new species of *Arisaema* – a genus related to the arums – which was described in 1881 from Père David's specimen by the German botanist Adolf Engler, who called it *Arisaema franchetianum* in Adrien Franchet's honour.[47] Arisaemas grow from under-

ABOVE

Schizophragma integrifolium in summer at Kew Gardens.

ground tubers and the leaf stalk is made up of false leaves or cataphylls that never open but serve to support the true foliage leaves and the spathe-tube. In the case of *A. franchetianum*, the cataphylls support a single large tri-partite leaf, with an over-sized lower section that is much bigger than the two lateral sections. The shiny bright green leaf is certainly handsome but it is the spathe-tube, with its curved hood, that draws the eye, as both are strongly marked with thin vertical chestnut-brown and white stripes. This curved hood ends in a long filament, and bears such a marked resemblance to a hooded snake that it is easy to understand why *Arisaema* species are sometimes called cobra lilies. *A. franchetianum* is a very striking species, and its impact is certainly increased by the decidedly sinister impression created by the cowled flower sheath.

Hongshanding

On 27 July, Père David set out on his second attempt to climb Hongshanding, but he had learned his lesson and this time took two of his hunters as guides, as well as plenty of provisions. He planned to spend a couple of days on the mountain and sleep in the house of a 'renegade' Christian who lived on the slopes. As he climbed, he noted how the vegetation changed, from conifers like the *thié-sha* at lower levels, to trees and shrubs, including rhododendrons, at higher altitudes and then, above 3,600 m (12,000 ft), he found that woody plants gave way to a rich herbaceous flora and higher still, at around 4,500 m (15,000 ft), the exposed slopes were covered with a medley of alpines, including gentians, saxifrages and aconites. Père David noted that the climb to the upper slopes of Hongshanding, even at midsummer and with the help of guides, was still hazardous:

> whether because of the distance that has to be traversed in one day or because of the steepness and practical impassability of the paths, always wet, muddy and sometimes close to frightening abysses... one has to be accustomed to dangers so as not to tremble and lose ones' equilibrium in some of the crossings, the mere recollection of which makes me shiver.[48]

The upper slopes of Hongshanding gave Père David his first clear view of the towering mountain ranges that rear up westwards towards Tibet and he estimated the heights of some of the peaks that stretched away to the south-west to be between 8–9,000 m (26–29,000 ft). He was one of the first Westerners to visit the Tibetan borderlands and his observations provided important information about the region.

Prominent on these high meadows were the upright flower spikes of various *Pedicularis* and during his time in Baoxing, Père David discovered five new species, including *P. davidii*, with dramatically coloured white and purple flowers. The louseworts, as they are commonly known, are very beautiful hemi-parasitic plants that colonise the roots of neighbouring species and flourish among the dense matrix of plants that form high upland meadows.[49]

Père David spent two or three days exploring the mountain and when he returned to the seminary he brought back several young plants of *Rhododendron orbiculare*, which he hoped to send to France.[50] Once he resumed his exploration of the lower forests, he began to come across large masses of imposing herbaceous plants with handsome leaves and panicles of tiny flowers borne in tall plumes. These were species of *Rodgersia*, which flourished in the damp sheltered glades and along the stream banks. Rodgersias are closely related to astilbes and Franchet originally identified one of Père David's finds as an astilbe, before realising that it was actually a new rodgersia, which he called *R. pinnata*. The specific epithet was suggested by the pinnate structure (in which the leaves are made up of a number of smaller leaf sections that are set either side of a mid-rib) of the large leaves, although this characteristic is very variable and the leaves can often appear to be palmate (that is, with sections radiating from a central point). The flowers vary in colour from cream through to various shades of pink.[51]

At the time, only a single rodgersia, *Rodgersia podophylla*, was known from Japan and Franchet thought that this was the other species Père David had found. However, in 1885, the Russian explorer Grigory Potanin collected specimens of the same rodgersia in eastern Gansu and northern Sichuan, and when

Photo Credit: Harry Jans

ABOVE

Pedicularis davidii in summer in the Wolong valley, Sichuan, 2,590 m (8,497 ft).

these were examined in 1893 by the Russian botanist Aleksandr Batalin, he recognised that they belonged to a new species, which he called *R. aesculifolia*. As with *R. pinnata*, the specific epithet was suggested by the form of the leaves, which in this case bear a strong resemblance to those of horse chestnuts (*Aesculus*). These large palmate leaves with their crinkled surfaces are always handsome, and forms such as 'Irish Bronze' are enhanced by copper or bronze young foliage. The feathery flower plumes are usually white and very tall, and *R. aesculifolia* makes an impressive ornamental plant for those with a wettish site and plenty of room.[52]

Buddleja davidii

In August, Père David was investigating the banks of some of the rushing melt water streams when he was attracted by the racemes of purple flowers belonging to an arching shrub with long pointed leaves, which Franchet later named *Buddleja davidii*.

It is a widespread species found throughout central and southern China and was subsequently collected on numerous occasions by other missionary-botanists, including Père Soulié, who sent seeds to France in 1893 (see chapter 14, p.197).

By the end of August, Père David had still not managed to replace all the skins that had been destroyed earlier in the year by poor preparation and pests. He would have liked to go elsewhere to find new specimens but recognised that it was probably not possible:

> The difficulty, expense, and fatigue of moving are such for me that I cannot think of it unless it is absolutely necessary. Here there are no stage coaches, carriages, or any animals, no hotels, no inns where a stranger can stay whenever he wants to and pay his expenses.[53]

It is interesting that he mentions fatigue – an indication that his illness in June and the months of exhausting physical exertion had taken their toll of his health. He was unwell again, suffer-

ing from severe headaches, fever and some of the other symptoms of typhus. This was very worrying, as he wanted to return to Chengdu in time to catch the Bishop's couriers who left for Wuhan at the beginning of October, as this provided 'the only good chance of the year' to get his precious collections safely on their way to Paris.[54]

Visit to Chengdu and illness

An improvement in his health enabled him to carry out his plan and he arrived in Chengdu on 5 September, having walked the whole way despite his recent illness: as so often, Père David's indomitable will and unshakeable faith helped make up for any physical frailties. The effort required took its toll on his limited strength and two days later he was struck down with swelling and 'almost intolerable pain' in his left leg. The pain was so severe that he remained in bed for twelve days and little could be done for him, although he eventually found some relief from 'poultices made with ginger and onions moistened with

LEFT

Rodgersia aesculifolia in summer at Wakehurst Place, West Sussex.

ABOVE

Buddleja davidii in its native habitat in a river valley at Jiuzhaigou, north of Songpan in north-west Sichuan.

brandy'. The Chinese called his illness 'bone-typhus', and it might have been septic arthritis or osteomyelitis. Père Mihière, one of Père David's old friends from Beijing, who had arrived in Chengdu shortly after he became ill, cared for him and then helped him finish packing up his cases. Père Mihière took the cases back with him to Chongqing, as the Bishop's couriers did not go that year.

Return to Baoxing and preparations for final departure

Père David was well enough to set out for Baoxing by the end of the month, although his leg had not completely healed and he had to be carried for most of the way. He managed to walk for the last two days but agonising pains in both legs returned in early October and he was forced to retire to bed, suffering with severe headaches and kidney pain. The local doctor recommended 'the fat of a leopard' to ease the symptoms. In his diary Père David commented that what he minded most was not the suffering but the loss of time, and he realised that

he would not be able to make any more lengthy expeditions into the neighbouring mountains. By 20 October, he was strong enough to celebrate Mass again and four days later, he made an excursion into the valley to test the strength of his legs. He packed up five more chests of specimens – which he had made himself, as none of the local people possessed the requisite carpentry skills.[55]

Winter was now beginning to set in and his hunters, who could no longer work in the fields, brought in many specimens including those of migrating birds, which he hurried to prepare as he intended to leave later in the month. He also made several further excursions himself, in spite of frequent snowfalls, and spent a final day on Hongshanding. The whole territory was now in an uproar because the local prince, who had refused to obey a Chinese order summoning him to Chengdu, had gone into hiding and instructed the local inhabitants to defend the borders against Imperial troops. This meant that all the villagers were involved in patrolling the mountains and none were available to transport Père David's baggage. This threw all his plans into disarray. Luckily, there was a most opportune delivery of rice from Sichuan and Père David managed to hire the rice-porters for the eight-day return journey to Chengdu. The weather was now bitterly cold and he worried that the rhododendrons and other young plants that he had given them to carry would not survive. He himself left the seminary at Baoxing on 22 November, after a stay of almost ten months, worn out from the 'hardships, difficulties, privations and illnesses' that had plagued him.[56]

Brief excursion to the north

Nevertheless, once back at the Bishop's residence at Chengdu, Père David decided to make one last expedition to the north before leaving Sichuan, and he made plans to travel to Koko Nor (Lake Qinghai) in north-east Qinghai. His confrères were aghast as they could see how weak and in need of rest he had become but he was determined and he left on 26 December. After travelling north-west through the mountains for twelve days, he found himself in a wooded valley that he thought was situated at the eastern end of the lake. In this he must have been mistaken as Lake Qinghai is much farther than twelve days' journey from Chengdu and it seems that he had, in fact, only reached the border with Gansu, a few days north-east of Long an fu. This uncharacteristic mapping error on Père David's part was perhaps because he was seriously ill with cholera, after travelling through territory that had recently been ravaged by the disease. He was unable to explore or make collections himself but the trip was not a complete write-off, as his servants brought him several zoological specimens, and he came across a large new conifer, *Keteleeria davidiana*. Keteleerias are closely related to the silver firs (*Abies*) and do best in hot dry climates, so they are generally better suited to warmer parts of America and to Australia than to Britain. *K. davidiana* was introduced to Kew in 1888 by the Irish plant collector Augustine Henry and to the Arnold Arboretum in 1907 by E. H. Wilson.[57]

Return to France

Père David returned to Chengdu at the end of March 1870 and spent a few weeks resting, preparing his specimens and packing, before setting out on the return journey down the Yangtze to Shanghai, which he reached on 18 June. He was physically exhausted and anxious to reach Beijing, but his ship was delayed in Shanghai and when it reached Chefoo (Yantai) on 25 June, the passengers learned that there had been a massacre at Tientsin (Tianjin) on 21 June and many Christians had been killed. When the ship arrived off Tianjin, Père David heard that all the French Christians, including the nuns of the Little Sisters of Charity (also Lazarists) had been murdered and the cathedral, nunnery, school and orphanage had been demolished and set alight – indeed, the buildings were still on fire. The situation

was judged too dangerous for the ship to dock and she returned to Shanghai.

Père David was broken-hearted and decided to leave for France immediately. There was more bad news when his ship reached Sri Lanka and he learned of the outbreak of the Franco–Prussian war; then, when he landed at Marseilles in September, he heard that the Prussians were besieging Paris. It was a sad homecoming.

Horticultural Developments

Camptotheca acuminata

C. acuminata has long been used in Chinese medicine and it has now been discovered that it produces an alkaloid called camptothecin, which has considerable potential as a cancer treatment. For this reason, it is also sometimes known as the cancer tree. As *C. acuminata* is now endangered in the wild in China, it is as well that it has been so widely planted as a valuable ornamental, and that it is also likely to be cultivated for medicinal purposes.[58]

Primula obconica

P. obconica has now fallen out of favour, as the primin in the hairs on its leaves causes allergic reactions in many people and has led to its being labelled the 'poison primrose'. Breeders have, however, now developed varieties that lack the allergen and perhaps these will help restore its popularity.

Helleborus thibetanus

The Japanese plant collector Mikinori Ogisu was determined to find this plant again and, in October 1989, he showed farmers in Baoxing a herbarium specimen of *H. thibetanus*. They took him to a clump of plants that were half buried under snow with only a few blackened leaves still visible. Ogisu could see that the leaves matched his herbarium specimen and knew that he had found Père David's hellebore. He returned the following year to see the plant in flower and to obtain seeds; and in 1991 seeds collected from plants near Baoxing were sent to the West from the Chengdu Institute of Botany. This introduction led to *H. thibetanus* becoming established in cultivation, and it is now commercially available. Gardeners were quick to recognise its merits as an ornamental plant for early spring when its lovely pink-flushed cup-shaped flowers can be properly appreciated, before its leaves die down and it is submerged by other plants.[59]

LEFT

Keteleeria davidiana at Shennong Sacrificial Templenear Shennognjia Forest Nature Reserve, Hubei.

Acer davidii

A. davidii and cultivars such as 'Serpentine' have become so popular because, in addition to their eye-catching bark which is attractive at every season, they have winged red seeds, which appear in summer and hang in pendulous clusters, and well-veined foliage that turns brilliant red in autumn. Seeds were first collected in 1879 by Charles Maries for the London-based nurseryman Harry Veitch, but the resulting tree did not flower and so was never properly identified. It was only after E. H. Wilson sent seed from Hubei province to Veitch's nursery in 1902 that *A. davidii* was introduced to cultivation. George Forrest, the Scottish plant-hunter who explored north-west Yunnan, also collected a form of *A. davidii* and his introduction now seems to be the one most frequently cultivated. Several recent seed collections of *A. davidii* have been made in different parts of China, and some of the best of the resulting plants have been named and are now widely available. *A.* x *conspicuum*, an excellent hybrid between *A. davidii* and North American *A. pennsylvaticum*, which arose in cultivation, is the parent of some very popular garden varieties including 'Silver Vein' and 'Phoenix'.[60]

Deutzia longifolia

D. longifolia is a beautiful garden plant in its own right and Victor Lemoine was quick to take advantage of its ornamental qualities in his deutzia-breeding programme. He crossed it with the closely-related *D. discolor*, (see chapter 8 p.105) and created the splendid pink-flowered *D.* x *hybrida* group of cultivars, which include such well-known garden plants as 'Mont Rose' and 'Strawberry Fields'.[61]

Corydalis flexuosa

C. flexuosa was rediscovered in 1986 by the nurseryman Reuben Hatch of Washington State, in the Wenchuan Wolong Panda Reserve north of Baoxing, and the form he collected is now known as 'Blue Panda'. In May 1989, the Compton, Darcy and Rix plant expedition explored the valleys of Wolong and Baoxing and found three more forms of *C. flexuosa*, which are now known as 'Père David', 'China Blue' and 'Purple Leaf'. As their flowers are of the various striking shades of 'gentian blue', which most gardeners find wholly irresistible, these *C. flexuosa* cultivars have quickly become ornamental favourites and are now some of the most attractive choices for shady situations.

ABOVE

Deutzia longifolia in early summer.

BELOW

Helleborus thibetanus in early spring.

Photo Credit: John Fielding

The original introductions were micro-propagated and widely distributed, and, as the species hybridises freely and further introductions have since been made from Sichuan, many different forms of *C. flexuosa* are now available.[62]

Lilium davidii

Père Farges, a missionary-botanist of whom we will hear more later in this narrative (see chapter 11, p.147) collected a floriferous form of *L. davidii* in north-east Sichuan in 1892 and he sent seeds to the French nurseryman Maurice de Vilmorin in 1894 or 1895. Vilmorin raised lilies that flowered in 1897 and sent seeds to Kew Gardens, where plants flowered in 1899. This variety of *L. davidii* was later named var. *willmottiae* after Ellen Willmott, a wealthy English plant enthusiast who had received bulbs collected by E. H. Wilson in Hubei in 1908–09 and exhibited the species in 1912. Wilson also sent bulbs to the Arnold Arboretum, and var. *willmottiae* has since been so widely distributed that it has superseded *L. davidii* itself in general cultivation.[63] This tangle of introductions and identifications was far from unusual as botanists and horticulturalists tried to work out relationships between plants they were seeing and growing for the first time.

L. davidii var. *willmottiae* was extensively used by breeders to develop hybrids that were later important in the creation of the Asiatic lilies, which were the first to have erect flowers, thus endearing them to both florists and gardeners.[64] In the 1970s and 1980s, Dr Christopher North used *L. davidii* and the excellent hybrid 'Maxwill' to create the North hybrids, which are

now among the most coveted of all lilies for the garden (see chapter 8 p.119). The way breeders have used Père David's lilies to develop new lily varieties is an example of the continuing role played by several of the missionary-botanists' discoveries in the development of a whole range of fine ornamental plants that today are often much more familiar and widely grown than the original introductions.

Lilium duchartrei

Sir Frederick Stern (1884–1967) grew *L. duchartrei* in his chalk garden at Highdown in Sussex in the 1920s and 1930s where he discovered that it spreads by stolons (underground rooting stems), which means that it tends to reappear some way from the original planting site. This wandering habit is now known to be a marked feature of the species. *L. duchartrei* is not often available commercially and in spite of the best efforts of those who do acquire it, has a depressing tendency to die away after a few years. It is one of the moisture-loving sub-woodland alpine species that are such a feature of the high mountains of western China and that are at once so covetable and yet so hard to establish in cultivation.[65]

Cotoneaster salicifolius

C. salicifolius is a variable species and three or four separate forms are recognised by botanists. At its best, it is a graceful evergreen or semi-evergreen shrub and many good selections have been made over the years, including forms with red, orange and yellow fruits.[66]

Rodgersia pinnata

There are several garden forms of *R. pinnata*, including white flowered 'Alba' and the excellent cultivar 'Superba', which has deep red-pink flowers and bronze-tinted foliage.[67]

Photo Credit: Julian Sutton

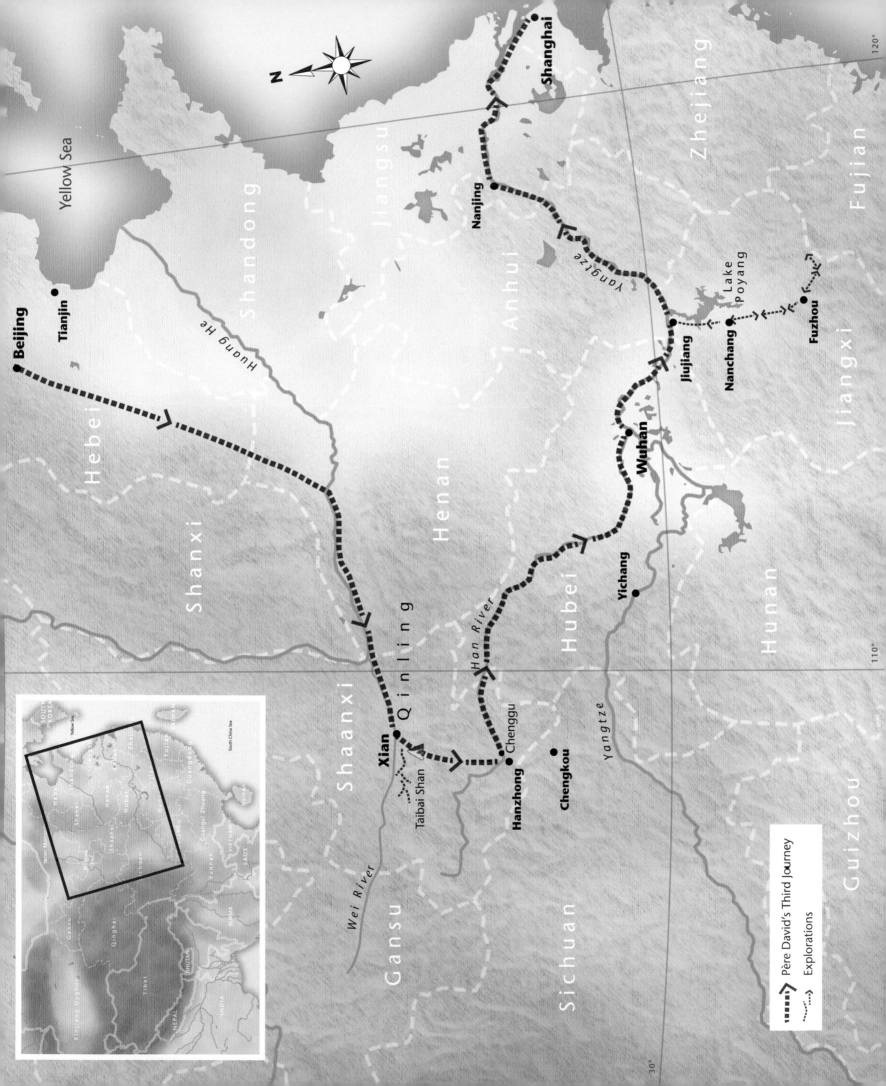

CHAPTER 5

The Qinling

—

[Père David] has ventured into unexplored regions, exposed himself to every danger, endured exhaustion and the vicissitudes of the seasons, and overcome immense difficulties to transport his fragile collections through the wildest country in order to advance our knowledge of creation. How can we fail to be touched by such commitment to the cause of science?

É. BLANCHARD, 1871[1]

—

Convalescence and return to China

The news of the siege of Paris by the Prussians meant that Père David could not return to the Lazarist mother-house as he had intended, so he decided to go instead to Savona in northern Italy where he had taught at the college. He spent several very happy months in the area, staying with his confrères, meeting old pupils and reading voraciously in the well-stocked local library, where he was particularly interested in the scientific discussions that had followed the publication of Charles Darwin's *Origin of Species* in 1859. This pleasant restful interlude did much to restore his health and he returned to Paris in June 1871, having recovered much of his old enthusiasm and energy. He was further delighted to discover that none of his specimens at the Muséum had been damaged during the siege. He spent August at home in Espelette, by which time his health was so much improved that he began to believe that he would soon be able to return to China to continue his natural history explorations. By mid-September he was again in Paris, where his superiors agreed to allow him to go back to China to make a third exploratory journey. He consulted the specialists at the Muséum and made a rapid visit to England – perhaps to meet the naturalists at the British Museum – before leaving France early in the New Year.[2]

Père David arrived in Shanghai in March 1872, where he met the Jesuit zoologist Pierre-Marie Heude and later the German explorer Baron Ferdinand von Richthofen. He also visited Ningbo to see the British ornithologist Robert Swinhoe, whom he had long wanted to meet. Swinhoe had been struck with paralysis down one side but his infirmity did not prevent him showing Père David his own collection of specimens, among which were two new bird species. Père David must have found it enormously rewarding to converse with these men who shared his own deep interest in Chinese zoology, geography and ornithology, and who also had first-hand knowledge of their subjects. When his confrères in the district learned of his presence, they pressed him to visit them and as the countryside around Ningbo was so beautiful, he decided on the spur of the moment to prolong his trip and spend a couple of weeks investigating the area. He had not brought his collecting equipment with him and rain prevented him making as much use of the time as he had hoped, but he accomplished enough to be pleased that he had made the effort to investigate the area. Robert Fortune had botanised in the same region very profitably twenty-five years earlier and Père David saw some of the roses and azaleas that Fortune had found, as well as a great deal of wisteria, which had been introduced to Britain in 1816. During his various journeys, Père David found wild forms of several plants that had long been grown in Chinese gardens and were already cultivated in Europe, including *Begonia grandis* subsp. *evansiana*, *Jasminum nudiflorum*, *Trachelospermum jasminoides* and *Kerria japonica*.[3]

Third journey of exploration to the Qinling

Père David returned to Shanghai and embarked for Tianjin, where he arrived on 31 May 1872. His mind was full of memories of his abortive visit in June 1870, two days after the massacre, when the torched buildings had still been burning. Once back in Beijing, he began to prepare for his next journey, which he intended to last two-to-three years. The continuing rebellion in Gansu still prevented travel in the north-west and so he planned to explore the forests of the Tsinling (Qinling) Mountains in the south of Shaanxi province in central China. The city of Xi'an, the old Imperial capital, is located in southern Shaanxi. In previous centuries the province had been at the heart of Chinese life but it was no longer of much political or economic importance.

Père David set out on 2 October with two young Beijing Christians and they travelled south-west through Hebei province towards the Huang He. Their luggage and equipment was carried in two carts and the journey, which took three weeks, was slow and difficult, as the roads were atrocious and they were hampered by storms and flooding. The weather improved as they approached the river and Père David noted that the flora and fauna began to assume a more southerly appearance.

They reached the Huang He on 24 October and the carts and mules were loaded onto the large ferry, but the boatmen, who Père David thought from the outset looked a ruffianly lot, anchored half way across and refused to continue unless they were paid more money. Père David and his assistants always carried their guns in full sight to deter brigands, but the boatmen knew that a French priest would not shoot them and had ignored the guns, assuming that they were safe from any retribution. They had picked the wrong man for their attempt at extortion as Père David, in spite of his gentleness, did not give in to bullying. He looked around for some way to coerce the boatmen and the sight of their 'marmite' or cooking vessel simmering on the brazier gave him an idea. He knew that if anything were to happen to it, the boatmen would go hungry: and so he aimed his gun at the pot and threatened to shoot it full of holes. This clever threat – which Père David later described as '*marmiticide*' – hit home, as the boatmen had no fancy for raw rice, and they grudgingly took the ferry to the opposite bank. Once ashore, Père David, who did not usually involve officials in his difficulties, was so incensed by the boatmen's racket that he informed the local authorities and they ensured that all the money the boatmen had prised out of the other unfortunate passengers was returned.[4]

Arrival at Xi'an in Shaanxi

Père David and his party turned due west along the southern bank of the Huang He, until they reached the border with Shaanxi, where the route continued alongside the Wei River.

Everywhere they passed signs of the destruction caused by the Muslim rebels, and the few surviving inhabitants were ragged and very poor. On their left, the Qinling Mountains stretched away to the south-west. They reached Xi'an on 3 November, where they stayed with the Italian Franciscans who oversaw missionary activities in southern Shaanxi. At the episcopal residence, Père David was amused to meet a French priest who had spoken nothing but Chinese for so long that he kept forgetting words in French.

The Qinling

With the help of the Franciscans, Père David hired a cart and three mules and was able to set out on 11 November, although the roads were in such a poor state that his party made slow progress. They travelled south-west and their destination was the small village of Inkiapo at about 1,400 m (4,593 ft) in the Lao-yu valley, where there was a Christian community. Here, Père David established a base from which to explore the northern slopes and valleys of the Qinling Mountains. The Qinling is one of the major mountain ranges in China and runs east–west, forming the southern border of the Wei River valley. Elevations

BELOW

Pinus armandii in summer at Westonbirt National Arboretum, Gloucestershire.

range from 1,300 m (4,265 ft) to over 3,700 m 912,340 ft), and the high peaks form a natural barrier between north and south China, with the northern slopes subject to harsh cold winters and the broad southern flanks enjoying a much warmer wetter climate. It was quite the wrong time of year for botany, as the north-facing valleys and slopes were already gripped by frosts and there were regular snowfalls, so Père David concentrated on his zoological collections. These grew apace, as local hunters brought him whatever they caught and he found that, as at Baoxing, once the men were released by the winter weather from work on the land they were happy to devote their time to hunting for him. He made excursions to nearby Christian villages and whenever he returned to Inkiapo he was kept busy preparing all the mammal and bird specimens he had acquired. He was also very pleased to acquire several new species of salamander (lizard-like amphibians).

As he explored the region, he saw that while large numbers of people had been killed during the recent rebel incursions, migrants from Sichuan had begun moving into the depopulated valleys. Areas that had long been inhabited were already deforested, but the flood of settlers wanted new houses and trees were now being felled on the higher slopes to meet the growing need for timber. Everywhere Père David went, he met long columns of porters carrying large planks and blocks of pinewood down from the mountains. He was considerably upset by the destruction:

> It is melancholy to see how rapidly these old-growth forests are being destroyed. In the whole of China only vestiges remain and they will never be replaced. With the great trees also disappear a multitude of shrubs and other plants, which can only flourish in their shade, and also all the animals, great and small, which need the forests to live and reproduce… and sadly, what the Chinese are doing in their own country, others are doing elsewhere.[5]

The wanton plundering of the natural world is a theme that occurs repeatedly throughout Père David's diaries, and he fervently hoped that one day men would stop abusing their environment and learn to live in harmony with their surroundings; although he was well aware that by the time this happened, many species would already have gone for ever. His belief in the sanctity of God's creation made him a conservationist *avant la lettre*.

Most of the trees that were being felled in the Qinling were conifers that Père David already knew, but one of the species was new to him. This was a tall straight pine that he referred to as *Pinus quinquefolia* in his diary because its needles were clustered in groups of three or, more commonly, five; but Franchet decided to call it *Pinus armandii* in his honour and this is the name by which it is known today.[6] *P. armandii* is found across central China from south-east Tibet to Hubei and, like many species with an extensive natural range, it is very variable. At its best, it makes an attractive ornamental tree with wide spreading branches, but not all the forms that have been collected over the years have proved garden-worthy. *P. armandii* was introduced to France in 1895 from north-east Sichuan, and the Irish plantsman Augustine Henry sent seed from the south of Yunnan province to Kew in 1897.[7] Père David also collected another conifer, which was later named *Abies chensiensis*. It was introduced from the Qinling in 1907 by E. H. Wilson and from Yunnan by Forrest, but it is still rare in cultivation.[8]

In the leafless woods, Père David had recognised walnuts, chestnuts, hazels, birches and oaks as well as rhododendrons and *Berberis*; and he could see that the flora was a mixture of northern and southern species. From what he could observe in the middle of winter, he thought that there might be much to interest a botanist at a more favourable season. As he explored the Lao-yu valley before Christmas, he noticed the interlaced stems of a wild vine growing all over the rocky outcrops and he thought they must belong to a new species as they were covered with stiff prickle-like bristles. He speculated that the plant might belong to a new genus, for which he proposed the name *Spinovitis*, but the species has since proved to belong to the same genus as all vines and is known as *Vitis davidii*. In March the following year, he came across another prickly-stemmed vine, and he managed to collect some shrivelled grapes from both species to send back to France. He was later given some of the local wine made from these grapes, which he thought pleasant enough but low in alcohol.[9]

On 23 January, Père David moved his base farther west to the village of Yen-kia-tsoun and continued to hunt and prepare specimens as before; but now he had occasional glimpses of the distant mountains that lay to the north-west in Gansu, and even though Bishop Chiais had confirmed that the borders were closely guarded, his old ambitions were rekindled and he determined to make an effort to reach the province. He left on 11 February with

RIGHT

Vitis davidii as *Spinovitis davidii*.

Photo Credit: *Revue Hort.* (1885) Fig.10, p.55

minimum equipment but, even so, he still needed porters to carry his luggage. The farthest Christian village with which he had contacts lay on a north-west spur of the Qinling. When he reached it, he could see that the mountains lay farther away than the two-to-three days he had been told, which meant that the expedition would have to last longer than he had envisaged. This was a set-back; but he then found it quite impossible to hire any porters, as everyone knew that the Gansu border and all the usual approach roads were closely guarded. As no one knew of any other routes, Père David reluctantly had to give up the idea and return to Yen-kia-tsoun. He had now done all he could north of the Qinling, and he prepared to cross the mountains and explore the southern flanks of the range. He knew that west of Mt. Taibai, the highest peak in the range at 3,767 m (12,369 ft), there was a pass leading south and he resolved to follow it. The local priest persuaded the villagers to act as his porters and Père David and his party set out on 18 February.

South of the Qinling

The party crossed the range at about 1,900 m (6,234 ft) and, once he began to descend the southern slopes, Père David saw a change in the vegetation as palms and other warm-climate plants became common. On 26 February, they met a Christian who was able to direct them to the Christian village of Ouang kia Ouan, about 12.9 km (8 miles) north-east of Hanzhong. To Père David's disappointment Padre Vidi, the Italian priest in charge of the area, was away and so he had no one he could ask for reliable information about the area. He had also been looking forward to having someone to talk to, and it is very apparent from his diaries just how lonely Père David found his solitary expeditions. He was a sociable man and he missed his friends and the conversation of fellow Europeans. He also found it extremely wearing having to contend with what he saw as the deceitfulness and trickery of so many of the Chinese people he met; sometimes even his servants tried to dupe him. It was, he said, 'hard to live as if constantly in the midst of enemies'.[10] The comments he makes about the Chinese people during the course of his third journey are much less charitable than formerly, and it seems that the horrors perpetrated at Tianjin in 1870 coloured his opinions to such an extent that he judged the motives and actions of all the Chinese people he met afterwards much more harshly.

The mountain crossing, coming as it did after months of hard work and poor food, made Père David very tired and this, together with the unexpected absence of Père Vidi, left him in such low spirits that he was tempted to abandon the rest of his planned explorations and hurry straight back to his confrères at Jiujiang, where he had stayed in 1868. However, once established at Ouang kia Ouan he began to feel more cheerful and resumed his natural history investigations. The weather

had also improved considerably as Père David had left the bitter winter behind when he crossed to the southern flanks of the Qinling and had exchanged the cool temperate weather of the northern slopes for a much warmer climate. He followed his usual procedure of making short excursions to neighbouring districts, while involving the local hunters in his search for birds and animals. During March, he collected specimens of a dense holly that he had seen growing throughout the range. The foliage was distinctive as the small sharply-pointed leaves had two or three pairs of strong spines along each edge, and were crowded along the branches. Père David collected some of the foliage but, as neither flowers nor fruit are present in March, his specimen was not sufficiently complete on its own for Adrien Franchet to describe; but in 1883 when Franchet examined a collection made by Père Perny during the early 1850s in Guizhou province, he recognised specimens of the same holly that Père David had found.

Père Paul-Hubert Perny of the Missions Étrangères de Paris had arrived in China in 1845 before concessions allowing missionaries to live openly in the countryside were properly enforced. In order to reach his post in Guizhou province he had travelled through Jiangxi province disguised as a beggar. His determination was recognised when he was made Bishop of Guizhou in 1850. He had an interest in natural history and, in 1850, sent 500 living cocoons of the Chinese oak silk moth to the natural history museum in Lyon. They hatched successfully and the moth was named *Antheraea perny* after their introducer. When Père Perny returned to France in 1857, he took with him a natural history collection, some living plants and several thousand botanical specimens, carefully labelled and annotated, which he presented to the Muséum. He went back to China in 1859 but had to wait a year in Guangzhou before being allowed to go on to Guizhou. He was transferred to Sichuan in 1862, where he remained until 1868, when he left

Photo Credit: Fonds Iconographique des Missions Étrangères de Paris

LEFT
Père Paul Perny in 1859.

China for good. On his return to France, Père Perny published a two-volume French–Latin–Chinese dictionary with an appendix on natural history. He left the Missions Étrangères in 1872 and settled near Paris.

Once back at home, Père Perny visited the Muséum and was very disappointed to find that his large herbarium collection had been neglected, and that many specimens had been lost and most of the labels thrown away. One that remained intact, though, was a specimen of the same holly that Père David had found; moreover, Père Perny had included both flowers and fruit so Franchet could provide a full description of the new holly, which he called *Ilex pernyi*. The bright spiny leaves are persistent, the red berries cluster tightly along the branches and the holly assumes a narrow pyramidal shape, and although it can reach over 6 m (20 ft) in the wild, it is usually much shorter in cultivation. *I. pernyi* was collected by E. H. Wilson in 1900 for Veitch's nursery, and later for the Arnold Arboretum.[11]

To Père David's delight, Padre Vidi returned on 23 March and was able to give him a great deal of useful advice and information about the area. Padre Vidi must have been equally delighted to have a European visitor as he lived the isolated life shared by most of the Catholic missionaries who served in China and whose only contact with fellow Europeans were annual meetings with some of their confrères. Père David's loneliness after a few months on his own was nothing compared to the solitariness of priests like Padre Vidi who spent their lives in remote Chinese villages, surrounded by those with whom they had nothing in common but the faith they had brought them.

It had warmed up considerably by the beginning of April, with temperatures averaging around 18°–23 °C (64.4°–73.4 °F), and although Père David thought that the flora was relatively slow to come alive, some of the plants were already in flower, including a spreading shrub with blue and white pea-like flowers and delicate foliage, which was later named *Sophora davidii*. It is widely distributed across southern and central China and flourishes on poor dry soils. Augustine Henry sent seed to Kew in 1898 and *S. davidii* first flowered there in 1902.[12]

Père David wanted to make one more excursion before leaving the area but his health would not permit him to travel, so he had to send his assistants instead. They let him down and returned early, and he commented bitterly in his diary that the Chinese had no understanding of the importance of his natural history work.

Shipwreck

Père David was now ready to leave and Padre Vidi found porters and four horses for him so that he was able to set out on 14 April for Chenggu on the upper Han River. Here, he took passage in a local cargo boat, which left on 17 April for Wuhan on the Yangtze; but disaster struck five days later when the boat hit a rock and foundered in the rapids. Père David and his assistants managed to haul most of his baggage onto the sandy shore before she broke up, but the chest of salamanders was swept away and he knew that the soaking everything had received would have ruined most of the specimens in the surviving cases. As these were nailed up, he could not check immediately and could only trust that the damage was not irreparable. His assistants went off to hunt downstream for any of their effects that might have been washed up on the bank, while Père David remained, drenched, bruised and barefoot, guarding their heap of luggage and trying to dry out as much of their soaked bedding and clothing as possible. It was 3 p.m. before

BELOW

Ilex pernyi in spring.

BELOW

Sophora davidii in spring at Deqin, Yunnan, 3,200 m (10,498 ft).

ABOVE

Boats on the upper Han River.

ABOVE

A river freight boat on the upper Han River.

they managed to find another boat to take them on board. They transhipped again on 30 April, this time into a much cleaner vessel, and continued their slow journey down the Han River. They eventually arrived at Wuhan on 7 May, where Père David, heartily sick of Chinese junks by then, was happy to see the big European steamers moored at the jetties.

He was welcomed at the same mission station where he had stayed on his way upriver in 1868. He was delighted to break what he called his 'eight month silence' by talking to the resident Italians and visiting Dutch, Belgian and French missionaries, and catching up with the news of France and the rest of China. He met the explorer François Garnier again and examined with great interest his book on the expedition up the Mekong, which the disruption caused by the Franco–Prussian War had prevented him seeing during his last visit to France. Père David also learned from his confrères that he himself was famous, as Professor Émile Blanchard of the Jardin des Plantes had written a series of articles in the *Revue de Deux Mondes* recounting his journey to Baoxing and the discoveries he had made there, which had been reprinted in Shanghai. There had also been similar articles in the British papers. The immediate consequence of his newfound fame was that the officers of the English steamer he boarded to take him down to Jiujiang gave him free passage and the best cabin on board – an agreeable luxury for one who had spent weeks travelling in the local boats.

Expedition to north-east Jiangxi

On 11 May 1873, Père David was back with his confrères at Jiujiang. His joy at being reunited with old friends was somewhat tempered by the shocking state in which he found his cases. Most of the ornithological specimens were ruined, although he managed to save some, and the fruits of many months of work

had been lost. Dispite this disappointment, he felt so much better now that he was back among Europeans that he decided to carry out further explorations in Jiangxi, even though he knew, from the summer he had spent in the area in 1868, just how bad the climate was likely to be for his health. He left Jiujiang on 22 May but did not feel strong enough to walk and so was carried in the traditional sedan chair used by Chinese officials. He made first for Nanchang and then continued south-east to Fuzhou, where the priest told him about the constant persecutions suffered by the local Christian community. He intended to spend the summer at the Lazarist seminary at Tsitou but his journey there was disrupted when one of the chairmen fled after stealing some of his money, which meant that in spite of his weakness he had to walk the last part of the way.

Père David arrived at Tsitou on 30 June 1873 and was met by Père Rouget, head of the seminary and its thirty students. He went about the familiar task of setting up his base and engaging the local hunters. Their help would be more important than usual, as he knew that he was not strong enough to undertake much fieldwork himself. As he travelled through the countryside he concluded that there were not many floral novelties to be found, and indeed his sojourn in Jiangxi did not produce much in the way of new plant species.[13] It was not long before his servants fell ill with a fever, and by the end of the month he was ill himself with headaches and fever. His discomfort was made worse by the oppressive heat for the temperature never fell below 30 °C (86 °F), even at night, and the land was parched with drought. His symptoms reminded him of the illness he had suffered at Baoxing and by the middle of August he was in a bad way, racked with fever, unable to face food and weak from lack of sleep. Throughout it all, he continued to prepare the specimens he was brought, but by 21 August he was almost delirious and could no longer work. He was forced to take to his bed. After a fortnight or so, he began work again but the fever quickly returned. It was apparent that he and his servants had contracted the marsh fever that was endemic in the area, which they were unlikely to shake off while they remained at the seminary. They would have to leave but, rather than return to Jiujiang, Père David decided to explore the mountains along the Jiangxi–Fujian border. Given the parlous state of his health, it was not a wise decision.

He and his servants left on 30 September, and spent four days at a small hamlet on the crest of the range separating the two provinces. By the time they arrived at the Christian village of Koaten, Père David was seriously ill, feverish, fainting and so exhausted after walking over mountainous terrain for two days that he had had to be towed with a rope through his belt for the last six hours of the march. A Spanish Dominican priest was stationed at Koaten and he did what he could for his guest, but Père David refused to rest. As soon as he was a little stronger, he insisted on carrying on as he usually did in any new location, instructing the local hunters, preparing specimens, and writing up his journal. His condition continued to worsen and he developed a bronchial inflammation and a persistent painful cough. By 9 November, he was so close to death that he received the Last Rites. He recovered very slowly, eventually managing a trip outside supported by his Spanish friend. Remarkably, given the state of health, he suceeded in collecting some plants, including a new gentian, which Franchet named *Gentiana davidii*. It impressed Emil Bretschneider when he saw it, but *G. davidii* is no longer cultivated.[14]

It was now apparent, even to one as dedicated to the pursuit of natural history studies as Père David, that the climate was disastrous for his health and that he would have to leave. He returned to Tsitou in December and left for Jiujiang in the New Year, arriving on 1 February. When he looked back on the last few months, he was pleased with the collections he had made, although he much regretted the long illness that had prevented him from exploring the areas where he had stayed himself. He said that it had been the torment of Tantalus to see the pristine countryside stretching before him, knowing it was full of wonders and yet being wholly unable to explore it. When he reached Shanghai, the French doctor examined him and then informed the Lazarist superior that Père David would not survive if he stayed in China. Regretfully, his superior began to make arrangements for Père David's departure and Père David had to accept that his great natural history explorations were over. He sailed from China on 3 April 1874, knowing that his fragile health would not permit him to return.[15]

Once back in France, Père David's health gradually improved, especially after a visit to Algeria. When he returned to the Lazarist mother-house in Paris, he worked on his ornithological specimens with colleagues at the Muséum and, in 1877, pub-

lished a detailed and beautifully-illustrated work on Chinese birds.[16] He made short visits to Tunisia in 1881 and Turkey in 1883 but otherwise lived quietly in Paris, a source of inspiration and advice for all those travelling to China. He died on 10 November 1900 aged seventy-four.

Horticultural Developments

Ilex pernyi

It was soon apparent that Perny's holly makes a distinctive ornamental plant and breeders made great use of it to develop new garden-worthy hollies, crossing it with *I. cornuta*, and also with *I. aquifolium*, to produce the *I.* x *aquipernyi* hybrids. The resulting progeny have since been used in various combinations to produce the complex series of attractive Perny holly hybrids, many of which are very suitable for small gardens.[17]

New vines

News of Père David's discoveries interested the vine specialist Frédéric Romanet de Caillaud of Périgneux, near Lyon, who wrote to Bishop Chiais at Xi'an asking for more seeds. He received these in 1880 and distributed them to various botanic gardens in France. Kew received material in 1885. Both species germinated quickly and the young vines proved vigorous, soon reaching a good size and producing fruits. In 1883, Romanet de Caillaud was able to describe the sweet black grapes produced by *Vitis davidii*, and the way its shining heart-shaped foliage turned a brilliant red in autumn. Three wild varieties are now known including var. *cyanocarpa*, which has fewer prickles and blueish fruits and was collected by Wilson for the Arnold Arboretum in 1907. Wilson noted that it was common in the mountains of western China but, although he saw it cultivated near Jiujiang, he thought the grapes had a 'harsh' flavour.[18] The second vine was named *Vitis romanetii*. It was described by the great plantsman W. J. Bean as, 'the finest of the true vines' but is not much cultivated today.[19]

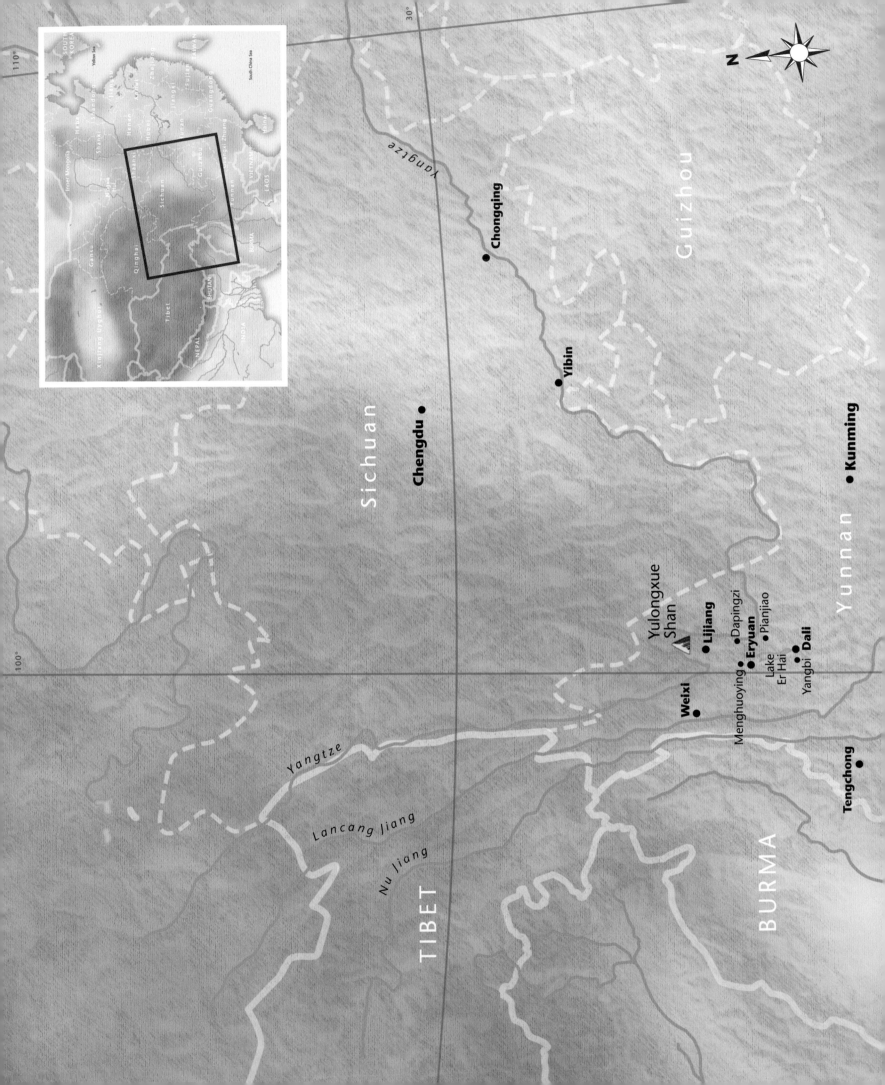

CHAPTER 6

A Botanical Eden

—

Each ravine has some peculiar plant, and this is the feature of Western China,

the astonishing richness of the flora; each new valley and range yields some

new species?

AUGUSTINE HENRY[1]

—

A S WE HAVE SEEN, THE ATTACK ON **Père Jean Marie Dela-vay** at Leizhou in December 1868 occurred shortly after his arrival in China, and although he was badly hurt and only reached the safety of the bishop's residence in Guangzhou the following May, his convalescence there was of short duration and just a few weeks later he was ready to resume his missionary duties. He was ordered to the small island of Weizhou 290 km (180 miles) west of Guangzhou and he set out in a small boat with two Chinese Christians at the beginning of July 1869. They were nearing the island when two pirate junks attacked the boat, killing the captain and forcing the crew to surrender. Père Delavay and his two Chinese companions were taken onto one of the junks, but the pirates were rather nonplussed by their unexpected captives and were not at all sure what to do with them. Some of the gang wanted to throw the prisoners overboard without further delay, but the majority thought it more prudent to keep them safe. After a couple of days' desultory sailing, they came across a fishing boat and the pirates immediately seized their opportunity and decanted the unwelcome prisoners. When Père Delavay and his companions learned that they were now not far from Shangchuan Island where there was a mission station, they paid the fishermen to take them there. Adverse weather delayed them but on 21 July they finally reached the island where they were made welcome by the priest. Shangchuan Island was famous as the burial place

of St Francois Xavier, the first Christian missionary to the Far East, who died there in 1522; and Père Delavay was so pleased to be have been given a chance to visit the Saint's shrine that he wrote to his brother saying that he almost felt grateful to the pirates who had inadvertently provided him with such an unexpected opportunity.[2]

Père Delavay was later transferred to the mission station at La-fou (Malu, Guangxi), close to the border with Tongking (now north Vietnam). There he was moved by the plight of Annamite women from Tongking who had been captured by pirates and sold into slavery along the borders of Guangdong and he spent much of his time writing letters to devout communities in France to raise funds to buy back their freedom. He was also involved in negotiations resulting in various treaties with Tongking.

Return to France

Père Delavay had begun his missionary service in 1867 with the robust constitution of a hardy young mountaineer but by 1880, after spending

Photo Credit: Fonds Iconographique des Missions Étrangères de Paris

LEFT

Père Jean-Marie Delavay in 1878.

thirteen years in the heat and humidity of southern China, his health had broken down completely and he had to return to France. The delight at seeing family and friends again cannot have made up for the despair he must have felt at having to abandon his missionary vocation; but once back in the mountains of his youth, he gradually began to recover his strength and vitality and, with them, the burgeoning hope that he might one day be able to return to China. He spent the winter helping out in a friend's parish before going on a pilgrimage to Rome and, by February, the improvement in his health was so great that he asked his superiors to send him back to China. He knew, though, that he could not live again in sub-tropical Guangdong and he requested a post somewhere with a more forgiving climate, such as the mountainous province of Yunnan in south-west China.[3] His wish was granted, but it was decided that he should spend the summer in France to complete his recovery and return to China in the autumn.

Meeting with Père David

Before Père Delavay left at the end of November 1881, he visited Paris where he met Père David. During his years in the southern province of Guangdong, Père Delavay had collected in a desultory way and had sent a few specimens to Henry Hance at Whampoa. Père David encouraged him to put his collecting on a much more organised footing and suggested that rather than sending the plants of his new district to Henry Hance, he should send them back to the botanists at the Muséum d'Histoire Naturelle. Unfortunately, we know nothing more about this meeting, but it turned out to be one of the most important in the whole history of botanical discovery and exploration.[4] Père Delavay was so inspired by Père David's words and by the example he had set that, on his return to China, he immediately set about making a comprehensive collection of the plants that surrounded his new home in north-west Yunnan, eventually sending over 200,000 specimens back to the Muséum.[5] As the carefully prepared specimens began to arrive in Paris, it became clear to Adrien Franchet and his fellow botanists that Père Delavay had discovered a remarkable new flora in western Yunnan that appeared at times to surpass even the plants found by Père David at Baoxing. Franchet's descriptions of the new plants were devoured by botanists and gardeners alike and within a couple of decades, plant-hunters from Britain and America had arrived in Yunnan to search out the diverse plants that Père Delavay had discovered, with a view to introducing them to Western gardens. Modern plant collectors still visit the areas that Père Delavay was the first to explore.

ABOVE

A street in Dali in 1903 – a scene that would have been very familiar to Père Delavay.

ABOVE

Primula poissonii flowering at 3,180 m (10,434 ft) Zhongdian, Yunnan.

The task that Père David had given Père Delavay could not have been more congenial to him. Botany was very popular at the time, especially among the clergy, and his own love of the subject was first inspired by the varied flora of the French Alps. He had spent much time during his boyhood collecting on the high slopes above his home at Les Gets.[6] The enthusiasm with which Père Delavay collected the plants of northwest Yunnan was no doubt fired by the fact that there he once again found himself in the midst of a cool temperate flora, which was much more to the liking of one who had grown up in the mountains, than the flora of his previous Chinese home in sub-tropical Guangdong. Even while training for the priesthood at the Grand Seminary at Annecy, the young Jean-Marie Delavay had found plenty of time to botanise and had discovered several plants that were previously unknown in the area. By the time he left the seminary, he was thoroughly acquainted with the characteristics of the main plant families, possessed a good knowledge of the local mountain flora, and had also learned how to collect and prepare plant specimens to a high standard.[7] Franchet later praised the admirably chosen specimens that he received from Père Delavay, describing how well they displayed flowers and fruit, and frequently the roots; and each came with a meticulous label giving the collection number and the location, together with some indication of terrain, altitude and flower colour.[8] Even before he went out to China, Père Delavay had had an opportunity to put his newly acquired botanical skills to good use, as he had collected specimens for the first complete account of the flora of the Savoie region, which was being prepared at the time.[9] His involvement in this project taught him the value of making repeated and comprehensive collections in a specific area in order to build up a complete picture of its flora.

When Père Delavay left France for the second time, in November 1881, he was forty-seven and although his health had recovered, he never completely regained his previous strength. Nevertheless, his desire to continue working as a missionary-priest in China was as strong as ever and he arrived in Shanghai in early January with high hopes for his new mission in Yunnan. Soon afterwards, he began the long journey westwards up the Yangtze, reaching Yichang in Hubei province in March. He spent a few days there and collected some plants, including the wild form of *Primula rupestris*, discovered in Yichang in 1879 by Thomas Watters, one of Henry Hance's correspondents. By the end of April, he had reached Sui fu (Yibin) where he left the river and continued overland to Tchenfongchan in north-east Yunnan, the first mission station in the province.[10] He botanised for the short time he was there and among the plants he collected was a new species of *Berberis*, *B. acuminata*, which Franchet described as one of the most beautiful species of the genus. It has handsome foliage and creamy yellow flowers that are larger than in most other *Berberis* but despite these attractive features it does not seem to be in cultivation in the West today.[11]

Arrival in Yunnan

Père Delavay then travelled westwards to Yunnan-fu (Kunming), the capital of Yunnan, where the Missions Étrangères had its provincial headquarters, and it was not until June that he arrived at Tali (Dali), after a journey that he calculated had taken six months and ten days, over half of which had been spent on the Yangtze.[12] Dali, the missionary centre for western Yunnan, lies some 250 km (155 miles) north-west of Kunming in the narrow plain between Lake Erhai and the Cang Shan mountain range. It had been the focus of the long-running

ABOVE

Dapingzi, Yunnan.

First plant discoveries

Dali is situated at 1,980 m (6,496 ft) at the foot of the snow-capped Cang Shan range, where the average height of the peaks ranges from 3,500 m (11,483 ft) to 4,122 m (13,524 ft), and Père Delavay must have been delighted to find himself once again in the mountains. He arrived in June when the main spring flowering period was over but there were still enough plants in flower to give him an idea of the area's floral riches and he could see that many of the plants which had already flowered were unknown to him. One can imagine his growing excitement as he explored farther afield and gradually realised that he had reached a region that appeared to contain many new species. His first find was *Primula poissonii*, then common throughout Yunnan and flowering in swathes that coloured large areas of damp meadows velvety-pink or red.[14] He sent seeds back to the Muséum and seeds were sent to Kew Gardens from the Jardin des Plantes in 1890. The usual form is bright purplish-pink but the red form was introduced to the West in 1992.[15] This primula with its tall tiered flowering stems has since proved itself a colourful garden plant and one that it is not difficult to maintain, provided it never dries out; but Reginald Farrer was not an admirer: 'There is something cold and clammy about the whole plant; the flowers are of an acrid and chilly magenta, and the

Muslim rebellion in Yunnan, which had only been suppressed seven years earlier after a lengthy siege of the town, and the whole area still showed signs of the devastation wrought by the Imperial troops during their campaign against the rebels. Much of the countryside remained deserted and the walled town of Dali itself was only thinly inhabited with parts still lying in ruins. Famine, malaria, cholera and a type of bubonic plague that was endemic in some of the valleys ravaged the population. It was, in the words of one observer, 'a frightfully poverty-stricken land'.[13]

Père Delavay was to spend a few weeks at Dali, familiarising himself with the work of the Mission Étrangères in Yunnan and learning the local dialect, which was different from the one spoken in Guangdong, before setting off for the village of Tapintze (Dapingzi), where he was to make his home. Père Jean Leguilcher, who now ran the mission in Dali, had lived in Yunnan since 1853 and had previously been based at Dapingzi, and so was able to give Père Delavay a great deal of information about his new district. While Père Delavay prepared himself for life in his new community, he began to botanise around Dali.

flopping smooth leaves with their pallid midrib are flaccid and unpleasant as a corpse's fingers.'[16] This touch of melodrama, though typical of Farrer, is unduly harsh: *P. poissonii* may not be the most attractive of Yunnan's primulas but many of the others require such specialised growing conditions that its very ease of cultivation counts for a great deal in its favour.

By July, Père Delavay was investigating the more accessible lower slopes on the eastern flanks of Cang Shan. He was too late for most of the rhododendrons, which have usually finished flowering by June, but on 4 July, on the slopes above Dali, he was able to collect specimens of herbaceous *Thalictrum dela-vayi*, the first and most garden-worthy of the six new species of *Thalictrum* he discovered.[17] He collected it on subsequent occasions and also sent seeds back to the Jardin des Plantes, which in turn sent seeds to Kew in 1889. As soon as the young plants flowered, it was apparent that *T. delavayi* was highly ornamental. It forms a tall framework of airy stems (up to 1.8 m/6 ft), which produce delicate panicles of tiny pendulous lilac and yellow flowers that flutter gracefully in the wind from midsummer onwards. The flowers last for some time and are followed by deep purple seed-pods.[18]

Life at Dapingzi

By the beginning of August 1882, Père Delavay was living in his new home at **Dapingzi**, about 65 km (40 miles) north-east of Dali. He now felt himself to be something of an expert on travel in Yunnan as he had covered around 595 km (370 miles) in the province since leaving the Yangtze in May, and he reported, in a letter to his Superior in Paris, feeling that 'the roads in Yunnan are abominable and the difficulties of communication defy the imagination...'[19] Even today, the old roads in the mountains of Yunnan leave much to be desired and are frequently rutted and pitted with giant potholes. Dapingzi might not have been far from Dali in terms of distance but it was 1,000 m (3,281 ft) lower and only about 15 km (9.3 miles) from the Yangtze, so the climate was very different, with high temperatures and much less rain. Although the climate of north-west Yunnan is essentially cool temperate, it varies with altitude and it must have been a disappointment for Père Delavay to exchange Dali's pleasant mountain climate for somewhere with summer temperatures that were reminiscent of his previous mission in Guangdong, even though it was generally drier and less humid.

The missionary district for which Père Delavay was responsible lay north-east of Dali and was called Houang kia pin (Huangjiapin), after a village situated south of Dapingzi and close to the border with the neighbouring Christian district, which was cared for by Père Pierre Proteau who lived at Pin kio (Pianjiao), some 40 km (25 miles) away. Père Proteau was fifty-one and had arrived in Yunnan in 1857, and although Pianjiao was a long day's march from the village of Huangjiapin, he was

A traditional house in Dapingzi today.

Père Delavay's closest fellow missionary and they became excellent friends. As Père Delavay became more familiar with his district, he estimated that about two-thirds of the population had died from plague and that half the remainder were Christians. The Christian converts came from the Lisu and Bai people, two of the ethnic groups living in Yunnan, and many of their villages throughout the area had Christian communities. Ministering to these scattered groups necessitated much travelling to and from his base at Dapingzi, which lay on the eastern border of the district, and Père Delavay used these trips as opportunities for plant collecting. He quickly discovered that the flora of Dapingzi, where summer temperatures reach 40 °C/104 °F and above, was sub-tropical, and he later told Franchet that he had found plants that he had first seen in Guangdong.[20] It was only when he reached the higher slopes of nearby mountains such as Mao Kou Shan (Maogushan) at 2,200 m (7,180 ft) that he was able to collect plants that were more likely to be suffiently hardy for temperate climates. This altitude-dependent temperature variation prevails throughout north-west Yunnan and has resulted in the evolution of an exceptionally rich flora that ranges from sub-tropical species at Dapingzi and in the depths of the steep valleys, to a multitude of alpines on the high slopes of Cang Shan and the interlinked mountains of the region.

Today, very little of the native flora that Père Delavay saw at Dapingzi remains, as the mountain slopes and hillsides have been almost completely denuded of their original vegetation while every scrap of flattish land around the village has been cleared for intensive agriculture. Farming was the principal occupation of the Lisu villagers that Père Delavay knew and, coming as he did from rural France, he would have found life in a small community governed by the seasonal rhythms of the agricultural year very familiar. Even today, despite all the changes that occurred in China during its turbulent twentieth century, there is still a great deal in Dapingzi that Père Delavay would recognise. Many of the two-storey houses have been rebuilt using modern materials, but some of the villagers still live in the traditional wattle houses, formed of a series of small windowless connecting rooms, with the upper storey used for storing grain. The floors are of beaten earth and there is little furniture: only a few low beds, a table and a scattering of diminutive benches. There are no chairs as people still squat on their haunches to eat and carry out everyday tasks. The smoky oil lanterns and tapers, which were all Père Delavay had to light these dim rooms, have now been replaced by electric light in the form of naked bulbs hanging from the ceiling, but the rooms remain shadowy. Bottled gas connected to hot plates is available for cooking, along with the traditional wood-fired ovens, but many still cook on an open wood fire in the middle of the floor, with the smoke seeping out through the ventilation gap at the top of the walls.

It was in such bare houses that the Catholic missionaries lived, sharing the hard lives of the villagers they had come to serve, subsisting on the same poor diet and subject to the same diseases. Père Delavay had already lived for over a decade in China so he knew exactly what was involved in the life he had chosen, but when one considers today the rigours of such an existence, one can only marvel at the devotion of the priests who subjected themselves to it. Western travellers who penetrated the remote areas occupied by the missionaries remarked on the extraordinary self-sacrifice of these men and their comments give us some idea of the hardships of missionary life. As the traveller A. J. Little observed, the members of the Missions Étrangères de Paris were drawn from 'some of the ablest men of the Catholic priesthood' and for these educated intelligent men life deep in the heart of rural China meant an intensely lonely existence, devoid of companionship and completely cut

BELOW

A valley in the Heishanmen range.

off from Western culture. They wore an artificial plait and Chinese clothes, spoke only Chinese and lived wholly Chinese lives, immersed in the communities they served. They could only work among the poor as 'no respectable Chinaman would ever admit a missionary into his home', which meant that they were unable to forge links with educated Chinese people and were deprived of the society of those who were their intellectual equals.[21] Isabella Bird writes of 'the anguish of loneliness which these Roman [Catholic] missionaries endure', and one priest she met told her that 'Madness would be the certain result but for the sustaining power of God, and the certainty that one is doing His work'.[22]

The priests were shunned by the ruling élite because missionaries were seen not only as preachers of a creed that they asserted was superior to the precepts on which traditional Chinese society was based, but also as representatives of a foreign power with threatening designs on China itself. Augustine Henry, an Irish plant collector who spent many years in China, saw this clearly: 'I doubt if the missionaries even ever get to be friends with their converts who are Chinese. The Chinese… detest us Europeans, as people they don't understand, and as people who are really their enemies, who come to rob them of their country.'[23] Chinese officials treated the missionaries with suspicion, if not outright hostility, and often turned a blind eye when missionaries or their property were attacked. Even during periods of relative calm, the threat of attack was never wholly absent and missionaries lived with the knowledge that at any moment the simmering mistrust might erupt into violence. The situation was much worse if the provincial authorities were headed by officials who actively resented the presence of the missionaries: in such circumstances, aggression towards them was encouraged or, at the very least, condoned. Père Delavay commented in a letter in March 1888 that the current precarious situation of the missionaries was caused by the present administration of Yunnan and that, although there was no active persecution at present, he was well aware that it could begin any day. Consequently, he and his confrères tried to live quiet and retired lives, so as not to provoke their enemies. Père Delavay understood Chinese fears and was ultimately forgiving, 'We are poor strangers in the midst of vast China and it is hardly surprising if, from time to time, we figure as Western devils and French barbarians.'[24]

The missionaries' obvious poverty further diminished their status in the eyes of respectable Chinese society. Their lack of funds meant that they lived frugal comfortless lives and the explorer William Gill described the priests he met as 'very thin men, with drawn features and sunken eyes'.[25] It was only their rock-like faith that enabled the missionaries to bear such an existence and one can understand how Père Delavay's overwhelming interest in the surrounding flora gave him an intellec-

tual pursuit that must have provided solace amidst the exigencies of his everyday life.

Heishanmen

Père Delavay's new home at Dapingzi was situated in a relatively flat area adjacent to the Yangtze, but this was exceptional as north-west Yunnan is mountainous, and his own missionary district was dominated by the Heechanmen (Heishanmen) range running north–south through its centre. It took Père Delavay about a day and a half to reach the western side of his district and the trip involved crossing Heishanmen, where the main pass lay at 3,000 m (9,800 ft). The flora of the mountain

ABOVE

The view north towards Heishanmen from the site of the priest's house at Menghuoying.

BELOW

The intensively cultivated plain at the foot of the deforested slopes of Heishanmen. Menghuoying can be seen in the centre of the photograph.

ABOVE

Rhododendron yunnanense in early spring.

Photo Credit: Roy Lancaster

ABOVE

Rhododendron fastigiatum growing on Cang Shan.

was exceptionally rich and Père Delavay called it the garden of Yunnan; he later reckoned that he had climbed it more than sixty times.[26] He came to know the range well and made several interesting discoveries there over the years. This shows the advantage that the missionary-botanists had over the professional plant-hunters who later visited the same areas: because they actually lived in the places where they collected plants they could easily visit locations year after year during every season, collecting seeds and fruit as easily as flowers.

In December 1884, as Père Delavay crossed Heishanmen, he collected the red berries of *Cotoneaster glaucophyllus*, a handsome variable shrub with greyish leaves, usually represented in gardens by var. *serotinus*, a floriferous variety that was originally introduced by George Forrest. In May 1886, again on Heishanmen, Père Delavay discovered a flowering specimen of *C. pannosus* and he collected seeds later in the year, which he sent to the Jardin des Plantes. Plants were raised in 1888 and distributed to Kew and other gardens in 1892. *C. pannosus* is now a familiar red-berried ornamental plant in warmer gardens, and has even become naturalised in parts of North America, especially in Californian coastal areas.[27]

Menhuoying

Across the Heishanmen range was the Bai village of Mo-so-yn (Menhuoying), about 14.5 km (9 miles) north of Lankong (now called Eryuan), where there was a sizeable Christian community.[28] The village had a church and a mission school, as well as a priest's house which Père Delavay used as a base for visits to the western areas of his district.[29] He described the location of Menhuoying in a letter to Père David as 'an area of great botanical interest. A wide high plateau [about 2,300 m/7,550 ft] surrounded by high mountains, partly wooded and partly covered with meadows'.[30] Over the years, he was able to explore

the area very thoroughly and he found several interesting species in the vicinity, one of the first of which was *Berberis pruinosa*, a handsome evergreen shrub with yellow flowers and blue-grey fruits, which he came across in February 1883.[31] He sent seeds to the Jardin des Plantes and in 1897 seed reached Kew.[32] In 1885, in the same area, he found the wild form of *Rosa banksiae*, which until then had only been known as a cultivated garden plant. On the highest slopes above the village, at about 3,000 m (9,800 ft), were rich pastures that were grazed by flocks of sheep in summer and it was here that he discovered fern-like *Corydalis cheilanthifolia*, the most decorative of the fifteen new species of *Corydalis* he found.[33] Today, as at Dapingzi, the natural vegetation of the plain around Menhuoying that Père Delavay explored with such delight has vanished and has been replaced by hundreds of intensively-cultivated small fields, connected by a network of paths and tracks. The diverse flora that once covered the mountain slopes has also disappeared, cut down for timber, firewood or animal fodder, but Père Delavay's specimens provide us with an invaluable record of the botanical riches that once flourished there.

Père Delavay had also begun to investigate the flora of Dapingzi and its immediate surroundings, and in the woods above the village in April 1883 he found *Rhododendron yunnanense*, a variable and often straggling species, which is usually very floriferous, producing sprays of white or rosy-pink flowers. It is one of the commonest shrubs in western Yunnan and can still be found in a wide range of habitats and at varying altitudes, tolerating even a relatively alkaline soil and flowering bravely in some extremely unpromising locations.[34]

Attacks on the missionaries

Père Delavay might have botanised whenever possible but he was first and foremost a missionary priest and in 1883, the situ-

ABOVE

Nomocharis pardanthina.

BELOW

Nomocharis aperta flowering in midsummer on Tian Bao Shan, Yunnan, 3,750 m (12,303 ft).

ation for missionaries in western Yunnan became much more dangerous. Only nine years earlier, Père Jean Baptifaud had been murdered at Pianjiao, where Père Proteau was now stationed, and anti-Christian feeling was again running very high in the province. On 28 March 1883, the church at Tchang-Yn near Yangbi had been besieged and Père Jean Terrasse, the priest there, murdered, along with fourteen members of his congregation. The violence then spread to other Christian communities in the vicinity, which were pillaged and destroyed.[35] Yangbi lay just west of Cang Shan so the threat to Dali and the other missions in the immediate area was very real. In such troubled times, it took considerable courage for solitary isolated priests to remain in their villages, surrounded by people who might turn on them at any moment, especially for one like Père Delavay who had already suffered at the hands of an angry mob. Yet, when he refers in his letters to the difficulties faced by the missionaries, he does so in general terms and usually only because something has occurred to curtail his botanical activities.

As a consequence of the destruction at Yangbi, the small daughters of the murdered Christians were sent to the school Père Delavay set up in Dapingzi, where they stayed until old enough to return home.[36] Making these arrangements and establishing himself in his new district took up his time, but he still managed to fit in several plant collecting excursions. At the beginning of June 1883, he set out to climb Cang Shan but, after an exhausting attempt, he was forced to turn back when only 300 m (984 ft) from the summit as he realised that he had taken completely the wrong route.[37] On the way up, however, he managed to collect a number of plants, including *Rhododendron fastigiatum*, a low spreading species with lilac flowers.[38]

New genus

One of the most remarkable of his Cang Shan discoveries was *Nomocharis pardanthina*, a bulbous plant that he found on 2 June, which turned out to be not only a new species but also a representative of an entirely new genus.[39] The genus *Nomocharis* was established by Franchet when he realised that Delavay's find lay halfway between lilies and fritillaries and, in fact, species of *Nomocharis*, which are high altitude plants, require the same cool, damp acid conditions required by the woodland lilies. They have distinctive pale open flowers that are often spotted or blotched and are unusual and attractive choices for those with cool gardens that do not dry out in summer. *N. pardanthina* is scarce in cultivation but in areas with suitable conditions, such as central Scotland, it can make a tall vigorous plant. George Forrest collected seeds in the Cang Shan range in 1906 and 1910, but he does not seem to have introduced it, and *N. pardanthina* actually came into cultivation from seeds collected by Père Édouard Maire (see chapter 9, p.124). This led to it usually being cultivated in gardens as *N. mairei*, although

N. pardanthina is the correct name.[40] The confusion caused by the wrong name becoming attached to widely cultivated plants was perhaps inevitable when several individuals were involved in the collection and identification of new and wholly unfamiliar species. In June 1889, Père Delavay discovered a second species near Dali called *N. aperta*, with flowers in varying shades of rose spotted with crimson, which Franchet first identified as a lily but E. H. Wilson later realised was a *Nomocharis*.[41] *N. aperta* was introduced by George Forrest.

Père Delavay returned to Dapingzi soon afterwards and continued botanising in the area. In September, he discovered *Jasminum polyanthum*, a rampant climber with fine heavily-scented flowers that needs warmth to flourish – not surprising, considering Dapingzi's hot climate. It was subsequently found in many places in south-west China but never higher than 2,200 m (7,218 ft), indicating that this species will not tolerate very low temperatures. In 1906 Forrest introduced *J. polyanthum* from the Dali valley and he collected it again in 1931 in Tengchong. His companion on the latter trip was the wealthy American plantsman Lawrence Johnston, who brought the new jasmine back with him and grew it very successfully in his garden at Mentone on the French Riviera, where it was perfectly suited by the climate.[42]

Franco–Chinese War

During the winter of 1883–84, Père Delavay devoted much of his time to drying and labelling his specimens and eventually packed up over 1,000 in two cases, ready to send off to Adrien

BELOW

Primula malacoides growing in the courtyard of Joseph Rock's house (now a museum) at Nguluko (Yuhu) near Lijiang.

BELOW

Clematis chrysocoma on Cang Shan above Dali.

RIGHT

Primula flaccida as *P. nutans.* (*Bot.Mag.* (1917) No. 8735)

Photo Credit: Roy Lancaster

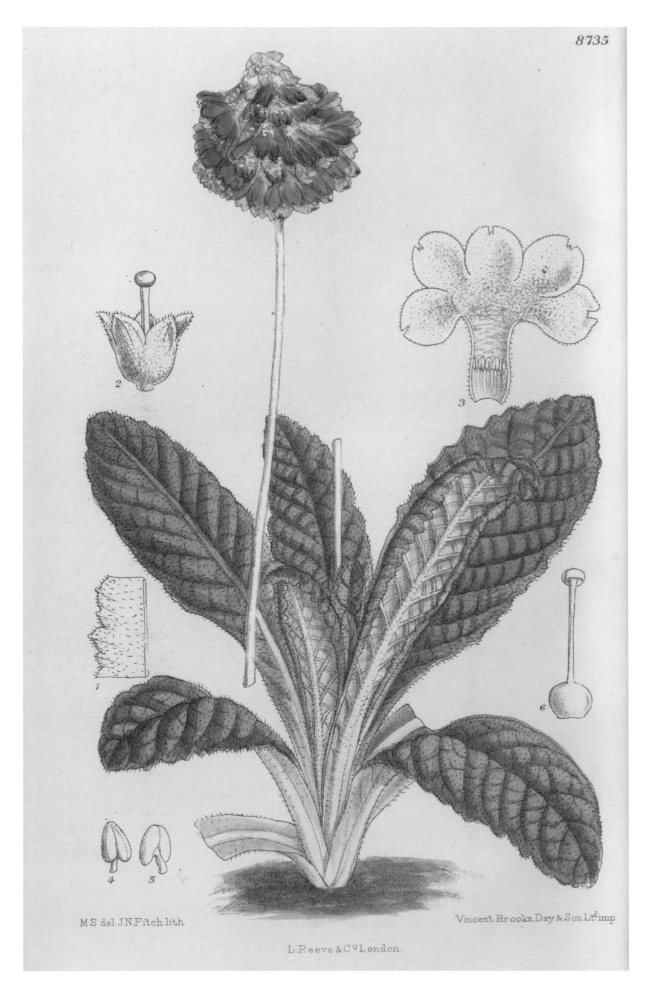

8735

M.S.del.J.N.Fitch.lith.

Vincent Brooks,Day&Son Lᵗᵈimp.

L.Reeve&Cº London.

Franchet in Paris. To have collected and prepared so many specimens in such a short time indicates both the exceptional richness of the flora in the area he had begun to explore and his remarkable industry and diligence as a collector. Once the collection was ready for despatch, he had to find some way of getting his cases to Yibin, where they would be transferred to a boat and transported down the Yangtze to Shanghai. In a letter to Père David at the end of May 1883, he had complained about the difficulties of sending sizeable consignments over the 'abominable' roads, and also about the expense, as he had discovered that the cost would amount to between Fr.180 and Fr.200. He comments, 'It is only a small sum but I don't have it. I have never felt as poor as I do here in Yunnan.[43] Nevertheless, by the spring of 1884, he managed to find the means to buy a horse and hire a man to take his chests on the forty-day overland journey to Yibin.

It was at this juncture that news reached Yunnan that France

BELOW

Gualopo village in the Heishanmen range.

and China were at war over Annam (Vietnam), a state which had always paid tribute to China but which France had annexed in 1874. As a result no official would give Père Delavay, who was seen as a representative of the enemy, a travel permit for his chests and so he was forced to store them, until it was once again safe to despatch them. This was a set-back but Père Delavay had already begun sending little packets of specimens back to Paris using the efficient Chinese postal service, and as soon as he heard that they had arrived safely he felt confident enough to start sending off small packages quite regularly.

Collecting in 1884

As the first signs of spring appeared in March 1884, he was back in Dali where he discovered annual *Primula malacoides* with its rose-lavender flowers, which was then a common sight in ditches around the fields. Seeds were sent to the French nurseryman Maurice de Vilmorin in 1895 but none germinated, and it was not until George Forrest introduced *P. malacoides* from the Dali area in 1908 that it reached the horticultural trade. Vilmorin acquired more seeds and exhibited flowering plants in

Photo Credit: Ken Cox

1911. Seed firms in Europe and America soon developed new forms and colours, which quickly became popular greenhouse and conservatory plants. *P. malacoides* can be grown outside in warmer climates, where it is now a familiar annual or biennial garden plant.[44] It is unusual among *Primula* in that it is monocarpic (dies after flowering and setting seed). Although abundant as a field weed around Dali until the 1930s, *P. malacoides* is no longer to be found in the area. It was under pressure even in the 1880s and Père Delavay reported to Franchet in July 1888 that, although he had looked for *P. malacoides* in several places where he had seen it before, he could not find it as the sites had all been cleared for agriculture.[45]

The following month, on Maogushan above Dapingzi, Père Delavay discovered another species of *Primula* that is now called *P. flaccida*, although it was known for many years as *P. nutans*. It is robust with lavender blue flowers and has proved a fine ornamental plant, being described as 'one of the most outstanding garden primulas.'[46]

In April, Père Delavay collected flowers from a low scrambling *Clematis*, which Franchet recognised as one of the *Mon-*

tana group and named *C. chrysocoma*. George Forrest thought it a beautiful species 'with its soft rose-coloured flowers, and golden, glistening foliage' and, as *C. chrysocoma* is always found flourishing in the driest sunniest places, it is not surprising to discover that it needs the protection of a glasshouse in all but the mildest and most sheltered gardens. Some of the very first seed pods that Père Delavay collected at Dali in November 1882 were from *C. chrysocoma*, and its seeds may well have been among those he sent to the nurseryman Maurice de Vilmorin, as it was cultivated in his nursery outside Paris and he sent it to Kew in 1910. Several seed introductions have been made in recent years and *C. chrysocoma* is once again available to gardeners. Shortly afterwards, Père Delavay was collecting in the vicinity of Menhuoying when he found another *Clematis*, which was so similar to *C. chrysocoma*, apart from its covering of very fine silky hairs, that Franchet identified it as a variety that he called var. *sericea*. It was introduced in the 1890s under that name but it was, in fact, the same as *C. spooneri*, which Père David had found at Baoxing in 1869.[47]

Père Delavay visited the northern part of his district in May and explored the environs of the village of Koua la po (Gualapo), situated at 3,660 m (12,000 ft) on the high pass above the town of Hokin (Heqing). This area was to prove a fine source of new plants, particularly rhododendrons. These included: *Rhododendron. lacteum* with primrose yellow or pink-tinged flowers, which formed dense woods on the surrounding slopes; *R. cephalanthum*, a beautiful small species with pink or white daphne-like flowers; *R. trichocladum*, another small shrub with greenish-yellow flowers; and *R. polycladum* (syn. *R. scintillans*), a fine compact blue-flowered species, which has

BELOW

Primula sonchifolia var. *sonchifolia* at 3,400 m (11,155 ft) on Laojun Shan, Yunnan.

BELOW

Cypripedium margaritaceum in summer, Yunnan.

Photo Credit: Harry Jans

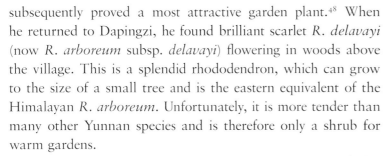

Photo Credit: Roy Lancaster

ABOVE

Rhododendron haematodes growing on Cang Shan where Père Delavay first found it.

ABOVE

Rhododendron decorum subsp. *duabrepes* in summer.

RIGHT

Yulong Xue Shan – the Jade Dragon Snow Mountain, north-west Yunnan.

subsequently proved a most attractive garden plant.[48] When he returned to Dapingzi, he found brilliant scarlet *R. delavayi* (now *R. arboreum* subsp. *delavayi*) flowering in woods above the village. This is a splendid rhododendron, which can grow to the size of a small tree and is the eastern equivalent of the Himalayan *R. arboreum*. Unfortunately, it is more tender than many other Yunnan species and is therefore only a shrub for warm gardens.

In June, Père Delavay climbed Cang Shan again and this time managed to reach the highest slopes, which he found covered in dwarf rhododendrons and low alpines. These high meadows are common throughout the mountains of northwest Yunnan, where pine forests give way to dense thickets of tree-sized rhododendrons at about 3,660 m (12,000 ft), which in turn peter out into the alpine prairies where Père Delavay was to make some of his most interesting discoveries. On his first foray to the summit, he collected more primulas, including delicate *Primula amethystina* and beautiful *P. sonchifolia* with ice blue flowers, the first of the 'blue' primulas to be discovered. *P. sonchifolia* is closely related to Père David's *P. moupinensis* but is distributed across a much wider area and can be found in shady places on the highest slopes in Yunnan and Sichuan where it flowers as the snow melts around it. This primula is unusual in that it overwinters as a large exposed resting bud. In 1930, when it had proved impossible to introduce to cultivation through seed-collection, the British Governor of Burma arranged for some exposed buds to be collected from the Hpimaw Pass close to the Burma–China border. These were then packed in bamboo tubes before being shipped back to Britain in a refrigerated hold, as a gift for George V. The plants raised from these buds flowered in 1931 and were exhibited by the Royal Horticultural Society in February that

year.[49] A good colour form of *P. sonchifolia* has recently been introduced from Dali.

During this exploration of Cang Shan, Père Delavay found a curious slipper orchid, *Cypripedium margaritaceum*, with a bulbous greenish flower that emerges on a very short stalk from the centre of two large ribbed leaves, which are borne horizontally like a platter. The flowers and the leaves, which are sometimes so wide they appear almost circular, are marked with dark maroon spots. Père Delavay found several other slipper orchids including purple-pink *C. yunnanense*, white *C. plectrochilum* at Dapingzi, and yellow *C. flavum*, which Père David had discovered at Baoxing. *C. margaritaceum* is the best known of the Yunnan slipper orchids, but it is now endangered in its native habit as human pressure on the land continues to increase. New slipper orchids are still being discovered in western China and we must hope that any species that are still unknown survive long enough for us to find them.[50]

Père Delavay also collected a purple-pink ground orchid, which turned out to be *Pleione bulbocodioides*, another of the species Père David had discovered at Baoxing (see chapter 4, p.41). Other ground orchids were common, including *P. yunnanensis*, which was long confused in gardens with *P. bulbocodioides*, and only correctly identified when botanists were able to examine living plants raised from seed collections made in the 1980s. *P. x taliensis*, a natural hybrid between these two species, has recently been described. These botanical puzzles, which are still being resolved, indicate the complexity of plant relationships in western China.[51]

Rhododendrons

Alpine primulas and various orchids were exciting enough, but it was among the Cang Shan rhododendrons that Père Delavay made some of his most ornamental discoveries. As well as *Rhododendron fastigiatum*, there was yellow-flowered *R. brachyanthum*; *R. campylogynum*, a charming dwarf species with several colour forms; and *R. haematodes*, a red-flowered species that many consider one of the handsomest of all the Chinese rhododendrons. Another red-flowered species that Père Delavay found on Cang Shan was *R. neriiflorum*. George Forrest, who later introduced many of these species from the same range, described finding the rhododendrons:

> in the gullies and gorges [where] one comes on them at every turn, either as solitary specimens or in groups, sheltered and supported by a jungle of mixed scrub and cane. In shady places such species as *R. neriiflorum* and *R. haematodes* eclipse everything in beauty with their cherry-coloured and deep crimson blooms, whilst again, beneath their shade is found the dwarfest of all known species from the region, *R. campylogynum*, only a few inches in height, with pendulous, dark plum-purple, bell-like corollas.[52]

Père Delavay subsequently discovered several other rhododendrons on Cang Shan, including excellent pink-flowered *R. rubiginosum*, a vigorous shrub that can reach the height of a small tree, and *R. taliense*, which was named after Dali (then written as Tali) at the foot of the range.[53] He also found *R. decorum*, discovered at Baoxing by Père David, which is still very common throughout the region and often found growing alongside *R. yunnanense*. In Dali, the white flowers of *R. decorum* are cooked with beans to make a palatable soup.

Seed collection

Père Delavay knew that botanists would be eager to see living material and, once he had collected specimens of most of the area's flora, he had more time to collect seeds and fruit. He managed to send back many packets of seeds, but not all the plants raised were successfully grown on, as it took time for the cultivators at the Jardin des Plantes to learn their individual requirements. For example, few rhododendron seedlings seemed to have survived long – perhaps because they were

grown in hothouses.[54] The unwillingness of gardeners to take risks with these completely unknown and very precious young plants is understandable but actually their hardiness depended on the altitude at which they were collected. Père Delavay estimated that everything growing at or above 3,000 m (9,845 ft) would prove hardy enough to withstand cold winters.[55] This assumption has generally proved correct, as was illustrated by two rhododendrons successfully raised at the Jardin des Plantes. These were *Rhododendron scabrifolium* with white or pink flowers, and low-growing pink-flowered *R. racemosum*, both of which Père Delavay had found while crossing Heishanmen; but *R. scabrifolium*, which is somewhat tender, is not found above 2,600 m (8,530 ft), whereas hardy *R. racemosum* grows up to 3,500 m (11,483 ft). *R. irroratum*, which Père Delavay had discovered in woods above Menhuoying, was another species successfully raised at the Jardin des Plantes, although it is usually found below 3,000 m (9,800 ft) and does best in warm gardens. He sent seeds back in 1890, which were shared with Kew and their plants flowered in a green-house in 1893.[56] The Jardin des Plantes was always quick to distribute living material to other botanic gardens, as growing introductions in several separate locations was the best way of ensuring the survival of new plants.

LEFT

Primula secundiflora in summer at Sir Harold Hillier Gardens, Hampshire.

BELOW

Picea likiangensis forest at the eastern foot of Yulong Shan.

ABOVE

Primula chionantha subsp. *sinopurpurea* at Bai Ma Shan, Yunnan, 4,100 m (13,451 ft).

Photo Credit: Roy Lancaster

ABOVE

Meconopsis delavayi flowering at 3,100 m (10,170 ft) above Lijiang, in the area where Père Delavay first found it.

First visit to Lijiang

In July 1884, Père Delavay visited Lijiang, which lay north of his own district, where he explored snow-capped Yulong Xue Shan or Jade Dragon Snow Mountain, a compact mountain range whose highest peak reaches 5,596 m (18,359 ft). These mountains are bordered to the west by the deep gorge of the Yangtze as the river flows south-east, before it swings north and then east on its long journey across China to the sea. In north-west Yunnan and eastern Tibet, the Yangtze runs parallel to the Mekong (Lancang Jiang) and the Salween (Nu Jiang), which also drain south-east from the Qinghai–Tibetan plateau. The parallel gorges and slopes of the steep mountain ranges separating the three great rivers are home to an extremely rich and

diverse flora. A few days on the limestone slopes of the Yulong Shan were enough to reveal this to Père Delavay. George Forrest, who later spent months exploring the range, was similarly impressed and described the area in glowing terms:

> There are shrubs in endless variety, with the ledges of the cliffs and surfaces of the boulders clothed with masses of Primulas, Saxifrages, beautiful Crucifers, several species of Meconopsis,… Along the Pine belt at intervals there are large, sheltered openings, covered with rich limy pasture; these are, floristically, the 'crème de la crème' of the range. There are found huge masses of the finer Primulas… several species of Androsace… also quite a number of Liliums… and a host of other rare and interesting species.… In the Alpine pastures are many Alpine forms of Primula, with innumerable Gentians, Corydalis, Anemone, Meconopsis… [57]

Père Delavay was only able to make a short visit in 1884 as he could not leave his district for too long, but the flora was so rich and diverse that he still had enough time to collect a host of new species. The largest of these was a fine spruce, *Picea likiangensis* with glaucous foliage and dark red young cones, which has since become well known in cultivation.[58] To his intense frustration, Père Delavay discovered that some of the finest of the alpines clung to completely inaccessible ledges; but among the new species he did collect were primulas such as *Primula secundiflora*, with nodding bell-like magenta flowers, which proved a fine addition to cool gardens when introduced by Wilson in 1905; and *P. chionantha*, which he found next to the Lijiang glacier.[59] *P. chionantha* does well in gardens with the cool wet conditions it requires and the form most usually seen in cultivation is the purple-flowered subspecies *sinopurpurea*, but there is also a very beautiful white form, subsp. *chionantha*, which is sometimes called the snowflower or snowflake.[60]

Meconopsis

As well as the primulas, Père Delavay also found various species of *Meconopsis*, including yellow-flowered *Meconopsis integrifolia*, which Nicolai Przewalski had discovered in Gansu in 1872.[61] He also collected a delightful little species that Franchet called *M. delavayi* in his honour. It is smaller than some other *Meconopsis* species and a true alpine with violet-purple flowers. It is a scarce plant that is only found in north-west Yunnan, chiefly on the eastern slopes of the mountains above Lijiang, where Forrest describes it as one of the most beautiful species, and where Christopher Grey-Wilson, a member of the 1987 Sino–British expedition, saw it a century after its discovery:

ABOVE

Paeonia delavayi in early summer.

Meconopsis delavayi is a dainty delight to see in the wild… High above the Lijiang Valley are steep, almost inaccessible mountain slopes covered with fir and rhododendron that give way in places on the eastern flanks to steep grassy meadows interspersed by scree. These meadows are full of treasures: corydalis, lilies, globeflowers and primulas. Amongst all the joys we found scattered colonies of *M. delavayi*. Many plants bore one or two flowers, though up to eight were counted. The colour ranges from pale violet-blue to violet-purple, the flowers nodding harebell-like or with the petals spreading widely apart to reveal the stamens beneath. The joy and excitement of seeing such a splendid little beauty so perfect in the wild can never be matched in cultivation. Long may it thrive on those distant slopes.[62]

M. delavayi was originally introduced by George Forrest but even though several seed collections have been made in Yunnan in recent years it is extremely rare in cultivation, as it requires very specific growing conditions. It has been grown successfully in Scotland, which has the cool wet summers and cold winters that it demands.

Fritillaria delavayi was another scarce species found by Père Delavay on the scree slopes of Yulong Shan, and members of the 1987 expedition rediscovered it, still flowering high on the inhospitable screes.[63] It is an odd-looking plant with mushroom-brown flowers that poke out of the broken rocks on short stems but, even though it is has a wide distribution that stretches

Osmanthus delavayi in early spring.

across into the Himalayan massif as far west as Bhutan, it has never been in cultivation.[64] Lower down, he found herbaceous perennials such as *Delphinium delavayi*, which has attractive long-spurred blue and white flowers.[65]

Tree peonies

On the same slopes, on 9 July, Père Delavay discovered the tree peony *Paeonia delavayi*, one of the best known of all the plants that bear his name. It is a medium-sized shrub with divided foliage and cup-shaped flowers that are usually dark crimson or maroon red, although the colour can be very variable. Forrest described the best forms of *P. delavayi* as having flowers of a 'deep rich red such as one may see when the sun shines through a glass of Burgundy'.[66] Père Delavay had already found a yellow-

flowered tree peony on the slopes above Dapingzi in October 1882, and in May the following year he found the same yellow peony as he crossed Heishanmen on his way to Dali.[67] Franchet described these yellow-flowered specimens as a new species called *P. lutea*. Père Delavay sent seeds of these peonies back to the Muséum and, when the plants flowered in 1892, they were exhibited by Maxime Cornu, the Professor of Horticulture.

P. delavayi is now known to be very variable but, before this was realised, the differences between specimens collected on different occasions and from different places led botanists to conclude that they were dealing with several new species, as Franchet had surmised when he examined the yellow-flowered specimens collected by Père Delavay that he called *P. lutea*. It was some years before even Père Delavay, who had plenty of living material to study, realised that only one variable species was involved and it is now generally accepted by botanists that all these different 'species', including *P. lutea*, are merely forms of *P. delavayi*. The yellow-flowered peony is now *P. delavayi* subsp. *lutea*.[68]

When Père Delavay left Lijiang, he did so with a heavy heart, knowing that he had barely scratched the botanical surface of Yulong Shan, and he resolved to return at the first opportunity. There was still, though, much left to discover in his own district, including *Osmanthus delavayi*, another of the excellent ornamental shrubs that bear his name, which he found while exploring the slopes near Eryuan. This fine evergreen has since become a garden favourite as it tolerates a variety of conditions and produces clusters of fragrant pure white flowers in early spring. Père Delavay sent seeds to the nurseryman Maurice de Vilmorin, who shared them among several colleagues. The only seed to germinate was one at l'École d'Arboriculture in Paris in 1890, but this was enough for professional plantsmen who were able to propagate it by grafting it on to stock from privet and other shrubs. It was soon available from several nurseries and most cultivated plants derive from this original introduction. Kew acquired a plant from Lemoine's nursery at Nancy that first flowered in 1912. Maurice de Vilmorin was impressed by the relative size of its flowers and, when he described *O. delavayi* in 1904, he remarked that he thought it would be a good small rock garden shrub for warm areas.[69] This shows just how new *O. delavayi* was to cultivation. Vilmorin had never seen a fully-grown plant and had no idea that O. delavayi is so far from being a compact low-growing species that it can reach 2.5 m (8ft), sometimes more in suitable conditions, although it can be pruned to keep it shorter.

In September, while Père Delavay was exploring the slopes at the northern end of Lake Erhai, he collected *Metapanax delavayi*, a close relative of *M. davidii*, which Père David had found at Baoxing. *M. delavayi* also forms a rounded evergreen shrub but its leaves are more delicate and more deeply divided, and it produces blue-black fruits in autumn. These two species are the only members of their genus.[70]

First consignments despatched

By February 1885, after what Père Delavay describes as 'the terrible alarms' of the previous December, the situation in western Yunnan was calm enough for him to obtain the requisite permits for the large consignment he had prepared, which now comprised four cases containing around 1,500 specimens.[71] Realising that it would be unwise to put a European address in European letters on the cases, he sent them to the Missions Étrangères' house in Shanghai and asked the Procurer there to forward them to Paris. Père Delavay was right to be wary, as a little later in the year two officials came to the village and accused him of raising a detachment of 2,500 heavily-armed men to march through Yunnan to Tongking. He managed to convince them that the whole story was nonsense, but not before his community had had a considerable fright.[72]

Horticultural Developments

Thalictrum delavayi

T. delavayi is very variable in the wild and there are four recognised varieties, including var. *decorum*, which is smaller than the typical variety with more delicate foliage and flowers and var. *acuminatum*, which has finely-dissected foliage. Both these varieties have recently been introduced to cultivation. Some fine long-flowering selections have been made over the years, including 'Album', a lovely white variety, and 'Hewitt's Double', with a froth of tiny double lilac-pink flowers. *T. delavayi* flowers after many other garden perennials have finished and has become a valuable feature of late summer displays.[73]

Paeonia delavayi

Paeonia delavayi was the first wild tree peony to be discovered by Westerners and its introduction led to a flurry of peony-breeding that continues today. The missionary-botanists' primary impulse might have been scientific but there is no doubt that horticulturalists have benefited enormously from their discoveries. Plantsmen such as Louis Henri of the Jardin des Plantes and Victor Lemoine immediately realised that they could develop a range of yellow-flowered ornamental tree peonies by crossing the yellow-flowered *P. delavayi* with cultivated *P. x suffruticosa* varieties. In 1897, Louis Henri introduced the first of the new hybrids called 'Souvenir de Maxime Cornu', and ten years later, Victor Lemoine introduced a fine yellow tree peony called 'L'Espérance'. It was the first of an excellent range of hybrids developed by the nursery that are now classified as *P. x lemoinei*. These included 'Alice Harding' (1935), which has pollen compatible with herbaceous peonies, leading to the creation of today's Intersectional Hybrids. In America, Arthur Saunders (1869–1953) used *P. delavayi* to develop dark-flowered 'Black Pirate' and several other fine hybrids, which have been described as 'some of the world's most stunning garden plants'.[74]

Photo Credit: Roy Lancaster

CHAPTER 7

The Riches of Yunnan

—

These mountains bear a flora which, apart from its intense botanical interest, is

composed of those natural orders and genera most rich in plants of horticultural

and economic value.

GEORGE FORREST[1]

—

Père Delavay had been in north-west Yunnan for only twenty-two months but he had already collected so many specimens that he was desperately in need of a botanical listing of the plants he had already sent to Paris. He asked Père David to let him have:

> the names of all the plants that I send; this would enable me to orientate myself amongst the large number of genera and species that I have to sort. It would also be very useful when I am collecting as such a classification would help jog my memory, so that I could recall more readily the places that I need to visit and the species that I need to collect, and whether they should be in flower or in fruit.[2]

He went on to say how delighted he would be if Franchet asked him for replacements for any unsatisfactory specimens and specified exactly what botanists wanted, as this would help him to concentrate on specific items and would 'stimulate his zeal which sometimes flagged'.[3] Even for a collector as dedicated as Père Delavay, it was not easy to maintain his enthusiasm or to work with continued energy and fervour in the complete isolation in which he lived. He himself was well aware of this and in

March 1885 he wrote to Professor Bureau to thank him for all the support he, Franchet and David had provided because, as he said, isolated collectors always needed a great deal of encouragement. This admission reveals something of Père Delavay's loneliness and intellectual solitude. It is hard for us today, who take for granted the instant access to the outside world provided

Photo Credit: David Rankin

ABOVE

Omphalogramma delavayi flowering in midsummer at 3,800 m (12,468 ft) on Cang Shan. Franchet originally classified it as a new type of primula.

LEFT

Peak above the main ridge of Cang Shan.

by the the web, social media, mobile 'phones, television and radio to appreciate the extent to which missionaries like Père Delavay, living deep in the interior of China, were cut off from the West. He had no reference books or journals and there was absolutely no one with whom he could discuss his finds.

The large collections he had made had shown him that certain plant genera were particularly strongly represented in the local flora. He had collected some twenty different *Primula*, for example, and he knew that most of them would prove to be new species; it was the same with other genera such as *Rhododendron*, *Gentiana* and *Pedicularis* – but why were there so many new species and why did so many of them belong to just a handful of mountain genera? He was already familiar with most of these genera from the mountains of his home but, in Europe, there were very few species from these groups – only three species of *Rhododendron* for example – and yet here there were dozens, which was remarkable considering the rela-

BELOW

Sarcoccoca hookeriana var. *digyna* in late winter .

tively small area he had been searching. He must have been longing to discuss his discoveries with fellow botanists, especially as he realised that such a preponderance of species in one area would have important implications for plant geography: but he was quite alone and all his queries had to be bottled up. His frustration must have been intense. He was making some of the most exciting botanical discoveries of the century, but the only outlet he had for his ideas were letters – with five to six months' delay before he could expect any answers.

Franchet's first papers

Père Delavay's small packets of dried plants began to arrive in Paris from September 1883 onwards, and Franchet, who had recently completed his descriptions of Père David's plants, was able to devote himself to studying and describing them. They proved a revelation. Prior to their arrival, the flora of Yunnan was unknown in the West but Père Delavay's assiduous collecting revealed unsuspected and extraordinary botanical riches. On 28 November 1884, after a year of work, Franchet delivered the first of what was to be a long and remarkable series of papers at one of the regular meetings of the Société Botanique de France.[4] In it he informed his colleagues that, of the twelve species of *Gentiana* so far collected by Père Delavay, two were already known from the Himalaya but the other ten, although related to Himalayan species, were completely new and showed many different forms. Ten new species collected in such a relatively small area was remarkable enough but, a decade later, when Franchet had finished examining all Père Delavay's *Gentiana* specimens, the total number of new species stood at thirty-two, which made *Gentiana* one of the most prominent plant genera in the flora of north-west Yunnan. Unfortunately from a horticultural point of view, although a few of Père Delavay's gentians have been cultivated in the past, including *Gentiana rigescens* and *G. cephalantha*, none have become established in cultivation, perhaps because many of them are autumn-flowering and rarely set viable seed in the West, making propagation more difficult.[5]

By January 1885, Franchet was ready to share some of his preliminary ideas concerning the newly discovered flora with his colleagues. He told them that, as had been seen with *Gentiana*, the specimens he had examined appeared to have a great affinity with the flora of the Himalayan massif. Franchet also remarked on the high number of new species and the predominance of certain genera. These were patterns that would become very familiar to botanists in the coming years, but even at this early stage it was apparent – as Père Delavay had also realised – that the discovery of so many new species in one small area shed new light on contemporary ideas of plant geography.

In July, Franchet informed his colleagues at the Société that 300 specimens had now been received at the Muséum, of which more than *two thirds* appeared to be new species. This was remarkable, but even more astonishing was his news that, so far, only a small fraction of Père Delavay's specimens had been seen as the majority, collected in 1883–84, had been delayed by the war and were still on their way.[6] The excitement among the botanists at these meetings must have been palpable, as they discussed the new discoveries and speculated on what future collections might have in store. The contrast with Père Delavay's intellectual isolation could not have been more poignant.

New theories

It was at one of the July meetings that Franchet described the new species of *Primula* Père Delavay had found, including one with deep purple flowers discovered on Cang Shan that Franchet called *Primula delavayi*.[7] *P. delavayi* was unusual in that it produced its flowers before its leaves and had winged seeds, characteristics that Franchet had never seen in other primulas.

This meant that he had to modify the way in which botanists defined species of both *Primula* and the related genus *Andro-sace*, and create a new subgenus or group within *Primula* that he called *Omphalogramma* to accommodate the new species.[8] As more discoveries were made by the missionary-botanists, Franchet later revised this initial assessment and decided that *Omphalogramma* should be treated as a distinct genus.

Further discoveries caused botanists to reassess other accepted ideas. Ever since Sir Joseph Hooker had discovered some two dozen new species of *Rhododendron* during an expedition to Sikkim in the Himalaya in 1848–51, botanists had believed that the centre of the genus lay in the Himalayan massif: but in April 1886 Franchet read a paper that turned such thinking on

BELOW

Rubus tricolor growing on the slopes of Cang Shan where Père Delavay first found it.

its head. He calculated that Père David at Baoxing and Père Delavay in the mountains around Dali had discovered a total of thirty-six new species of *Rhododendron* between them, with only *R. decorum* common to both regions. Franchet added that he thought Père Delavay would find more new rhododendrons once the war was over and he could travel again. If Père David's discoveries had cast doubt on the theory placing the centre of *Rhododendron* in the Himalaya, those made by Père Delavay clearly indicated that the heartland of the genus lay farther east, in the mountains and deep river gorges of western China.[9]

Several of the rhododendrons discovered by Père Delavay were introduced to cultivation by George Forrest and have become valued ornamental plants, including *R. campylogynum*, *R. fastigiatum*, *R. haematodes* and *R. neriiflorum*. As we have seen, Père Delavay found several of them on Cang Shan

but the range was also rich in other woody species and it was there that he found *Vaccinium delavayi*, also introduced by Forrest.[10] *V. delavayi*, which is closely related to Père David's *V. moupinense*, is a low-growing evergreen shrub that colonises rocky places and cliffs. It can be slow to produce its white flowers when young, but this is more than made up for by its red-flushed new growth, which is most attractive. The berries, when they appear, are blue-black.

Difficulties in 1885

Père Delavay was investigating a limestone ravine near Menhuoying in January 1885 when he came across another small evergreen shrub that has since become an unpretentious but extremely useful garden plant. This was winter flowering *Sarcococca ruscifolia* var. *chinensis* with shiny narrow leaves,

BELOW

Magnolia delavayi in summer.

scented white flowers and dark red berries; and in the woods above Dapingzi in July 1886, Père Delavay found *S. hookeriana* var. *digyna*, another, more ornamental member of the genus, which also has polished shiny leaves and highly-scented white flowers, but differs in its purplish stems and black berries.[11]

The excursion to Yulong Shan in 1884 had been so promising that Père Delavay had hoped to make a further exploratory journey to Lijiang in the summer of 1885, but the continuing war made the situation too uncertain. The situation had deteriorated so much by March that he thought it possible that the missionaries might be expelled from the province altogether and, as he could not travel far, he resolved to explore the most accessible areas of his own district as thoroughly as possible.[12] He did manage to get to Dali in June, and during an excursion to Cang Shan, discovered *Rubus tricolor*, a distinctive low-grow-

BELOW

Dipelta yunnanensis in spring.

ing bramble that is unusual, as it has neither spines nor prickles, although the stems are covered with fine reddish hairs. Wilson collected it in 1903 for Veitch and then for the Arnold Arboretum in 1908. *R. tricolor* still grows in profusion on Cang Shan alongside the roads and paths that wind up the mountain.[13]

One of the specimens Père Delavay collected near Dapingzi in July belonged to a tree that Franchet was unable to assign to an existing genus. He therefore had to create a new genus, *Delavaya*, to accommodate it and Père Delavay's discovery, *Delavaya toxocarpa*, is still the sole species.[14]

Ill-health

In the event, it was not only the unsettled state of the country that prevented Père Delavay from venturing farther afield but also his poor health. He was already beginning to feel the ill effects of the deleterious climate of Dapingzi and these, together with the meagre diet and the spare comfortless existence, sapped his strength, leaving him too weak to do more than visit his Christian communities whenever his fragile health permitted.[15]

Discoveries in 1886

The cooler winter months allowed Père Delavay to recuperate and by the beginning of 1886 he was feeling stronger. On 10 January he received a letter from Franchet that did much to restore his spirits.[16] It contained a list of all the new species that he had sent so far and this enabled him to see where there appeared to be gaps or where better specimens were needed. Now that he knew his collections were being examined so promptly and were making such an important contribution to botanical knowledge, Père Delavay returned to his botanical studies with renewed enthusiasm. By 4 April, he had prepared enough specimens for four large packages, in addition to the small packets of seeds and specimens he regularly sent by post.[17] His health had improved and he was well enough to resume plant collecting during his regular visits around the district. Crossing Heishanmen in May, he discovered a noteworthy crab apple, *Malus yunnanensis*, which colours well and produces fine red fruits in autumn. It was introduced by E. H. Wilson in 1900.[18]

Père Delavay visited Eryuan and found another of the magnificent plants that now bears his name. This time it was an evergreen tree, *Magnolia delavayi*, with handsome glossy foliage and creamy flowers. The cup-shaped flowers are produced sporadically from late summer onwards but they are short-lived and the permanent glory of the tree is its shining leaves, among the largest of any evergreen that can be grown in cool temperate climates. When the staff at Kew acquired a young sapling from Veitch's nursery in 1902, they thought it would need considerable protection and it was first grown under glass. However, although *M. delavayi* is tender, it will withstand considerable frost when established in a sheltered spot and Kew now has a fine specimen planted outside in the walled garden. E. H. Wilson recognised *M. delavayi* as one of his first really good finds when he came across it 1899 and it is a fine ornamental choice for warmer gardens. It is important that *M. delavayi* continues to be planted in parks and gardens as it is now endangered in the wild. The Veitch nursery first exhibited it in October 1912, along with other recent introductions including *Viburnum davidii*, *Ilex pernyi* and *Malus yunnanensis*. Visitors must have been fascinated to see so many newly-introduced Chinese plants already available for them to try out in their own gardens. The wealth of new plants on show illustrates how the eagerness of professional nurserymen like Veitch to exploit the commercial possibilities of recent discoveries ensured that they reached the horticultural public with as little delay as possible.[19]

Père Delavay received a very welcome letter from Franchet in May, enclosing the latter's photograph. He was delighted to have a picture of the man with whom he was beginning to develop a close intellectual partnership, but he was over-

whelmed by the 2,000 francs Franchet had sent, protesting that it was far too much. However, such a sum removed any financial hurdles that might have prevented further plant collecting. Professor Bureau and Franchet had obviously taken to heart Père Delavay's description of the loneliness in which he worked and realised that he needed tangible signs of the Muséum's appreciation. In addition to the financial subsidy, Professor Bureau had also arranged for him to be awarded the prestigious title of Officier d'Académie.[20]

At the end of the month, Père Delavay was exploring Yentzehay, a pass across the Heishanmen range where he had had found *Berberis yunnanensis* the year before. *B. yunnanensis* is an ornamental shrub with large leaves and fruits that turns a glowing scarlet in autumn, and was one of the very few plants he had managed to collect in 1885. This time he discovered *Lonicera setifera*, a shrubby honeysuckle relative that produces small pale pink flowers on bare branches at the very end of winter. A few days later, he came across *Dipelta yunnanensis*, a graceful shrub with unusual tubular flowers that are followed by papery pink-tinged winged fruits. Franchet recognised its potential as an ornamental garden plant and, at the end of his description, included the hope that it would soon be introduced; however this was not realised until 1910 when Forrest sent back seeds.[21]

Shortly afterwards, Père Delavay came across *Rhododendron bureavii*, a handsome slow-growing evergreen shrub that can reach the size of a small tree in the wild. It is chiefly valued in cultivation not for its white flowers but for its superb foliage. The young stems and the undersides of the glossy green leaves are covered with a thick pelt of eye-catching wool-like orange hairs and the new foliage is held upright in neat candelabra-like clusters. Franchet recognised that it was something out of the ordinary and named it in honour of Professor Bureau, who had done so much to encourage the discovery of the Chinese flora.[22] It was introduced by Forrest in 1904 and has become a highly-desirable garden plant.

Blue poppy

As well as discovering shrubs at Yentzehay, Père Delavay also found a small alpine species of *Meconopsis* with deep purple flowers. It was named *Meconopsis lancifolia* but, like *M. delavayi*, its relative from Yulong Shan, it has never been successfully cultivated.[23] In July, Père Delavay came across another species of *Meconopsis* at Gualapo, but this time it was not a diminutive alpine species but a sizeable blue poppy. Blue poppies have since become some of the most easily recognised and

LEFT

Rhododendron bureavii at Edinburgh Botanic Garden.

RIGHT

Meconopsis lancifolia in midsummer at 4,535 m (14,878 ft) on Hongshan, Yunnan.

Photo Credit: Harry Jans

most celebrated of all flowering plants and today, even if we have never seen living plants, we have all admired photographs of their beautiful flowers and can understand something of the impact they make on those who see them for the first time. In 1886, however, the blue poppies were virtually unknown in the West and Père Delavay, who had only seen small alpine *Meconopsis*, was wholly unprepared for his first sight of blue poppies flowering on the slopes above the pass. It is not hard to imagine his delight. There is something about their limpid colour that is always breath-taking, no matter how often one sees it. Franchet named Père Delavay's discovery *M. betonicifolia*; but although Père Delavay sent back whatever seeds he could find, he did not succeed in introducing the new species.[24]

Abortive second visit to Lijiang

Père Delavay had begun writing to Franchet every month and at the beginning of September he reported that his trip to Lijiang that year had to be cut short as constant rain meant that he could not dry his specimens properly. Another problem was the inhabitants' belief that his collecting forays would offend the tutelary mountain spirit, which would then exact revenge on their village. This had made them so afraid that they had asked him to leave.[25] He saw enough of Yulong Shan to recognise that he had barely skimmed the surface of its rich flora and he believed that many new species lay awaiting discovery, including new rhododendrons and primulas, which he thought he was now unlikely to find in his own district. He also realised that a thorough investigation of the range would take several months.[26] His assessment was correct as later plant-hunters, particularly George Forrest and Joseph Rock, spent months at a time exploring the mountains around Lijiang and found them a rich source of new species. In just one year, for example, Forrest found forty new types of *Rhododendron*.[27]

BELOW

Meconopsis betonicifolia on Laojun Shan.

Photo Credit: Harry Jans

Plague

During the year, Père Delavay had been able to send over 600 specimens back to the Muséum but, on 11 November 1886, he told Franchet that unrest in the province once again prevented him from despatching large consignments of specimens, although he could still send small packets. He apologised for not having more specimens ready to send, explaining that he had been struck down with the bubonic plague that had long ravaged his district and that it had left him so weak and exhausted that he hardly had the strength to leave his bed. He went on to say that he had no idea when he would be able to get back to work.[28] Père Delavay seems only to have mentioned this shattering episode so that Franchet would excuse what he might otherwise have interpreted as laziness, once he noticed that the supply of specimens and seeds from Yunnan had tailed off. In fact, although Père Delavay does not dwell on his illness, he had been lucky to survive, as the plague epidemic had torn through his community, killing thirty-two of the forty-eight people who had fallen ill.[29] His health never fully recovered and he remained frail and rheumatic for the rest of his life.

Horticultural developments

Sarcococcas

Both Père Delavay's sarcococcas were originally introduced from western Sichuan by E. H. Wilson in 1908, and under their common name of sweet or Christmas box have since become familiar shrubs for shade. Their delicious scent is the real glory of these small unassuming evergreens and, on a winter's day, it can overwhelm passers-by with its promise of spring. The American plantsman Dan Hinkley collected *Sarcococca hookeriana* var. *digyna* north of Dali in 1996, and describes it as: 'a sensational plant, with light green foliage contrasting well with the dark purple stems, and a steady progression of intoxicatingly fragrant flowers from Christmas through February; indeed, we deliberately planted this along our driveway… for a direct olfactory hit while driving past.'[30]

An attractive form of *Sarcococcas. ruscifolia* var. *chinensis* now called 'Dragon Gate' was collected by Roy Lancaster near Kunming in 1980.[31] It is exciting that species and varieties first discovered by missionary-botanists over a hundred years ago are still being introduced to cultivation by today's plant-hunters.

Meconopsis betonicifolia

The history of *Meconopsis betonicifolia*'s introduction to the West has been confused with that of another *Meconopsis* called *M. baileyi*. In 1913, Col. F. M. Bailey collected a blue meconopsis in south-east Tibet, which was given the name *M. baileyi* when it was examined at Kew. The plant-hunter Frank

Kingdon-Ward collected seeds in 1924 from the same area and successfully introduced *M. baileyi* to cultivation. Further seed collections were made along the Tibetan and Burmese borders and *M. baileyi* soon proved itself a robust garden plant, quickly becoming a favourite across the temperate world with names such as Bailey's blue poppy, or the Tibetan or Himalayan blue poppy. As further specimens were collected and cultivated living material became available for examination, it became clear that Tibetan *M. baileyi* was very similar to Chinese *M. betonicifolia* and botanists decided that they were the same species. According to botanical rules, the first name given to a plant has priority so, in this case, as the Chinese species had been named first, *M. betonicifolia* was taken as the correct name for all the plants hitherto grown in gardens as *M. baileyi*.

Recent studies have shown sufficient differences between the Tibetan and the Chinese plants to indicate that they really are separate species. One important difference between them is their growth habit, as *M. betonicifolia* spreads by underground stolons or runners and thus forms colonies, whereas *M. baileyi* is clump-forming and is generally a much more robust plant than its Chinese relative. As they appear to be distinct species, they should rightly be known by different names, and this view has already been accepted by horticulturalists. Consequently, as virtually all the plants grown in gardens during the twentieth century under the name *M. betonicifolia* came from Tibetan seed collections, they should properly be called *M. baileyi*, the name of the Tibetan species.[32]

Now that plants long-cultivated as *M. betonicifolia* have been identified as *M. baileyi*, it has become apparent that the Chinese species is not in cultivation. It is difficult to understand how this has come about. Admittedly, *M. betonicifolia* is limited to a relatively restricted area of north-west Yunnan and seems to be much less common in the wild than Tibetan *M. baileyi*, but both Forrest and the American plant-hunter Joseph Rock managed to find it and collect seeds. Unfortunately, all the seedlings raised from these collections soon died. In 2000, the Alaska Rock Garden Group expedition collected seeds of *M. betonicifolia* from Jianchuan, just west of Heqing, and the plants they subsequently raised in Alaska are still thriving. Other seed collections have been made recently and perhaps they will also result in healthy plants of *M. betonicifolia*, so that one day we might hope to see Père Delavay's blue poppy in general cultivation.[33] It is encouraging that this species is still found growing in some of its native territory, as it is unlikely that many plants survive around Gualapo where Père Delavay first found it. That area is now intensively cultivated and grazed: the original rhododendron forests have been destroyed; and the slopes alongside the road down from the high pass to the town of Heqing are now disfigured by gravel pits, landslips, and proliferating spoil and slag heaps.

LEFT

Meconopsis baileyi in summer at Wakehust Place, West Sussex.

Photo Credit: Nick Macer

CHAPTER 8

Our Dear Project

—

As regards botany… this region [Yunnan]… is, I imagine, the most interesting in

the world. It is evidently the headquarters of most of the genera which are now

spread all over Europe and Asia…

AUGUSTINE HENRY[1]

—

IN SPITE OF HIS PESSIMISTIC LETTER to Franchet, Père Delavay had regained sufficient strength by the middle of November to undertake a few light tasks and begin working again on his specimens. He made good progress and among the first that he had ready to send off was the beautiful poppy from Gualapo, which he described in a letter to Franchet as 'a Meconopsis with large clear blue or violet blue flowers'.[2] Later that month, in spite of communications being disrupted by unrest along the Yangtze, he received a parcel of periodicals from the Muséum, including one containing Franchet's 1885 paper on his *Primula* finds. This proof that the results of his diligent collecting were being accorded serious attention must have been enormously encouraging.

Plant distribution puzzles

Père Delavay might have been physically weak but his intellect was as sharp as ever, and in his next letter he set out his detailed comments on Franchet's *Primula* identifications, with which he sometimes disagreed, and these reveal how deeply he thought about the plants he collected. One of his principal concerns was to understand why there was such a multitude of species in only a handful of mountain genera and how they were connected to American and European floras. He did not think that climatic differences between regions were enough on their own to explain the great proliferation of species that he had uncovered, although he agreed with Asa Gray's theory that prolonged periods of ice would wipe out the flora of an area, perhaps leaving only a few plants to survive in isolated pockets. This theory explained the pattern of disjunct distribution revealed by comparing the floras of eastern North America and western China, but Père Delavay also speculated that North America and Yunnan had once been joined by a chain of mountains that had since been submerged by the Pacific, thus isolating the two floras.[3] He was right to suspect that dramatic changes in the physical landscape had something to do with the species-rich flora that he was uncovering, although his ideas about the linking mountains were wrong. The origins of the narrow mountain ranges and steep valleys he was investigating actually lay in the movements of the plates forming the earth's crust, and as these were only understood with the development of the theory of plate tectonics during the twentieth century, it is not surprising that Père Delavay and his contemporaries found it so hard to come up with geological explanations for the remarkable landscape that they found in western China.

Through the study of tectonic plate movements, geologists have now been able to establish that, forty to fifty million years ago, the northwards drift of the Indian plate led to a collision

LEFT

Schefflera delavayi in the foreground (with *Rhus chinensis*) in its native habitat.

with the Asian plate and over the slow millennia, as the Indian plate pushed north, the whole Asian landmass folded and buckled, giving rise to the Himalayan massif, the Tibetan plateau and the mountains of western China. Once these immense mountain ranges appeared, they began to attract warm wet winds that blew across southern China, keeping temperatures higher than in the ice-bound areas farther north, which meant that much of the original flora survived. As the mountains rose ever higher, they produced a variety of new habitats for these plants to colonise and the simultaneous formation of deep narrow valleys created barriers that kept individual plant colonies apart, so that they evolved quite separately. Over millions of years, this resulted in the evolution of entirely new forms of primula, rhododendron and other mountain genera.

As we have seen, when Franchet examined many of the plants that Père Delavay had sent him from Yunnan, he discovered that, although they were obviously closely related to the Himalayan flora, they frequently exhibited enough differences to warrant designation as new species and this eventually resulted in the identification of twelve new species of *Primula*, twenty-four new species of *Rhododendron*, and dozens of new species of *Gentiana*. When Franchet thought that one of Père Delavay's specimens was the same as a species already described from the Himalayan massif, comparison was easy if the Paris herbarium already had a specimen of the Himalayan plant; otherwise, he would have to borrow a Himalayan specimen from another herbarium – Kew, for example – to compare the two. More often than not, however, he would compare Père Delavay's specimen with the original published description of the Himalayan species: a process that was certainly quicker but less accurate than actually comparing physical specimens. This sometimes meant that when the new plants became better known and botanists came to re-examine Père Delavay's specimens, they disagreed with some of Franchet's original conclusions. For example, the outstanding red-flowered rhododendron, which Père Delavay had collected near Dapingzi and Franchet had identified as a new species he called *Rhododendron delavayi,* is no longer thought distinct enough from Himalayan *R. arboreum* to warrant species status, but is now merely classified under *R. arboreum* as subsp. *delavayi* (see chapter 6, p.76). Similarly, *R. crassum*, which Père Delavay found on Cang Shan, is now considered only a subspecies of the Himalayan *R. maddenii* and is known as subsp. *crassum*. Plants of *R. maddenii* subsp. *crassum* raised from seeds collected in Yunnan are generally hardier than other forms of *R. maddenii*.[4]

Nevertheless, in spite of all the geological and climate differences, some species do remain essentially the same across the huge expanse of territory that stretches across the Himalayan massif and into the mountains of western China. One of these is a scented white-flowered rhododendron that Père Delavay collected on Cang Shan in 1886, which Franchet identified as a new species he named *R. bullatum*. It is now considered to be the same as tender *R. edgeworthii* discovered by Joseph Hooker in Sikkim in 1851, and although plants of *R. edgeworthii* raised from seed collected in China are often hardier than forms from farther east, even these need shelter in cool gardens. Another species with a similarly wide distribution was a *Bergenia* discovered by Père Delavay on Cang Shan that Franchet originally named *Bergenia delavayi,* but which is now considered by many botanists to be the same as Himalayan *B. purpurascens*, a species found from Nepal eastwards into southern China, and *B. delavayi* is thus treated as a synonym. Père Delavay's discovery, with its rich red winter colour, has proved such a distinctive garden perennial that horticulturalists were unwilling to see it disappear under the *B. purpurascens* umbrella, and still distinguish it with a separate name: *B. purpurascens* var. *delavayi.*[5]

The flora of north-west Yunnan is closely connected not only to the flora of the Himalayan massif but also to the mountains of western Sichuan explored by Père David. This was brought home very clearly to Père Delavay when Franchet sent him a copy of *Plantae Davidianae* in 1887, and he realised that many of the species that Père David had found at Baoxing such as *Rhododendron decorum, Rodgersia pinnata* and the ground orchid *Pleione bulbocodiodes* were also common in his own area.[6] It is apparent from Père Delavay's letters to Franchet that he thought very highly of Père David, and that, as a result of their regular correspondence, he made zoological collections for Père David, sending him various specimens of insects and small mammals. His study of Franchet's descriptions of Père David's plants brought home to him just how much he needed such precise botanical information about his own discoveries and he repeated his request for copies of everything Franchet had published about his Yunnan plants.[7]

Collecting again

By the end of November 1886, Père Delavay, although still very weak, was well enough to venture outside and he even managed to collect a few plants, notably a new white-flowered species of *Camellia* called *Camellia yunnanensis*. This fine camellia flowers through the winter and, as it is resistant to camellia flower blight, it is one of the species being used by hybridists to breed new resistant cultivars.[8] He also collected seeds of several rhododendrons, including *Rhododendron arboreum* subsp. *delavayi* and *R. decorum*, as well as seeds of *Paeonia delavayi* and shrubby *Clematis delavayi*, which thrives in hot dry sites and was one of the very first plants he had collected after arriving at Dapingzi in August 1882.[9] He was very busy with his pastoral

Rhododendron arboreum subsp. *delavayi.*

SYNTYPE
of Nonelia insignis Franch.

N.º 2498.

EX HERBARIO
MUSEI
PARISIENSIS

HERBIER E. DRAKE

Nonelia
Nonelia insignis Franch.

EX HERBARIO
MUSEI
PARISIENSIS

Photo Credit: Kew Herbarium

Photo Credit: Roy Lancaster

duties during the next few months, as he helped his community come to terms with the tragedy that had befallen so many of its members. Despite this, he still found time for plant collecting and, in March, he was exploring close to Dapingzi when he found a most unusual shrub in the aster family. When Franchet examined the specimen he discovered that it did not fit into any known genus and so he created the new genus *Nouelia* of which this species, *N. insignis*, is the sole member. It is only found in north Yunnan and west Sichuan, where it is now seriously endangered as so much of its habitat has been destroyed. It is estimated that perhaps only 5,000 plants still remain in the wild and efforts are being made to conserve it.[10]

During this period, Père Delavay also worked hard at drying his specimens and by the beginning of May he had a further 500 ready to send off. By this time, he thought he had recovered sufficiently to investigate Yulong Shan again but he had only got as far as Menghuoying when an outbreak of typhus in the village prevented the men who were to have accompanied him from leaving their families, and this forced him to postpone the trip.[11] He botanised around Menhuoying for a while but had

to return to Dapingzi, as he felt unwell again and in need of further rest.

Père Delavay's map

On 13 August 1887, Père Delavay received two letters from Franchet that had been written in April, and in his reply he included a rough map he had drawn of the area, which showed the location of his most important collecting sites.[12] This sketch map provided Franchet with a vital guide to the geography of the area that was producing so many new species. Until he received it, the only map that Franchet would have seen was the map of China produced by the Jesuits in the early eighteenth century, and although the Jesuit map gave an overall view of the geography of Yunnan, it was not always accurate nor was it detailed enough to show the mountain ranges, towns and villages mentioned most often on Père Delavay's specimen labels. Knowledge of western China increased during the latter half of the nineteenth century, as Europeans began to travel through Yunnan and Sichuan and started mapping the provinces. These efforts were given considerable impetus by the British gov-

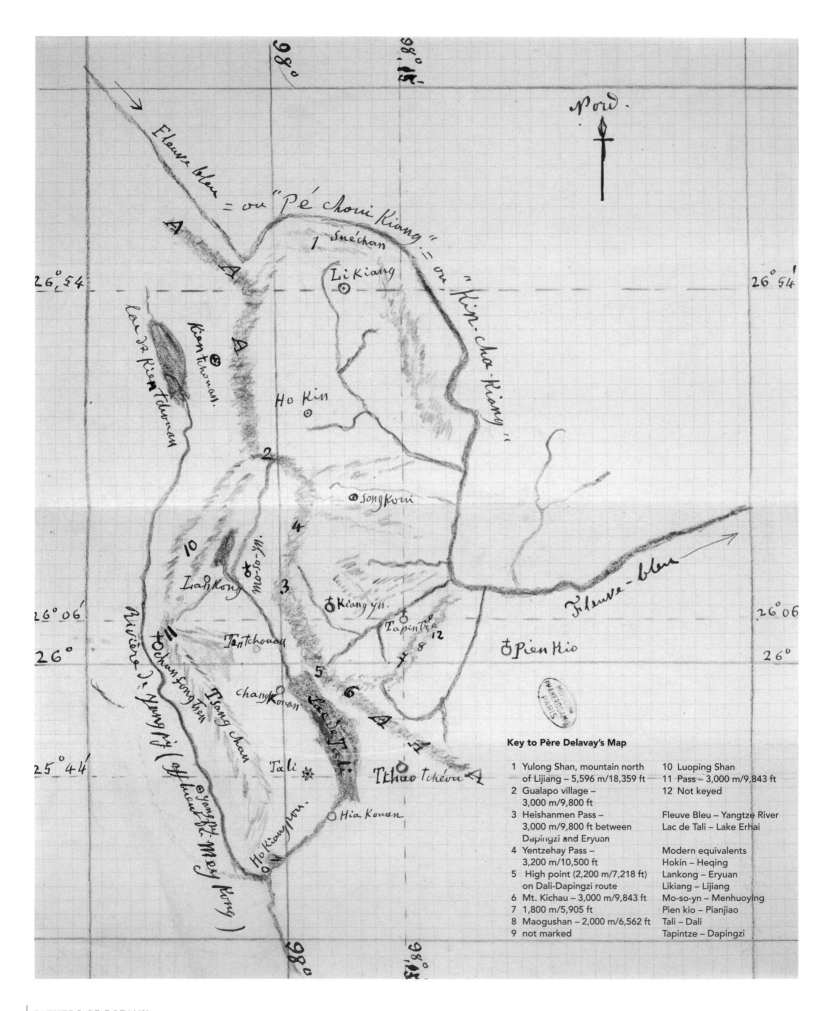

Key to Père Delavay's Map

1 Yulong Shan, mountain north of Lijiang – 5,596 m/18,359 ft
2 Gualapo village – 3,000 m/9,800 ft
3 Heishanmen Pass – 3,000 m/9,800 ft between Dapingzi and Eryuan
4 Yentzehay Pass – 3,200 m/10,500 ft
5 High point (2,200 m/7,218 ft) on Dali-Dapingzi route
6 Mt. Kichau – 3,000 m/9,843 ft
7 1,800 m/5,905 ft
8 Maogushan – 2,000 m/6,562 ft
9 not marked

10 Luoping Shan
11 Pass – 3,000 m/9,843 ft
12 Not keyed

Fleuve Bleu – Yangtze River
Lac de Tali – Lake Erhai

Modern equivalents
Hokin – Heqing
Lankong – Eryuan
Likiang – Lijiang
Mo-so-yn – Menhuoying
Pien kio – Pianjiao
Tali – Dali
Tapintze – Dapingzi

ernment, which was eager to discover a direct overland trade route from the British protectorate of Burma to Yunnan and the upper Yangtze. Consequently, they despatched officers from the India Survey team to map south-west China, and in 1909, Major Davies pulled all the available cartographic information together and published a comprehensive map although, in spite of his best efforts, certain areas of Yunnan and Sichuan were still unsurveyed and had to be left blank.[13]

Conifers

In his August letter, Père Delavay also mentioned that he was sending specimens of a dozen conifers, including four silver firs. He had collected one of the firs on the summit of Cang Shan in 1884 and Franchet, recognising that it represented a particularly fine species, called it *Abies delavayi* in his honour.[14] It is one of the most handsome of the silver firs, with violet-purple cones and eye-catching silver-white undersides to the needles. As might be expected from a mountain fir whose natural habitat lies above 3,000 m (9,845 ft) in mountain ranges that receive the full force of the monsoon rains, it requires long cold winters

ABOVE

Cone of *Abies delavayi* – the white substance is solidified sap.

BELOW

Abies delavayi growing on the west flank of Cang Shan where Père Delavay first found it.

Photo Credit: Roy Lancaster

and cool damp growing seasons to flourish. It is unfortunate that such an ornamental species is usually only seen in specialist collections and is rarely cultivated in gardens, even in the cool wet areas that might be thought ideal. When other specimens collected by Wilson and Forrest came to be examined, it was apparent that, although superficially the same, they were actually somewhat different from Père Delavay's silver fir. This led botanists to think that *A. delavayi* must be a very variable species and that the new specimens represented various subspecies or varieties.[15] Subsequent analysis has confirmed them to be completely separate, albeit closely related, species, and the correct name for the plant that Forrest collected on Yulong Shan, originally known as *A. delavayi* var. *forrestii*, is now *A. forrestii*. Wilson's specimens, collected in Sichuan and called *A. delavayi* var. *fabri*, are now known as *A. fabri*, as they have been correlated with a species that was first discovered in 1887 on Emei Shan in Sichuan by the German missionary Pastor Ernst Faber (see chapter 10, p.140). These taxonomic tangles show once again how difficult it was to differentiate between similar species and sort out relationships in a completely unfamiliar flora, and it is a tribute to Franchet's skill as a botanist that, faced with thousands of completely unknown specimens, he got so much right. Wilson later wrote that he believed the advantage of having seen living plants in the wild in all their differing stages of growth sometimes gave him a clearer insight into the identities of the various species and types involved than was possible for botanists examining dried species in the herbarium.[16]

Third visit to Lijiang

In October, Père Delavay reported that he had finally managed to get to Lijiang where he had found a few new plants, although the weather had generally been so bad that he had to return much sooner than he had hoped.[17] This was disappointing, as he did not think that he would find many more new plants in his own territory and, although he continued to collect plants as assiduously as ever to provide Franchet with extra specimens for comparison purposes, he began to focus more and more on seed collecting. He had already sent several packets of seeds to Paris but, as Franchet never mentioned them, he became despondent and began to fear that none of them had germinated. This fear turned out to be misfounded: he was quite delighted to receive a letter from Franchet at the beginning of November telling him that several plants derived from his seeds were already growing in the Jardin des Plantes.[18] This was most encouraging, as he always took particular care to collect seeds from plants that he thought were both ornamental and hardy enough to be grown outside in Paris.

This was certainly true of a *Philadelphus*, which Père Delavay discovered flowering above Menhuoying in June 1887 and

ABOVE

Philadelphus delavayi flowering on Cang Shan.

thought he recognised as *Philadelphus coronarius*, a familiar European species. He later sent seeds, which reached the Jardin des Plantes in March 1888 and when the resulting plant flowered in May 1890, Maxime Cornu duly exhibited it to the Sociéte Nationale d'Horticulture as *P. coronarius*. After further examination, Franchet realised that it actually represented a new species, which he called *P. delavayi*. It is a vigorous upright shrub with attractive deep green leaves and scented white flowers that has been described as 'one of the most beautiful flowering shrubs of any genus while it is in flower'.[19] Some forms such as var. *melanocalyx* and the cultivar 'Nymans Variety' have dark purple calyces, which are particularly striking.[20] *P. delavayi* is undisputably ornamental but it can reach 5 m (16 ft) and needs space to flourish.

Two years earlier, in June 1885, Père Delavay had found *Indigofera pendula* on the slopes above Menhuoying, and he sent back seeds at the beginning of 1888. This decorative long-flowering shrub with its pendent rosy-purple racemes is best in warmer gardens, although it will shoot back if cut to the ground during hard winters in cold areas. *I. pendula* proved the most garden-worthy of the eight new indigoferas that Père Delavay discovered and has recently been reintroduced.[21]

New deutzias

In 1889, the Jardin des Plantes received seed of another attractive Yunnan shrub from Père Delavay, but this time it was a *Deutzia* that Franchet originally identified as a variety of *Deutzia discolor*, although it is now considered a species in its own

right and is called *D. purpurascens*. The new shrub flowered the following year and was exhibited by Maxime Cornu at the Sociéte Nationale d'Horticulture on 22 May 1890. He gave stock to the nurseryman Marc Micheli and these plants flowered in April 1893. Seeds were also sent to the Arnold Arboretum, where the young shrub that was raised was overwintered in a cold frame as its Yunnan origin made the gardeners doubtful at first of its hardiness. *D. purpurascens* first flowered there in June 1893 and subsequently proved perfectly hardy outside.[22] The buds and the backs of the white star-like flowers were stained a deep purple-pink, which made a great impression on observers as, apart from the occasional pink-tinged *D. scabra*, all the deutzias known at this time had pure white flowers.

D. purpurascens was actually the second species of *Deutzia* that Père Delavay had discovered, but this was not recognised at the time as Franchet originally thought the first one, which had been found above Dapingzi in June 1883, was the same as Père David's *D. longifolia*. When E. H. Wilson and Alfred Rehder, his colleague at the Arnold Arboretum, came to re-examine Wilson's specimens of *Deutzia* during the course of their work on Wilson's Chinese collections, they realised that Père Delavay's first *Deutzia* was an entirely new species, which they named *D. calycosa*. In 1981, almost a century after its discovery, *D. calycosa* was reintroduced to the West from seeds collected on Cang Shan by the Sino–British expedition. It might, of course, already have been introduced by Forrest, but it would then have been grown as either *D. longifolia* or *D. purpurascens*.[23]

New *Incarvillea*

Père Delavay was also responsible for the introduction of a fine herbaceous species with brilliant petunia-pink trumpet flowers, which he had first found on Heishanmen in May 1883. He collected it again on several occasions during 1886 and 1887 and Franchet identified it as a member of the genus *Incarvillea*, named after Père Nicholas d'Incarville SJ, the first missionary plant collector in China (see chapter 2, p.14). Franchet called the new discovery *Incarvillea delavayi* and it seems fitting that the name of this outstanding plant should commemorate two such pioneering botanical collectors. Père Delavay sent seeds to the Jardin des Plantes and also to Maurice de Vilmorin, who raised healthy plants that flowered in 1891. Vilmorin immediately passed one of the flowers to Franchet for comparison with Père Delavay's dried specimens and, the following year, the Jardin des Plantes' own *I. delavayi* flowered.[24] The new introduction was illustrated and described in the *Revue Horticole* in 1893, when *I. delavayi's* merits as a garden plant were already apparent. This early horticultural promise has subsequently been borne out in gardens with the fertile well-drained soil in which it flourishes. *I. delavayi* grows from a fleshy tuber. In rich soil it makes a vigorous clump of large deep green leaves, which produce eye-catching pink flowers that rise up on tall stems in spring but in poor soil the plant will rapidly fade away. A form with paler pink flowers was introduced by Forrest from Cang Shan while he was collecting for Bees Nursery and this introduction is usually called 'Bees' Pink'. In 1889, Père Delavay discovered a second species, *I. lutea*, a very much larger plant

reflect the process of trial and error that nurserymen and the cultivators at the Jardin des Plantes, Kew and the Arnold Arboretum were forced to adopt. The plants they were dealing with were completely unknown and the only guides they had when it came to propagation and cultivation were instinct and experience. Over the years vast amounts of seeds were sent back from China but Charles Musgrave, a prominent British plant enthusiast and an expert propagator, took a pessimistic view of their probable fate and commented:

> the great majority of the seedlings raised from the Chinese expeditions have been lost because we did not realise that we were dealing with a new plant of which we knew nothing beyond the field notes, which rarely help in cultivation, and that our chances of success with them lay in the making of experiments.[26]

Such experiments were risky but sometimes paid off, as happened with an iris that Père Delavay sent to the Jardin des Plantes, which was first planted in a dry well-drained part of the garden – where it never grew and never flowered. In spring 1894, the cultivators decided on a change of tactics and dug it up, replanting it in a flooded hollow – whereupon it proceeded to grow apace, producing large purple flowers in June 1895. Quite by chance, they had discovered that this handsome vigorous species, which was named *Iris delavayi*, needs plenty of moisture to flourish. In suitably damp conditions, *I. delavayi* can reach 1.5 m (4.9 ft) and, like its close relative *I. chrysographes* with which it often hybridises, belongs to the group of water-loving Sibirica irises.[27] It says much for the skill and determination of all the professional plantsmen involved in the introduction of the flora of western China that they managed to grow and propagate so many of the new plants and introduce them to commerce within a few years of their discovery.

with pale yellow flowers that redden as they age. Although very beautiful, this species has never become firmly established in the West. Perhaps this is because it takes a long time to reach flowering size and although Reginald Farrer thought it grew 'like any cabbage', it has actually proved so very slow growing and so prone to sudden and mysterious collapses in cultivation that several gardeners have simply lost patience with it.[25]

Raising new plants

The differing fates of the plants raised from Père Delavay's Yunnan seeds, from the failures with many of the rhododendrons to the successes with the philadelphuses, deutzias and incarvilleas,

Death of Père Proteau

At the end of February 1888, Père Delavay learned that his great friend Père Proteau, who cared for the neighbouring missionary district, had died at Dali. He was devastated. The companionship and support of his neighbour had eased the loneliness of his solitary existence and from his letters it is clear that he was grief-stricken at the loss of his friend. Even though Père Proteau was not much interested in botany, he had facilitated Père Delavay's plant collecting expeditions by looking after his community during his absences and Père Delavay knew that his replacement would not be as likeable or as obliging. In the event, Père Proteau was not replaced immediately, which meant that Père Delavay had to look after the neighbouring district as well as his own. He was full of despair and even his botanical pursuits lost their appeal. As he wrote to Franchet in March, 'I need to work to cheer myself but work itself has become laborious'.[28] His spirits remained low and he was very grateful for Franchet's continued correspondence. He thanked him at the end of May, 'Your letters raised my morale a little as, since my dear neighbour Père Proteau died, I haven't had the heart for anything and have had to force myself to do a small amount of work, but your letters have given me new strength.'[29]

He had decided the previous December to further his investigations of the 'vast forests' that flourished at lower levels throughout the region and, in spite of his low spirits and continued ill health, he resolved to continue with this particular project. The difficulty with such exploratory excursions into sparsely-populated areas had always been to find lodgings for himself and the men accompanying him for the two or three nights that they would be away.[30] However, he knew that if his Yunnan collections were to be as comprehensive as possible, he had to extend his collections to include the plants growing at low altitudes in the hot narrow valleys, as they were very different from those that he had already found in the much colder alpine zones. This meant further exploration of the mixed deciduous and coniferous forests that clothed the middle slopes of the mountains and renewed investigation of the side gorges, which were choked with a dense undergrowth of shrubs.

It was as well that Père Delavay made his collections when he did, while much of the old-growth vegetation still remained because, even as he did so, the situation was changing. Once peace had been restored to Yunnan after the defeat of the insurgents in Dali, prosperity returned to the area and the population began to grow quickly. The lower slopes of the main valleys had already been denuded of timber and brushwood by the local inhabitants, but more land was now required for housing and for growing food, which put increasing pressure on existing farmland and on natural resources such as timber. To meet these needs, villages expanded and hamlets appeared in formerly uninhabited spots, and the previously untouched valleys and accessible slopes were cleared of vegetation and turned over to intensive agriculture. Higher up the mountains, where the terrain was too steep for cultivation, trees and shrubs were chopped down for timber, firewood and fodder, and the meadows were grazed by sheep and cattle, with herds of goats taking whatever was left. Only the flora on the highest slopes survived relatively intact.

Deforestation

The destruction is particularly apparent in some of the valleys and slopes of the Lopinchan (Luoping Shan), the northward continuation of the Cang Shan range to the west of Eryuan. After one of his visits, Père Delavay commented in a letter to Père David that, although he had collected some good plants, he found these mountains 'gloomy'.[31] If he found them 'gloomy' then, it is hard to imagine what he would say if he could see the same mountains today. The great forests have gone, pillaged for their timber, and the soil has been stripped of the grasses and herbaceous plants that held it together so that it now blows about in great yellow clouds, which cover the sparse young pines and struggling saplings in a thick pall of dust. This blurs the outlines of the stumps which are all that now remain of the

mature forests that once flourished here: only their great size indicting the impressive stature of the trees and rhododendrons that once covered the whole area. Several stumps have produced new shoots but these are thin and etiolated, flapping weakly in the wind; and straggling bushes of *Rhododendron yunnanense* cling to the shifting soil, although the colour of their pale flowers is barely discernible through the enveloping layer of yellow dust. It is very hard to imagine the teeming wonders of the original vegetation when confronted with today's ravaged and despoiled landscape.

Luoping Shan presents a particularly bleak example of deforestation but most of this part of western Yunnan has been affected and there are very few old trees to be found in the open countryside. An occasional mature shade tree survives in the villages but otherwise the vegetation is immature, consisting mainly of young pines, principally *Pinus yunnanensis*, various scrubby oak species and straggling rhododendrons, particularly

BELOW

Deforested slopes at Luopingchan, Yunnan.

Rhododendron decorum and *R. yunnanense*. Tree stumps are ubiquitous, broken branches litter the ground, and herds of goats forage on many of the hillsides, browsing on low shrubs and eating herbaceous plants and grasses right down to the roots. Nowadays, some areas are being replanted; but the species most often used is *P. yunnanensis*, originally found above Dapingzi by Père Delavay and chosen for its vigour rather than for any other attribute. At least, these young pines stabilise the soil and provide some shelter from the wind and, although visually dull at the moment, they will become more interesting in time. E. H. Wilson saw mature trees of *P. yunnanensis* and he described their bark as deeply-fissured and peeling off in red strips on the upper part of the trunk.[32] In effect, though, a lacklustre monoculture is now replacing the original diverse mix of species.

Today, it is only in the most remote areas and on the highest slopes that some of the plants first collected by Père David and Père Delavay, and by plant-hunters like E. H. Wilson and George Forrest who followed soon afterwards, can still be found. However, their dried specimens provide a record of the indigenous flora of the region and the living material they collected ensured that many plants, which might otherwise have been lost, survive in cultivation.

It was during a visit to Luoping Shan in 1884 that Père Delavay found one of the most attractive of his discoveries, an elegant shrub that Franchet named *Clethra delavayi*, a relative of the more familiar eastern American shrub *C. alnifolia*. *C. delavayi* has deep green lance-shaped leaves and, in midsummer, horizontal clusters of pink buds open to scented white flower spikes that are borne on short upright stems. It was introduced in 1913 by George Forrest. The great plantsman W. J. Bean considered it the finest of all the clethras; but it has not proved easy in cultivation and is not reliably hardy below −10 °C (14 °F).[33]

New genus

Two years later at Luoping Shan, Père Delavay discovered an anomalous primula-like species with large blue flowers resembling those of *Vinca major*, which Franchet named *Primula vinciflora*, and assigned to *Omphalogramma*, the same new subgenus of *Primula* that he had created for *P. delavayi* (see chapters 7 and 15). When he decided that the differences were sufficient to warrant the recognition of *Omphalogramma* as an entirely distinct genus, these erstwhile primulas became *Omphalogramma delavayi* and *O. vinciflora*.[34] *Omphalogramma* now contains around thirteen species, with nine species in western China and the rest in the eastern Himalaya and northern Burma.[35]

When Forrest saw *O. vinciflora* growing in masses of twenty to thirty plants, he was so delighted by its deep indigo purple flowers that — calling it by its old name of primula — he declared, 'None of the many primulas I have seen can compare in beauty with this unique plant growing in its natural habitat, which is sheltered grassy openings in pine forests at an altitude of 10,000–11,000 ft [3,200 m].'[36] Its habitat was already threatened in the 1880s and Père Delavay told Franchet that seeds from *O. vinciflora* were hard to come by, as the species grew high up in heavily-grazed pastures where there was scarcely a blade of grass left at the end of the season.[37]

Lilies

Père Delavay visited Luoping Shan again in the early summer, when he discovered a tall lily, *Lilium lankongense*, with pendent spotted turkscap flowers in varying shades of lilac and

BELOW

Omphalogramma vinciflora flowering in midsummer at 3,850 m (12,630 ft), Hongshan, Yunnan.

mauve. It is restricted to north-west Yunnan and adjacent areas of Tibet and appears to be the southern counterpart of *L. duch- artrei*, which Père David had found in Baoxing, although it does not have the same wandering habit (see chapter 4). It was introduced by Forrest in 1920 and flowered at the Edinburgh Botanic Garden in 1922.[38]

Père Delavay had discovered his first lily, *Lilium taliense*, at Gualapo in 1883. It is a tall handsome species, with large yellow-white turkscap flowers heavily marked with purple, which is restricted to the Dali–Lijiang area. Like Père David's *L. duchartrei*, which Père Delavay also collected, *L. taliense* has a wandering habit. It was not introduced until 1935 when Forrest found it on Cang Shan and it remains relatively rare in cultivation.[39]

Two of Père Delavay's other lily finds relate to *Lilium bake- rianum*, a species that was first found in 1888 by Colonel Col- let during an expedition to the Shan States of upper Burma. The first, which Père Delavay collected at Dapingzi in 1887, has long greenish white tubular flowers. Franchet originally thought it was a new species and named it *L. yunnanense*, although it is now recognised as *L. bakerianum*. He found the second lily above Menghuoying the following year and it had similar large greenish-yellow trumpet flowers, although these were spotted with purple or red, and Franchet called it *L. delavayi*. *L. bake- rianum* is remarkably variable with five different forms found throughout south-west China and *L. delavayi* is now consid- ered to be one of these variant forms and known as *L. bakeria- num* var. *delavayi*. Like *L. taliense*, its range is limited, stretch- ing from Dali to Lijiang and just into south-west Sichuan. It was introduced by Forrest in 1911, but tends to dwindle away in cultivation.[40]

Primula vialii

Père Delavay told Franchet in September 1888 that he was send- ing him twenty large packs, mainly of duplicates but with the occasional new species. One new species was a *Primula* he had found in April on the slopes above Eryuan and he asked Fran- chet to call it *Primula vialii* in honour of Père Paul Vial, one of his fellow missionaries.[41] In 1881 Père Vial had arrived at Chu-tung (Qutong), south west of Dali and part of the Yangbi missionary district, and the following year he had accompa- nied Archibald Ross Colquhoun, a Scottish railway surveyor and explorer, on a surveying journey to Bhamo in Burma.[42] Père Vial was fortunate that Qutong was two days' journey from Yangbi and that the anti-Christian riots in 1883, during which Père Terrasse was killed, did not spread to his own dis- trict. He remained at Qutong until 1885 and it was during this period that he and Père Delavay became friends. He was then

BELOW

Lilium lankongense in summer at Sir Harold Hillier Gardens, Hampshire.

BELOW RIGHT

Lilium bakerianum var. *delavayi* at 3200 m/10,498 ft near Litang, Sichuan.

Photo Credit: Harry Jans

transferred to the east of the province and, in 1887, he established the first mission to the Yi people, becoming an expert on their language and culture. He was severely wounded by thieves in 1892 and had to go back to France for treatment, but he returned to his mission in 1894 and subsequently published several works on the Sani Yi people.[43]

Archives MEP

Primula vialii was introduced by Forrest in 1906 but, as he did not realise it had already been named, he called it *P. littoniana* after G. L. Litton, the British Consul at Tengchong, a friend and plant-hunting companion who had died in 1906. *P. vialii* was the name first given to this species so it remains the correct name, but it is interesting that both Père Delavay and George Forrest thought the species handsome enough to be worthy of their friends. Reginald Farrer was also impressed, describing it in a flamboyant passage:

> It makes a tuft of upstanding oval foliage, ribbed and downy and soft; and then up shoots a tall powdered stem, terminated by a spike often 6 inches in length, of brilliant scarlet bracts, from which, as the stem grows taller, unfold the innumerable pendent little packed flowers of lavender-lilac or deep violet, till in mid-bloom the spikes seem tapering ghost-flames of blue aspiring to their long tips of crimsoned fire, making an unparalleled effect as they hold up their millions of tall steady candle-lights in the lush grasses of the Yunnanese alps...[44]

Primula. vialii is an attractive choice for cool damp areas, as are some of Père Delavay's other primula finds, including *P. poissonii*, *P. secundiflora*, *P. chionantha* and *P. flaccida*. These, and species for hotter climates such as *P. malacoides*, are fre-

quently found in cultivation, unlike most of Père David's woodland primulas which are scarcely grown even by specialists.

In April, while botanising around Dapingzi, Père Delavay discovered a handsome evergreen honeysuckle with sweetly-scented cream and yellow flowers, which Franchet thought was a new species and called *Lonicera delavayi*, although it is now considered a variety of *L. similis*, a species first discovered in central China.[45] Seeds were sent to Vilmorin in 1901 and *L. similis* var. *delavayi* first flowered in his nursery in 1904; he sent a plant to Kew in 1910. In spite of its early introduction this vigorous robust honeysuckle is not common, which is a

pity as it is ideal for covering large pergolas and arches, and although revelling in sun, as one would expect of a plant from Dapingzi, it is also reasonably hardy. Père Delavay was exploring near Dapingzi in June, when he collected a *Cotoneaster*, *C. coriaceus*.[46] The fact that there were still new species to be identified in the vicinity of Dapingzi, an area that Père Delavay had been exploring assiduously for five years, emphasises that repeated visits are necessary to establish a comprehensive view of the flora of a particular area. *C. coriaceus* was only introduced in 1993, but as it appears garden-worthy, it might eventually become established as an ornamental plant.[47] As we have seen so often throughout this narrative, the horticultural story of the missionaries' various discoveries is still very much a work in progress.

BELOW

Cotoneaster coriaceus in midsummer.

Photo Credit: Jeanette Fryer

Ill-health

Père Delavay had been extraordinarily diligent, in spite of low spirits, recurrent bouts of marsh fever and general ill health, as he had continued to collect and dry plants all through the summer of 1888. Not surprisingly, he was quite worn out by the autumn and he let Franchet know that he hoped to take a short break the following year. This is the first time that he mentions taking any sort of holiday, which gives an indication of the parlous state of his health: as well as fever and persistent anaemia, he was now suffering pains around his heart. He was completely exhausted and urgently needed to rest but he told Franchet in October that he still had a further twenty-five large packets to prepare: a task he describes as 'daunting'. In January, fifteen packs of the previous summer's plants still required attention and his desperation and exhaustion led him to write an acrimonious letter to Franchet, reproaching him for not producing any lists of the plants he had been sent from Yunnan and concluding that, as things were moving so slowly, in spite of all Franchet's fine assurances, he doubted whether he would ever live to see a report on his Yunnan collections.[48]

Franchet's publications

Then, at last, instalments of the botanical account of his plants that he had despaired of ever seeing began to arrive from Paris and Père Delavay understood immediately that Franchet had been as good as his word. When he saw the first parts of the *Plantae Delavayanae*, Franchet's enumeration of his discoveries, he was overwhelmed and told Franchet that he had not dreamed that it would be so magnificent.[49] The *Plantae Delavayanae* represented Franchet's third attempt at preparing a comprehensive description of Père Delavay's collections: in 1885, he had published the *Plantes du Yunnan*, an article that provided a complete account of the first plants received at the Muséum; the following year, he began the *Plantae Yunnanenses* and described ninety-eight new plants in the first few sections, before abandoning it, as the plants began to arrive too quickly to keep up with. He then published several papers on individual genera and species in 1887 and 1888, before making another attempt at a complete enumeration in 1889 with the *Plantae Delavayanae*. This too was abandoned in 1893, and he resumed publication of papers on specific plants. The flood of material arriving from Yunnan was simply too great for one man to work through quickly.

Identification of new species was the first task but, when several new species of a particular genus had been found, the whole genus usually needed revising to accommodate the recent discoveries. This involved checking dozens of specimens in the Muséum's collection, as well as borrowing specimens from other herbaria, and was a process that simply could not be hurried. Franchet was the acknowledged expert on the flora of the Far East, and although other specialists at the Muséum dealt with certain groups like the mosses, lichens and mushrooms, he had to do the work on most of Père Delavay's specimens himself. It was not as if Père Delavay's plants were the only ones arriving at the Muséum, either; other missionaries and plant-hunters in the Far East were sending back their collections and these also had to be determined and the new species written up. Sometimes Franchet could share the work with colleagues like Professor Bureau, but mostly he worked alone: it is quite astonishing that he was not completely overwhelmed and that for twenty years he was able to produce a steady stream of papers describing the latest discoveries.

On receipt of the printed results of Franchet's labours, Père Delavay was ashamed of his bad temper and wrote Franchet a heartfelt apology for his outburst. He spent a restful summer and was feeling a great deal better by September. Most importantly, the piles of specimens waiting to be sent off were diminishing and he believed that he could now see an end to the task he had set himself of making a comprehensive collection of the flora of his area.[50] His specimens were now mostly duplicates and, after six years' hard work, he believed that he had virtually exhausted the possibilities of finding new species.

The flora of north-west Yunnan was so rich, however, that there were still surprises, notably *Prunus serrula*, which he had come across in July 1889 on the slopes above Menghuoying at around 3,000 m (9,842 ft). This handsome cherry has become a favourite garden plant and is grown for its outstanding chestnut-red bark, which shines like highly-polished furniture. *P. serrula* is attractive all year round, especially in winter when the glossy bark stands out boldly in the bare landscape. It was introduced by E. H. Wilson, who collected it in the vicinity of Kangding, Sichuan in 1904 and again in 1908.[51] Later that year, in September, Père Delavay collected *Schefflera delavayi* in one of the gorges near Gualapo, but this fine ornamental shrub had to wait until the 1990s before being introduced to cultivation. It has already proved a handsome evergreen addition to warmer gardens, with its divided palmate leaves and upright panicles of tiny white flowers.[52]

Seed collecting

Père Delavay knew that the place that he really should investigate more thoroughly was Lijiang but he could not find anyone to look after his district for the fortnight that he would need to be absent. Now that Père Proteau had died, there was no one on whom he could rely for help, which meant that he was confined to his own district. Despite this, he was still able to take two or three days to visit Cang Shan and he made a successful expedition at the end of August to collect seeds, which had now become his main focus. He sent the results, including seeds of several species of *Primula* and of *Omphalogramma delavayi*, back to the Jardin des Plantes using the San Francisco postal route, which he had used before and found to be a quick way of getting fresh seed back to Paris.[53] What he really wanted, though, were seeds from some of Cang Shan's rhododendrons and he climbed the mountain at the beginning of December. The trip was a gruelling one: the climb took nine hours and, although he had intended to pass the night sheltering in a cave, it was bitterly cold and his assistants were frozen, so they had to make a rapid descent by moonlight. In the rush, the few seeds he had managed to collect during the climb were all jumbled up.[54]

In January 1890, Père Delavay sent Franchet more seeds from the evergreen *Rhododendron racemosum*, which he thought 'a very pretty little species that could easily be introduced in France'. He had already sent seeds of this species, which had germinated in 1889, and living material was soon sent to Kew and reached Veitch's nursery near London shortly afterwards (see chapter 6, p.81). When Veitch exhibited a flowering specimen of *R. racemosum* in 1892, it was immediately awarded a First Class Certificate by the Royal Horticultural Society, which recognised its potential as a garden plant. In the wild, *R. racemosum* is very variable in size and flower colour but the best-cultivated forms tend to be small compact shrubs with flowers in brilliant shades of pink. Forrest introduced several forms and he described *R. racemosum* as one of the most common and widely distributed rhododendrons on the mountains of south-

west China, often dominating the vegetation to the exclusion of everything else and occupying the same position as heather does in Scotland.[55] The cultivar 'Rock Rose' was raised from seeds collected by the plant-hunter Joseph Rock.

Another successful introduction was *Rhododendron rubiginosum*, a vigorous and free-flowering species, which Père Delavay sent back to Paris in 1889, and it was not long before it reached Britain. Some of the seeds he sent to the Jardin des Plantes were successfully cultivated and living material was then distributed to other botanic gardens but these forms did not usually find their way into the general horticultural trade, unless Maurice de Vilmorin also acquired some of the same

BELOW

Shining bark of *Prunus serrula*.

BELOW

Prunus serrula at Sir Harold Hillier Gardens, Hampshire.

seeds and introduced the plants through his nursery. Thus, most of Père Delavay's discoveries are now present in gardens in forms that were found by George Forrest or by other professional plant-hunters, who were often directly sponsored by nurseries. These plant-hunters were able to travel widely and usually had help. Forrest, for example, showed his team of local assistants how to collect seeds and, in the autumn, would send them back to the areas he had reconnoitred earlier in the year. This enabled him to acquire seeds of particular species from several different locations, which was very useful as forms from some areas, especially those from higher altitudes, proved hardier than those originally discovered by Père Delavay, or were

more suited to cultivation in other ways. Similarly, plant-hunters today, on the lookout for more floriferous or hardier forms of familiar plants, continue to collect seeds from different locations, thus expanding the gene pool available to cultivators and breeders, which in turn helps ensure the long-term future of a species, especially if it is under threat in its native habitat.

One of Père Delavay's finest introductions from a horticultural point of view was *Ligustrum delavayanum*, a handsome privet. He sent seeds to Vilmorin in 1889 and plants were raised that flowered in 1893. In 1900, Vilmorin sent a specimen to Kew and gave plants to the Paris nurseryman Georges Boucher, which he propagated and sold. In 1901, Alfred Rehder of the Arnold

BELOW

Rhododendron racemosum in spring.

Arboretum visited Vilmorin's collection at Les Barres and was able to examine the new privet for himself. *L. delavayanum* was well worth the visit, with neat glossy leaves and dense clusters of small white flowers with purple anthers, which are followed by black fruit; it makes a good hedging plant, although it is not for the coldest areas. Both Forrest and Wilson collected it and Forrest, who found it on the Yulong Shan, thought it one of the twelve best shrubs that he had introduced.[56]

Among the seeds that Père Delavay sent to Vilmorin in 1889, was a new rowan that was later named *Sorbus vilmorinii*. Only two of these *Sorbus* seeds germinated and although both saplings were soon lost, Vilmorin had already sent grafts to Georges Boucher, who managed to increase the stock successfully. One of the young trees flowered in 1902 but the fruits were poor. The following year the fruits were full of seeds and Vilmorin gave a grafted sapling to Kew in 1905. *S. vilmorinii*

BELOW

Sorbus vilmorinii in the author's garden in autumn.

is one of the most attractive of the rowans and is an excellent small tree for any garden that does not dry out in summer. It is attractive at every season with white flowers in spring, delicate almost fern-like foliage and excellent autumn colour, even on limey soils, when the leaves turn from red to tawny-gold. At the same time, its pendulous clusters of fruit change colour as they ripen, fading from crimson to pink and eventually to white. The long red leaf buds for the following year's foliage also appear in autumn and enliven the leafless winter silhouette. There are various similar trees in cultivation, resulting from seeds collected in different areas at different times, and these, particularly one with larger coarser leaves called *S.* 'Pearly King', are sometimes confused with *S. vilmorinii*. However, the original stock of *S. vilmorinii* seems to have derived from the seeds Père Delavay sent to Vilmorin.[57]

Maurice de Vilmorin was a dedicated plantsman who had a nursery near Paris and an arboretum at Les Barres, near Orléans in the Loire Valley, which had been founded by his grandfather. The collection of trees was extensive and constantly growing, as he had a worldwide network of contacts and also received specimens from the Jardin des Plantes, which was beginning to run out of space. In 1893, he set aside land alongside the arboretum for a fruticetum, where he could establish a collection of hardy shrubs. These collections gave him an opportunity to observe newly-introduced species as they grew to maturity; in many cases, he was the first to see living examples of plants that were known only from herbarium specimens. As he had plenty of room, he did not have to limit his collection to just one example of a particular species but could plant together individuals raised from seed collections made in different areas, which gave him several living examples for comparison. This was particularly useful when it came to variable species, or to species with distinct forms.

This was the case with some roses raised from Père Delavay's 1889 batch of seeds. When the young roses flowered, they were similar to the beautiful white Himalayan *Rosa sericea*, which is unusual among roses as it only has four petals. One of them was the same as a rose called *R. omeiensis* found two years earlier on Emei Shan in Sichuan by the German missionary Pastor Ernst Faber (see chapter 10, p.141). The second of Père Delavay's roses was a variety of *R. omeiensis* with the most remarkable thorns, and it is these that render this particular rose so striking. The thorns are very large with elongated triangular bases and they are translucent red when young, only turning woody in the second year. Sometimes these thorn bases are so large that the entire branch appears winged. This variety of *R. omeiensis* is known as var. *pteracantha* (which means winged spines), and it was much admired when first exhibited in London in 1905 by the nurseryman Arthur Paul, on behalf of Maurice de Vilmorin.[58]

Maurice de Vilmorin was generous with the plants he raised and sent plants to Kew Gardens and the Arnold Arboretum, thus giving their botanists a chance to study living material. Distributing plants among different gardens also helps ensure their survival, which is particularly important in the case of species that are threatened in the wild. One such species is *Corylus chinensis*, the Chinese filbert, which Père Delavay collected on one of the passes between Dali and Heqing. He thought it had the appearance of a lime but it is a relative of the hazel, although it is much larger, and in suitable conditions forms a

magnificent tree with flaking bark that can reach 36 m (118 ft). Père Delavay sent seeds to Vilmorin, which were successfully germinated and the resulting saplings were planted in the arboretum at Les Barres. They did well and began to flower and produce catkins regularly. Unlike a herbarium collection of dried specimens, an arboretum is a living collection and it changes and develops as it matures; and in a large arboretum like the one at Les Barres, where species from all over the world were grown together, such developments can be very interesting. It was there that, some time around 1911, wind-borne

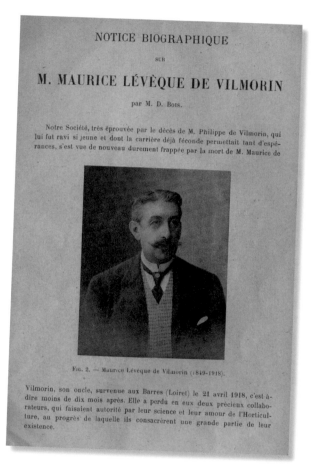

ABOVE

Maurice de Vilmorin.

RIGHT

The translucent red young thorns of *Rosa omeiensis* var. *pteracantha*.

Plate 541

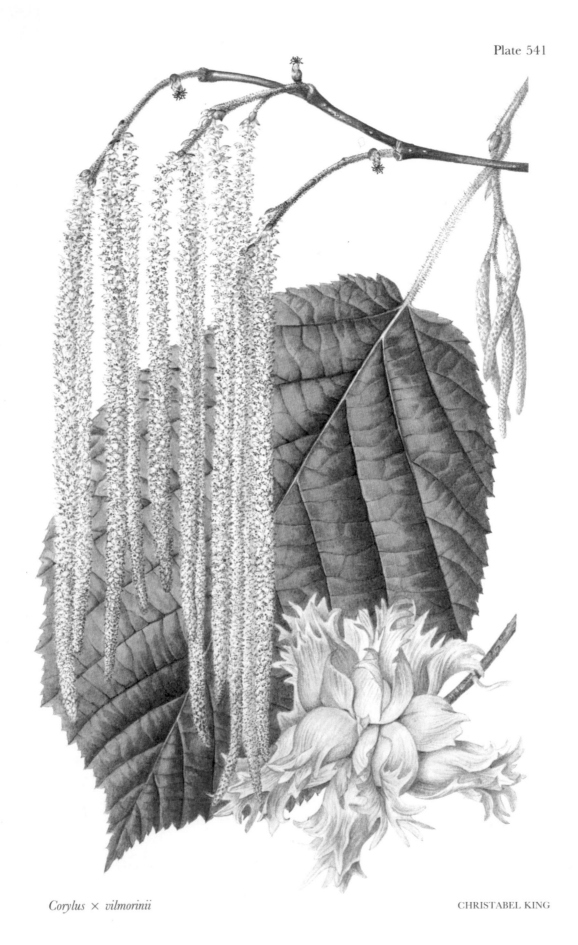

Corylus × vilmorinii

CHRISTABEL KING

LEFT

Corylus x *vilmorinii* by

Christabel King. (*Curtis's Bot.*

Mag. (2005) 22 4 Plate 541)

pollen from *C. avellana*, the common European hazel, reached female catkins of *C. chinensis* and fertile seeds were produced. The resulting hybrid, which is particularly ornamental when it develops its long pendulous catkins in late winter, was named *C. x vilmorinii* by Alfred Rehder in 1926.[59]

The successful germination of some of his seeds was encouraging, and Père Delavay now received publications and articles much more regularly from Franchet, including recent papers concerning some of his *Cotoneaster* and *Geranium* discoveries.[60] (Considering how many packages they posted to each other over the years, remarkably little seems to have gone astray.) Another paper described the thirty-five new species of *Pedicularis* which he had collected: a remarkable total for such a relatively small area. It is now known that over three-quarters of the 271 species of *Pedicularis* endemic to China are found in the mountains of north-west Yunnan and adjacent ranges.[61] More than 600 species are known worldwide, but the louseworts are rarely seen in cultivation as they are hemi-parasitic. The seeds germinate freely but the seedlings soon die, as they cannot exist without the matrix of host plants on which they depend. The only way to see established colonies of the species of *Pedicularis* discovered by Père Delavay or by Père David is to visit the high alpine meadows of north-west Yunnan and south-west Sichuan, although anyone who has seen louseworts flourishing in the mountains of Europe or North America will have a good idea of the brilliant picture they present in their native alpine meadows.

Convalescence

It was clear by the end of March 1890 that Père Delavay's health would not recover if he remained in Yunnan and he had to relinquish any plans to visit Lijiang again, as it had been decided that he was to go to the sanatorium run by the Mission Étrangères in Hong Kong, once he had finished up in Dapingzi. Nevertheless, in April he felt strong enough to do a little botanising on Ma'an Shan above Heqing and, to his great delight, he found a coltsfoot, *Tussilago farfara*, which he had last seen flowering in the Haute Savoie. It was anything but a rare plant but it was very familiar and it reminded him of home.[62] It was one of his last collecting excursions as he then concentrated on finishing as many packs as possible for Franchet, before leaving for the sanatorium where he arrived on 24 August 1890.

Horticultural developments

New deutzias

Maurice de Vilmorin, who was in touch with Père Delavay, received seeds of *Deutzia purpurascens* but, once again, it was Victor Lemoine who really made the most of the new *Deutzia*. He acquired seedlings and, having quickly recognised the breeding possibilities presented by its pink colouring, proceeded to develop some exceptional hybrids. The best known of these include early-flowering *D. x rosea* (*D. purpurascens* crossed with *D. gracilis*) and its cultivar 'Carminea'; and *D. x elegantissima* (*D. purpurascens* crossed with *D. sieboldiana*), which is now most frequently represented in gardens by its pink-flowered cultivars 'Rosealind' and 'Fasciculata'. In fact, its hybrids have virtually taken the place of *D. purpurascens* itself in gardens, which is a pity as the species is very attractive.[63]

Various colour forms of *Deutzia calycosa* ranging from white to pale pink and a large-flowered cultivar, 'Dali', are now available.

Lilium lankongense

L. lankongense is not easy in gardens but, in the 1960s, Dr Christopher North (1917–2004) of the Scottish Research Institute began a successful breeding programme, crossing *L. lankongense* with other Asiatic lilies such as *L. davidii* and its forms, particularly the hybrid 'Maxwill'. He eventually developed an excellent series of pendent-flowered lilies in a range of colours varying from deep red to pinks and apricot, which are now known as the North Hybrids. These have since become highly desirable garden plants.[64] *L. lankongense* is still being used to breed new ornamental lilies.

LEFT

Deutzia x elegantissima 'Rosealind' in the author's garden in early summer.

CHAPTER 9

Convalescence and Return

—

In the name of the love that we share for the dear plants that are one of the most

glorious manifestations of Divine power and beauty…

PÈRE BODINIER TO PÈRE DELAVAY, 1891[1]

—

Back to Yunnan

After a month's rest at the sanatorium in Hong Kong, Père Delavay felt a great deal better and ready to resume his botanical studies. When he wrote to Franchet, he made it clear that his poor health had nothing to do with plant collecting, 'My illness is not due to my expeditions into the mountains but to my lengthy eight year stay in the valley of Dapingzi in a hot and unhealthy climate.' He went on to say that he wanted to buy various botanical works and that he would be grateful for a subsidy from the Muséum, as he had used up all the funds previously sent. After the first payment in 1886, Franchet had arranged for him to be sent further sums for which Père Delavay was very grateful, although he did not always need the whole amount. In 1888 he asked Franchet to put half the money towards publishing an enumeration of his plants but in 1889 his health was so poor that he refused to take anything at all, as he did not feel that he could do enough work to justify a payment.[2]

In January 1891 he appeared well enough to return to Yunnan and he travelled by the Red River route through Tongking. When he reached Mengtze (Mengzi) in southern Yunnan, he made a plant collecting foray into the nearby hills, where he found two new rhododendrons. He asked Franchet to call one of these *Rhododendron vialii*.[3] As southern Yunnan is much hotter and more humid than the north-west of the province

with a sub-tropical flora that is generally too tender for cultivation in cool temperate regions, red-flowered *R. vialii* is only suitable for warm gardens. After a short stop, Père Delavay continued on to Kunming but by the time he arrived he was too exhausted by the long journey to go straight to Dapingzi, and had to stay with his confrères and rest. Nevertheless, he still planned further collecting expeditions and wrote and told Franchet on 24 February that, once the Muséum had sent him more funds, he had every hope of making a long trip to Lijiang and the Yulong Shan.[4] His desperation to continue with his plant collecting blinded him to the true state of his health: he was, in fact, very ill and shortly afterwards he took such a turn for the worse that he was given the Last Rites.

Return to France

Père Delavay survived but it was apparent to his confrères, if not to himself, that he would not recover if he stayed in China and arrangements were made for him to recuperate at the sanatorium the Mission Étrangères maintained at Montbeton, Tarn-et-Garonne, in south-west France, where he arrived towards the end of August 1891. He longed to go to Les Gets to see his family and to Paris to see Franchet but he was too ill; and he suffered an attack in October that left his limbs and head partially paralyzed. This further blow to an already shattered constitution would have overwhelmed most people, but the weakness of his body had not dulled Père Delavay's mind and his first priority was to find enough to occupy him during what he

saw as an inconvenient period of enforced idleness. He asked Franchet to subscribe to a botanical journal for him and to send him whatever publications he could to help pass 'the long sad days'; but by November he had decided that what he really needed was a new pastime and so he joined some of the other invalids who were learning Russian. There is then a long gap in the correspondence but a year later he was well enough to visit Franchet in Paris. It must have been an emotional occasion: the first meeting between two men who had worked together for so long on what Père Delavay referred to as 'our dear project', and whose partnership had led to an immense increase in the knowledge of the Chinese flora and to a related expansion in the understanding of plant geography.[5]

At the beginning of February 1893, Franchet reported that the Dapingzi herbarium had arrived at the Muséum in fourteen packages, but Père Delavay's relief at its safe arrival was tempered by worry as to how the 100 packets of specimens that he had left had been so considerably condensed. No doubt much was discarded that he himself would have preserved, but the care and patience that he would have given his own carefully chosen and dried specimens could not be expected from those who, however well-meaning, lacked his own extraordinary dedication to the cause of botany.[6]

By the spring, it was apparent that the dismal state of his health had been exacerbated by the long winter spent at Montbeton and Père Delavay realised that he was now too accustomed to the warmth of Dapingzi to tolerate the damp chill winters of his homeland. He resolved to go back to China before the onset of another cold season. The Missions Étrangères agreed to let him return to Yunnan, although there was now no question of him resuming his life as a missionary priest, and it was decided that he should stay at one of the missions in the north-east of the province. His departure was fixed for November and Père Delavay immediately wrote to Franchet, informing him that he would arrive in Yunnan at the end of the winter and would therefore be in time to collect the spring flowers. Now that he was too frail to undertake any missionary duties, he abandoned himself to his passion for botanical discovery, which became all-consuming. He stressed to Franchet how important it was that the specimens he sent to the Muséum were examined and classified without delay. He added a heartfelt plea, 'Nothing is more discouraging for collectors than to go for long periods without hearing anything of the plants they have collected and sent back.'[7] Yet, as we have seen, immediate classification of specimens was not always possible, especially when a plethora of species and forms was involved. Père Delavay had just read with the greatest interest Franchet's recent account of nineteen

LEFT

Rhododendron vialii at Glendoick Gardens, Perthshire.

of his *Senecioneae* discoveries, several of which he had collected in 1884, almost a decade earlier; and Franchet went on to describe a further fifteen species over the next four years.[8] In spite of the pleas of collectors in the field, the slow work of determination could not be hurried and the natural eagerness of collectors to see descriptions of their hard-won collections in print would always come up against the painstaking time-consuming work of the botanist in the herbarium.

Père Delavay visited Franchet again in April and then went to Les Gets to spend the summer with his family. His health improved a little in his native mountains and he felt able to face the long journey back. Before he went, he told Franchet that his friend Père Vial, whom he described as 'a man of great energy and a sound constitution' had returned to France to receive treatment for injuries sustained in a severe assault. He added that he was likely to make a full recovery and return to his mission in China.[9] The courage and dedication of the priests of the Missions Étrangères was remarkable. As for himself, he recognised that it would not be sensible to enter into a formal undertaking with the Muséum, now that he was so weak. As he acknowledged to Franchet, 'the investigations and the expeditions that I should have liked to continue with in Yunnan will have to be completed by younger and stronger men. After all, we must leave our nephews with something to do'.[10]

Return to China

On his way back to Yunnan, Père Delavay stopped for a day at Hong Kong where he met **Père Émile Bodinier**, a fellow-member of the Missions Étrangères and another eager plantsman with whom he had already corresponded. Père Bodinier was delighted to meet the famous collector but, as he later wrote to Franchet, although he was tremendously impressed by Père Delavay's extraordinary energy and courage, he thought that his constitution was now so weakened that the end could not be far off.[11]

On 20 February 1894, Père Delavay arrived at **Tchenfongchan**, 120 km (75 miles) south of Yibin in north-east Yunnan, which he had first visited twelve years earlier while on his way to Dali. The area was mountainous, with an oppressive humid climate where temperatures averaged about 20 °C (68 °F) and it rained every day, with almost constant mist and fog and very little sunshine. Tchenfongchan was the oldest mission station in the province and lay at 2,500 m (8,200 ft), in the midst of thick forests. Père Delavay could see that many of the trees surrounding the station were new species but he could not collect specimens as he had no way of reaching their flowers.[12] The humidity was an even more serious problem, as it made it extremely difficult to dry plants. Père Delavay kept finding that specimens he thought were completely dry had actually rotted in their papers and needed to be thrown away. He tackled the problem with his usual determination and by the end of September he had prepared 700 specimens, which represented an extraordinary level of industry for someone in such poor health working in such difficult conditions. He sent them off to the Muséum in six or seven packs, but he was not happy about their condition and suggested to Franchet that they should be swiftly unpacked and treated for insect infestations, before being left to dry out properly.[13]

Père Delavay also visited the mission at **Longki**, which lay 25 km (15.5 miles) to the west, and while there he collected flowers from a shrub that he also saw at Tchenfongchan, where he collected its fruits. Franchet named it *Stachyurus chinensis* and it is very similar to *S. praecox,* its Japanese relative, which also flowers early in the year, although *S. chinensis* is generally more vigorous and produces its long pendulous racemes of greenish-yellow flowers about a fortnight later. At the end of the summer, the buds appear on stiff polished purple-brown stems, where they are exposed to the full brunt of the winter weather, before opening in early spring. The shining stems are an attractive feature but *S. chinensis* really comes into its own when the tiny flowers open, which they do before the leaves, as the whole shrub then seems to be draped in a net of golden filaments. It is a popular shrub in North America, in areas where hot summers and mild winters encourage it to give of its best.[14]

It was at Longki that Père Delavay discovered two new representatives of the herbaceous genus *Asarum*: *A. delavayi* and *A. cardiophyllum.* Both these low-growing species produce brown funnel-shaped flowers that are hidden away at ground level beneath the glossy heart-shaped leaves. The foliage is very attractive and these asarums make good ground cover in damp shady woodland areas, as long as slugs can be kept at bay.[15]

Père Delavay found another herbaceous species in the woods around Longki, but this was a *Podophyllum*, a relative of *P. peltatum,* the may apple of North America. His discovery, *P. delavayi*, is a truly stunning foliage plant, emerging from the bare soil in tall spear-like buds, which open into large lobed velvety leaves that are strikingly marked with a tracery of mottled chestnut splashes. The pendent brownish-pink flowers, which are produced under the leaves and smell unpleasant to attract pollinator carrion flies, are certainly curious but they are hardly ornamental and *P. delavayi* is really cultivated for its remarkable long-lasting leaves. The species is slow-growing but, when planted in moist humousy soil and dappled shade, will eventually make a substantial clump about 30 cm (12 inches) high, and gardeners fortunate enough to possess the conditions *P. delavayi* requires to flourish will find it an enduring delight.[16] The species was originally introduced as *P. veitchii* by E. H. Wilson, who collected it during his 1903–05 expedition.

While at Longki, Père Delavay met **Père Édouard Maire**, a capable, energetic priest from Trondes, near Nancy, who was then forty-six. He had been in Yunnan since 1872 and had spent six years at Tchenfongchan before being moved to Longki in 1890, when he was made an assistant bishop and Superior of Lower Yunnan. In 1898, Père Maire was transferred to Kunming to oversee the building of a large new church; and while there he acted as interpreter for the Governor of French Indochina, who arrived in 1899 to finalise the route of the railway the French wanted to build from Tongking (North Vietnam) to Kunming. All the missionaries' property at Kunming, including the new church, was attacked and pillaged during the Boxer rebellion: but, once peace had been restored, the missionaries set about rebuilding all that had been destroyed. The railway was completed in 1910 and thereafter provided a direct link between the port of Hai Phong in Vietnam and Kunming, which made communications in the region a great deal faster. This was particularly useful for plant hunters like George Forrest who used it to get living plant material back to Britain and America as quickly as possible.

As assistant bishop Père Maire had many administrative duties, but he was a keen plantsman and he was inspired by Père Delavay's example to botanise whenever he could. In 1906, while based at Kunming, he discovered a diminutive allium, *Allium mairei*, that is well known to alpine garden enthusiasts as it makes an excellent subject for troughs.[17] Père Maire returned to Tchenfongchan in 1908 and continued to collect specimens for several years. He remained at Tchenfongchang until his death in 1932 aged eighty-four. His discoveries included an attractive early-flowering herbaceous peony, later named *Paeonia mairei*, and a variety of the Himalayan yew, *Taxus wallichiana*, now known as var. *mairei*.[18] Many of the plants he collected had already been found by others. One of these was a

RIGHT
Stachyurus chinensis in early spring.

LEFT
Père Édouard Maire in 1923.

ABOVE

Podophyllum delavayi in early summer at East Bergholt Place, Essex.

handsome oak *Quercus schottkyana*, first discovered by Augustine Henry, which has only recently been introduced to cultivation.[19] In 1912, Père Maire introduced *Pseudotsuga sinensis*, the Chinese Douglas-fir, to Chenault's nursery in Orléans. He thought it an exceptionally handsome species. Although *P. sinensis* needs a warm situation to flourish, attempts should be made to cultivate it as it is now under threat in the wild.[20]

Incarvillea mairei, the most famous of the plants named in his honour, had actually been collected on several occasions by Père Delavay, but the bright pink, short-stemmed flowers of *I. mairei* are so similar to those of *I. delavayi* that they had previously always been confused. Today it is known that the two species are quite separate, with *I. delavayi* found only in northwest Yunnan and west Sichuan, while *I. mairei* is widely distributed through western China, Bhutan and Nepal. *I. mairei* is now almost as common as *I. delavayi* in cultivation.[21]

Return to Kunming and final collections

As the winter came to an end, Père Delavay travelled back to Kunming. His sight was now beginning to fail but he contin-ued to botanise and one of the plants he collected in February was *Magnolia laevifolia* (formerly *Michelia yunnanensis*). He first collected this outstanding plant in June 1882 shortly after his arrival in Yunnan, and he later found it cultivated at Dali. When *M. laevifolia* was introduced to the West by the Kunming Botanical Garden in 1986, its ornamental qualities were quickly apparent and it is has recently been described as 'a superb, floriferous addition to horticulture'.[22] *M. laevifolia* is a moderately hardy evergreen shrub that does best in warm sheltered sites and produces creamy highly-scented flowers in spring and then sporadically throughout the summer.[23]

Père Delavay spent the summer at a mission station northeast of Kunming where, notwithstanding his failing health, he managed to collect double the number of specimens he had collected the previous year. Not surprisingly, given this level of activity, he fell ill again and his sight continued to deteriorate; but he still had a job to finish and, on 9 December, he sent the last seven packs of specimens collected over the summer to Franchet, along with more seeds. It was to be his final despatch. He was now sixty-one and seriously ill. On 29 December he was visited by Dr Deblenne, who was staying in Kunming with a French trade mission. The doctor could see at once there was nothing to be done but he prescribed laudanum, which relieved

Photo Credit: Roy Lancaster

Photo Credit: Gail Harland

Photo Credit: Keith Rushforth

ABOVE

Allium mairei.

ABOVE RIGHT

Paeonia mairei.

RIGHT

Taxus wallichiana var. *mairei.*

the pain and allowed Père Delavay a last few hours of peaceful sleep: for which, characteristically, he thanked the doctor.[24] The end came just before midnight on 31 December 1895.

Père Delavay was buried at the foot of the mountains near the college in Kunming run by the Missions Étrangères. When Père Bodinier, who had the warmest admiration for his confrère, visited the grave in March 1897, he was very moved to find *Gerbera delavayi* growing wild all around the site:[25] but Père Delavay's lasting memorial is in the thousands of specimens that he collected and prepared, which are now kept in the Herbarium of the Muséum d'Histoire Naturelle in Paris, with duplicates in herbaria throughout the world, and also in the numerous seminal papers describing the previously-unknown flora of north-west Yunnan published by Adrien Franchet and the other botanists who examined and identified his specimens over the years. His discoveries provide an enduring record of the native flora of north-west Yunnan and led to the expeditions of plant-hunters like George Forrest, who were eager to introduce his discoveries to Western gardens. As so many of the plants Père Delavay discovered have proved extremely ornamental, he now has living memorials in gardens all over the world. What is extraordinary is that he had accomplished so much in just ten years of active botanising, during which he had continued to fulfil his ecclesiastical duties and to administer to the scattered communities in his care, while surviving an attack of plague and suffering almost continuously from poor health.

Successors

Père Bodinier, during his visit to Kunming in early 1897, was able to inform Franchet that Père Delavay's spirit lived on in **Père François Ducloux**, the Superior of the seminary run by the Missions Étrangères at Kunming, as he shared Père Dela-

vay's love of mountain flowers. Père Ducloux, who was then thirty-three and came from Pélussin near Lyon, had been Père Delavay's successor at Dapingzi and had inherited his botany library. He had already accompanied Père Bodinier on several botanising expeditions around Kunming, and they had sent over a hundred specimens to the Muséum at the end of March. Many of these plants had previously been collected by Père Delavay, including *Magnolia laevifolia*, but there were discoveries such as tender *Vaccinium duclouxii*, which was common in the mountains of north-west Yunnan. *V. duclouxii* is larger than both *V. moupinense* and *V. delavayi*, but is only really suitable for the mildest locations. Seeds were first collected by Forrest in 1914–15, but the species has never become established in cultivation.[26] Père Ducloux described himself to Franchet as a botanical beginner but promised to collect more plants from the vicinity of the city, as he knew that this was an area that had not been searched by Père Delavay. As Père Bodinier commented, even if no new species were discovered, Père Ducloux's collections would help establish distribution patterns and contribute to knowledge of the plant geography of the region.[27]

ABOVE

Père François Ducloux in 1889.

RIGHT

Hedychium yunnanensis in midsummer.

LEFT

Incarvillea mairei.

Père Ducloux carried out his task with enthusiasm and by the end of 1899 had sent about 250 specimens to the Muséum, which included – in spite of Père Bodinier's pessimistic assessment – dozens of new species, particularly ferns. He had been thorough and his specimens were well prepared, comprehensively labelled and often accompanied by drawings. One of the first new species he discovered was *Hedychium yunnanense* with its fragrant creamy-white flowers and striking orange seeds. It is one of the hardier members of the warmth-loving tribe of ginger lilies and is therefore a good choice for cooler gardens.[28] Père Ducloux's botanising was temporarily halted in 1900 when the seminary and other mission buildings at Kunming were burned to the ground during the Boxer Rebellion, and he had to flee to Hong Kong. At the end of the year he went back to Tchenfongchan, where the Christians from Kunming had taken refuge, but he and his community were only able to return to Kunming in 1903. Once the upheavals of 1900–1901 had subsided, the situation became a great deal safer for the missionaries living deep in the Chinese interior and Père Ducloux was able to resume his plant collecting expeditions.

Photo Credit: Roy Lancaster

His collections provide a snapshot of the indigenous flora and provide a fixed reference point for later botanists. The importance of these first collections is illustrated by the example of a small yellow-flowered herbaceous plant that Père Ducloux collected in February 1906. When it finally came to be examined in 1998, after lying undisturbed in the herbarium for over ninety years, it was seen to belong to a completely new genus that was given the name *Paraisometrum*, and Père Ducloux's plant was called *Paraisometrum mileense*. *P. mileense* is found only in south-east Yunnan and is the sole known representative of the new genus. By the time it was described, it had not been seen for so long that it was considered extinct. Nevertheless, Chinese botanists rediscovered the plant in 2006 and its

BELOW

Ripened fruits of *Hedychium yunnanensis* in autumn.

identity was confirmed by checking it against Père Ducloux's specimen in the Muséum's herbarium. Seeds have been saved and botanists can now make a detailed study of this unusual and very rare species.[29]

One of Père Ducloux's most ornamental but least known discoveries was *Mahonia duclouxiana* (syn. *M. dolichostylis*), which he first collected in 1904, and again in 1905. Père Delavay had collected *M. duclouxiana* near Heqing in March 1885, but it was only described in 1908 when Père Ducloux's specimens were examined. It is an evergreen shrub with long pinnate foliage and large erect trusses of yellow-orange flowers that open from scarlet buds in mid-winter. Dan Hinkley considers *M. duclouxiana* to be the most splendid of all the flowering plants in his garden and has described it as 'the most handsome and least appreciated of the winter flowering mahonias'. Unfortunately, in spite of these outstanding ornamental qualities, it is very rare in cultivation and unavailable commercially.[30]

Père Ducloux's collections included many plants that had already been collected by others, notably Père Delavay, and already identified. One such specimen was of a tree related to a columnar conifer that was very similar to narrow upright *Cupressus sempervirens*, the familiar Italian cypress of southern Europe; indeed, when Père Delavay had found the same conifer above Menhuoying, Franchet had identified it as *C. sempervirens*. Maurice de Vilmorin, who had received seeds from Yunnan, introduced the new conifer to Europe in 1905 and this gave botanists an opportunity to compare living individuals of both the Chinese and European plants. These comparisons made it clear that, although closely allied, the two species were quite separate and it was decided to call the Chinese species *C. duclouxiana* in honour of Père Ducloux. The native habitat of *C. duclouxiana* is in the deep river gorges of the Yangtze, Lancang Jiang and Nu Jiang, and the Austrian botanist Heinrich Handel-Mazzetti saw several growing beside the track along the Lancang Jiang in 1915: 'gigantic sombre pyramids towering into the air, as regular and uniform as an avenue of poplars.'[31] Handel-Mazzetti noted that it was still common on the steep rocky slopes of the gorge, but it is doubtful if any natural old-growth stands of this species remain elsewhere in the region. It does survive as a cultivated plant, as Père Delavay had recognised when he found trees growing around human habitations, and *C. duclouxiana* is still extensively planted in Yunnan for its handsome ornamental qualities.[32]

Père Ducloux took charge of a newly-built seminary at Kunming in 1903 and was made assistant bishop in 1908. He continued to discharge his increasing administrative and priestly responsibilities until, in 1934, he gave up his diocesan duties at the age of seventy to live quietly at the seminary. He died in January 1945, aged eighty-one. He had lived in Yunnan for fifty-five years.

In one of his last letters, Père Delavay had told Franchet that his collections at Tchengfongchan and Longki represented barely half the local flora, but he knew that he could do no more himself and that the investigation of the area, and the wider province, would have to be completed by his botanical 'nephews'. He wondered somewhat dispiritedly if anyone would be interested in botany in the future: but he need not have worried.[33] The flora of Yunnan is so rich and diverse that its wonders have not yet been exhausted, and today's botanists and plant collectors are as eager to continue Père Delavay's investigations as were Père Maire, Père Ducloux and George Forrest, the first of the succession of botanical 'nephews' that followed so eagerly in his footsteps.

BELOW

Cupressus duclouxiana at Bedgebury Pinetum, Kent.

Horticultural Developments

Podophyllum delavayi

P. delavayi did not make much impact when first introduced but gardeners now value good foliage plants much more highly, and when the American plantsman Dan Hinkley reintroduced *P. delavayi* in 1986 from Emei Shan in Sichuan, demand was so great that he had to have it micro-propagated. Other seed introductions have been made recently and various forms are now available. During his 1903–1906 expedition, Wilson discovered another species of *Podophyllum*, *P. difforme*, which also has large dramatically-marked leaves and is closely related to *P. delavayi*, with which it has hybridised. An excellent *P. delavayi* hybrid called 'Spotty Dotty' retains its coloured markings throughout the season; this was raised for Terra Nova Nurseries in Oregon in 2005 and is now widely available.

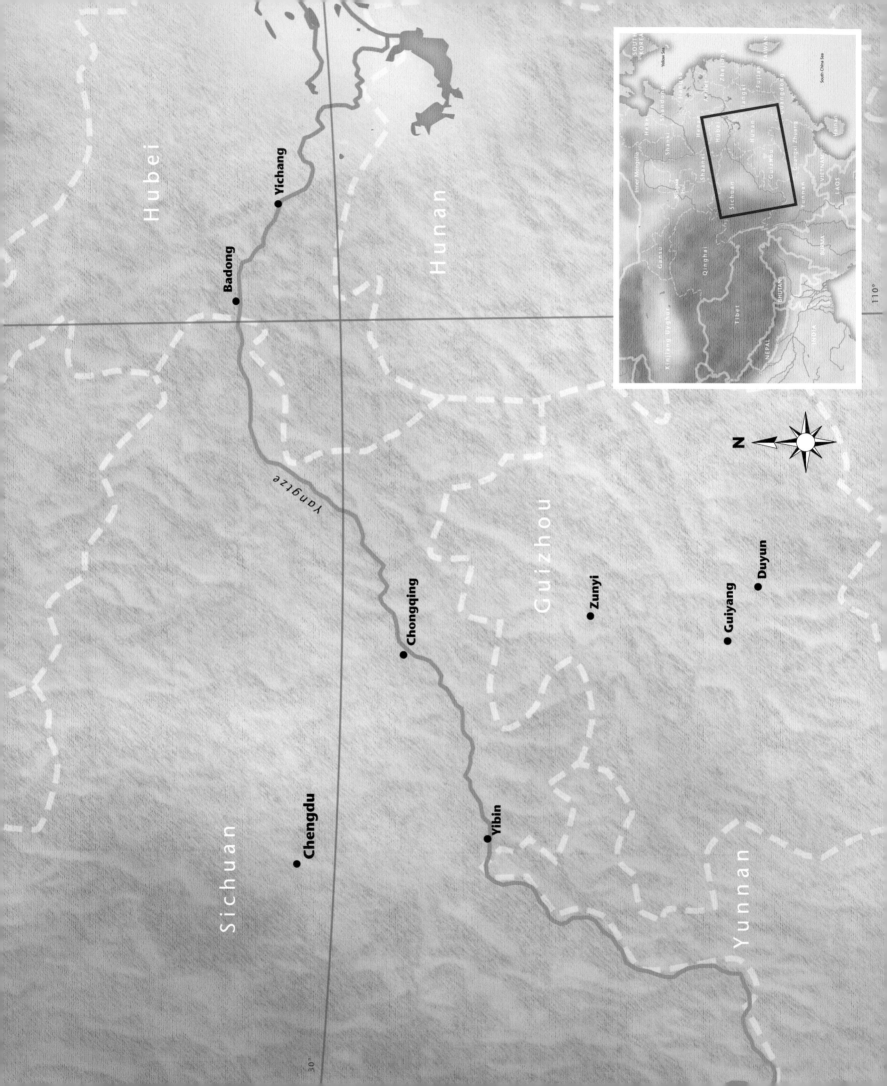

CHAPTER 10

A Most Interesting Flora

—

The opening of several new ports, and the annually increasing facilities for

penetrating into the interior of the Empire, encourage the hope that we may

soon acquire a far better and more comprehensive knowledge of one of the

most interesting Floras which can occupy the attention of botanists.

HENRY HANCE, 1878[1]

—

Guizhou

Père Émile Bodinier was able to bring his visit to Kunming to an end in 1897, when the missionaries were finally allowed to return to the province of **Guizhou**. Père Bodinier had first arrived there in 1865 as a young missionary of twenty-three, and he was delighted to be going back to his home province after a ten-year exile, even though he knew as well as anyone just how difficult life in Guizhou province had always been for the priests based there. As Père Delavay had told Franchet, the situation for the missionaries was always much worse where the provincial authorities were actively hostile, because the missionaries then had no protection in cases of attack which, when they were not punished, were frequently repeated. This had long been the case in Guizhou. Local officials and landowners encouraged their adherents to attack the missionaries and their communities, and the culprits got away with such assaults because the provincial Governor never pursued them with any vigour, in spite of the missionaries' vociferous demands for justice which were backed up by the French Minister in Beijing. The missionaries' safety was entirely dependent on the willingness of local officials to protect them and when such protection was withheld, the priests were effectively on their own. In such cases, the local authorities had to be forced to take action by robust instructions from Beijing, and it was to ensure that such instructions were issued that the French Minister in Beijing

expended his efforts. He had never had much success in Guizhou, though, and attacks were frequent.

Northern Guizhou was particularly unsettled and in 1869 the church and mission buildings at Tsen-y (Zunyi) were destroyed; then in May the following year, Père Bodinier was attacked and driven from his own mission at Erlang in the north-west of the province, after trying to protect his community. For several days afterwards, he had to hide in the houses of Christian converts, and spend nights in a cave or outside in the forest, before eventually getting away. Shortly after he reached safety, he and his confrères learned of the Massacre of Tientsin (Tianjin) on 21 July. Two years later, he was assigned to Zunyi with the task of rebuilding the church, re-establishing the school and orphanage, and re-assembling the shattered community. All was relatively quiet until 1884, when the outbreak

Photo Credit: Fonds Iconographique des Missions Étrangères de Paris

LEFT

Père Émile Bodinier.

of war between France and China inflamed latent anti-foreign feeling and once again angry mobs pillaged and burned churches and mission buildings across Guizhou. The missionaries, including Père Bodinier, were forced to leave, although the bishop commented that he did not believe that their lives were in danger as the mobs did not want to make 'martyrs' of the priests, but merely wished to destroy every outward vestige of a religion that they saw 'as foreign and even as French'.[2] Père Bodinier, who had been most closely affected by the destruction of Zunyi, went to Shanghai and Beijing to plead for justice but without immediate success; and it was not until early 1886 that the missionaries were able to return to Guizhou.

There, they found everything in ruins: their buildings razed, their possessions stolen and their communities dispersed. The unrest had died down and with their enemies momentarily quiescent, the missionaries resolutely began the slow work of rebuilding their congregations. It was at this point that Frederick Bourne, a British consular official, visited Guizhou:

> I was most hospitably received by the missionaries of the Missions Étrangères de Paris, as indeed wherever that Society is represented. The worthy Fathers forget all differences of nationality and religion in their cordial hospitality to a fellow-European. The Curé of Tsun-i, Père Bodinier, had only just returned after his expulsion during the Franco–Chinese war. Little did I think that, within three months the Father would be again a prisoner, with half the population shrieking for his blood and numbers of his converts murdered before his face.[3]

It was indeed so. Anti-foreign riots in Chongqing, an important river port that lay just north of Guizhou on the Yangtze, set off a fresh round of violence in the province in July 1887, and the Christians in Zunyi were attacked again. A mob, wielding knives, broke down Père Bodinier's door and he was only saved because some of his congregation were guarding the house. He fled and managed to reach the sanctuary of the local courthouse; but many of his community were wounded and killed, and every building connected with the missionaries was torched. He was luckier than Père Thomas Lin, a Chinese priest from one of the neighbouring mission stations, who had hurried to warn Père Bodinier of the impending attack but, on returning to his own church, was himself captured and murdered.[4] Once again, Père Bodinier had to flee the province; and his superiors, having appointed him Assistant Bishop of Guizhou, sent him to Beijing to argue the missionaries' case for compensation and reinstatement.

It was at this juncture that Père Bodinier really became aware of the important botanical contributions that some of his confrères were making, as Franchet had just published his account of Père David's plants, and his various papers describing Père Delavay's discoveries had begun to appear. Père Bodinier was himself a keen botanist, and he was particularly interested in the similarities he had noticed between the flora Père Delavay

BELOW
Caves are a typical feature of karst limestone formations such as these near Zunyi, Guizhou.

ABOVE

House close to the Loushan Pass in Zunyi County, Guizhou – modern additions include a satellite dish and electricity wires.

had found in Yunnan and the flora he knew from his years in Guizhou.[5] He had not previously thought to make collections himself, but he now seized the opportunity to explore some of the less well-known areas of the Beijing region and during the summer of 1888 he and two fellow plant enthusiasts, a Lazarist priest called Père Alexis Provôt and a Trappist monk called Frère François, explored the mountains about 160 km (118 miles) west of Beijing, where the Trappists had a monastery and ran a school of agriculture. They also investigated the high mountain chain that lay some 60 km (37 miles) northwest of Xuanhua. The flora in both these previously unexplored regions was abundant and they made extensive collections, eventually sending some 930 specimens to Père David to pass on to the Muséum.[6]

Père Bodinier's own experience of the unwillingness of the provincial authorities to assist the missionaries, in defiance of China's various treaty obligations, and his calculation of the large financial losses suffered by the pillaged missions and their communities, made him an angry and forthright advocate for the missionaries' rights in Guizhou. His fierce ardour did little to endear him to the officials at Beijing and after three years it was decided that success might be more likely if negotiations were left in the diplomatic hands of the French Minister. Père Bodinier was transferred to Hong Kong but he still firmly believed that his return to Guizhou was imminent and in August 1891, he wrote as much to Père Delavay, now in the sanatorium at Montbeton, with whom he had begun corresponding. Père Bodinier was too optimistic, and it was actually another six years before the missionaries were allowed to re-enter Guizhou.[7] In the meantime, Père Bodinier occupied himself in Hong Kong by preparing and publishing a dictionary

of the Guizhou dialect, and by collecting the island's plants. It was during this period that he finally met Père Delavay, when the latter was returning to Yunnan at the end of 1893, and he promised him that when he returned to Guizhou, he would collect plants for Franchet.[8]

Père Bodinier was a meticulous collector and he was able to make a comprehensive collection of the flora of both Hong Kong and Macau, thereby filling several gaps in the Muséum's collection.[9] He collected seeds for the Jardin des Plantes and Franchet began to send him botanical publications, both for himself and for distribution to other botanically-minded missionaries. Père Bodinier also sent some Hong Kong plants to Emmanuel Drake del Castillo, a wealthy private botanist who worked at the Muséum and who had sent him money, for which he was very grateful, as his stipend was limited and the extra funds allowed him greater freedom to collect and prepare plants.[10]

Return

The missionaries' return to Guizhou was eventually secured by the diplomatic efforts of Auguste Gerard, the French minister at Beijing, and Père Bodinier acknowledged the minister's help in a letter to Franchet, writing that he was 'grateful to my country and in my small way will do what I can to repay her in sending my plants to the Paris Muséum'.[11] These patriotic sentiments were common to the French missionaries who understood just how much they owed to France's determination to protect their position within China. Père Bodinier realised that he would not have the opportunities for botanising that he had had as a young man, and he lamented that his age (he was now fifty-five), and his responsibilities would limit what he could do.[12] He was now based in the Bishop's residence at Kouy-yang

BELOW

Houses east of Guiyang built in the style of the local Miao people.

(Guiyang), the capital of Guizhou.[13] He explained to Franchet that as he only had one day off a week and one month's holiday a year he did not have much time to devote to botany himself; nevertheless, his enthusiasm was such that he still found time to botanise during the first month in his new home.[14] He went on to tell Franchet that he had enlisted the help of those of his confrères who were also plant enthusiasts, including **Père Léon Martin** who was based farther south. As we have seen, Père Paul Perny had made extensive herbarium collections in Guizhou between 1850–1857, but most of his specimens had been lost or spoiled when they reached the Muséum in Paris, so it was as well that Père Bodinier had marshalled his confrères to provide replacements.

One of the most ornamental plants collected by Père Bodinier was *Callicarpa bodinieri*, commonly known as the beautyberry. It is a variable shrub valued by gardeners for the clusters of dramatically-coloured violet-purple berries, which accompany the bright autumn foliage and remain after the leaves have fallen, enlivening the bare winter garden. Père Bodinier first collected a fruiting specimen in October 1897, and Père Martin added a flowering branch the following June. It was collected again in Guizhou by several of Père Bodinier's confrères, including **Père Pierre Cavalerie** and **Père Joseph Esquirol**.[15]

ABOVE

Père Léon Martin aged twenty-five in 1891.

ABOVE

Père Joseph Esquirol.

Perhaps Père Bodinier's finest herbaceous discovery was *Epimedium leptorrhizum*, an evergreen species with large pale flowers suffused with lilac-pink, which was only described in 1933 and then not seen again until the 1980s, when the Japanese plant-hunter Mikinori Ogisu rediscovered it and introduced it to cultivation.[16]

In December 1897, Père Bodinier sent the fruits of his first season's collecting back to the Muséum, but he did not receive an acknowledgement from Franchet so he did not send his 1898 collection. By May 1899, he was so puzzled by Franchet's prolonged silence that he wrote anxiously asking if he had retired or if, as he had heard, the Muséum was overwhelmed by the arrival of so many new collections from China, as well as from other parts of the world. The fact that this story had reached as far as Père Bodinier in the heart of Guizhou gives some idea of the pressure on Franchet and his colleagues now that they were faced with what amounted to a tidal wave of new material. Franchet made haste to reply and as soon as Père Bodinier received his letters, he sent off three cases of specimens in November. It was to be his last consignment, as he had contracted an illness in September 1899 from which he never recovered. He died in February 1901.[17]

Père Bodinier's confrères continued to collect in the province, particularly Père Esquirol, who discovered a new species of *Camellia*, *C. costei*, in 1905, and **Père Pierre Cavalerie**, from Roussenac, Aveyron in the Midi-Pyrenées, who was based at Duyun in southern Guizhou and made extensive collections in his own district.[18] Many of the plants he collected had already

LEFT

Callicarpa bodinieri in autumn.

been found by others, including *Citrus cavaleriei*, a species of lemon hardy enough to grow outside in the British Isles, which was described from his specimens, although it had first been found by Augustine Henry. Among his discoveries was *Nothaphoebe cavaleriei*, which has since been hailed as 'one of the most beautiful of broadleaved evergreen trees... with glossy dark green leaves, glaucous grey-white below, and conspicuously veined.'[19] *N. cavaleriei* was championed in the 1990s by the late J. C. Raulston, director of North Carolina State University's arboretum, and is now available commercially in the USA. It is to be hoped that this fine species will soon be offered elsewhere.[20] Père Cavalerie's other discoveries included a new species of *Magnolia*, *M. cavaleriei*, and *Ypsilandra cavaleriei*, a hardy woodland member of the lily family. Pierre Cavalerie's later life was unsettled, as he left the priesthood and the Missions Étrangères in 1919 and retired to Kunming, where he set up as a rice merchant. He was murdered there by one of his servants in 1927.[21]

Index Florae Sinensis

In his correspondence, Père Bodinier remarked on various occasions that Franchet could produce the most wonderful Flora of China if he wanted, and Franchet was certainly the best qualified of all the botanists engaged in the study of the Chinese flora to undertake such a task. But Franchet was hard pressed just keeping up with the specimens arriving in the Paris herbarium and had no intention of getting bogged down in such an ambitious project. Sir William Thistleton-Dyer, Sir Joseph Hooker's successor as director of Kew Gardens, and Francis Forbes, a keen amateur botanist who had worked in China for the American firm of Russell & Co, and had himself made a large collection of Chinese plants, were convinced that a general checklist of all the plants found in China to date would be of such assistance to botanists that they instigated the preparation of the *Index Florae Sinensis*, which was to provide 'an

Photo Credit: Kew Archive

ABOVE

William Botting Hemsley.

ABOVE

Augustine Henry in later life.

enumeration of all the plants known from China Proper'. The project began in 1884 and was carried through by William Botting Hemsley, a botanist at Kew specialising in Chinese plants.

Hemsley's first section of the *Index Florae Sinensis* was published in instalments between 1886 and 1888, and listed all the Chinese plants that had been described to date, as well as new plants identified by Hemsley in the herbaria at the British Museum and at Kew (which had been considerably augmented by the acquisition of Henry Hance's vast herbarium in 1886).[22] When the first parts of the *Index Florae Sinensis* appeared in 1886, Franchet commented that it was a pity the authors could not have waited a few more months, as the work would then have been enriched by the 1,000 new species discovered by Père Delavay.[23] A few months' delay would not have made a great deal of difference, even though a few of Père Delavay's plants were included in later instalments of the first section, as publication of the *Index* had begun just as thousands of new specimens collected in previously unexplored areas of China began pouring into herbaria in Paris, St Petersburg and Florence. The *Index* was out of date almost before the ink was dry and the project was always beyond the efforts of one man: the modern *Flora of China*, which has been published during the last fifteen years, has involved teams of botanists from China, Europe and America and runs to twenty-five hefty volumes.

Collections in Hubei province

Quantities of Chinese specimens began to arrive at Kew from 1886 onwards, as the British had acquired a collector of their own. This was Augustine Henry, a young Irishman in the service of the Chinese Maritime Customs, who was appointed in 1882 as the medical officer at the Yangtze river port of Yichang in Hubei province. A new treaty in 1876 had allowed Britain to establish a consular office at Yichang, some 1,600 km

(994 miles) from the sea, and also to trade at three other river ports on the upper Yangtze and these concessions opened up the centre of China to British officials. They also considerably expanded the area accessible to botany enthusiasts and, by 1878, collectors were sending Henry Hance plants from these hitherto unexplored regions.

As well as his medical duties, Augustine Henry also carried out some customs duties, but he still had considerable leisure and it was really to occupy his free time that he began to collect plants around Yichang in November 1884. He quickly developed a real enthusiasm for the task and began to correspond with Hance, who gave him a great deal of encouragement. Hance was anxious to encourage Henry's initial efforts, as he recognised the value of a keen resident collector in an area that had never been properly investigated, and when Augustine Henry wrote to Kew in 1885 he found the botanists there equally encouraging. He sent his first collection totalling over 1,000 specimens to Kew early in 1886 and the response was everything he could have hoped for.[24] Still, he played down his own role, saying that the flora was so abundant and novelties were so plentiful that he hardly needed to go more than thirty miles from Yichang for the first three years. He was, he said, no more than 'a pioneer digger in a glorious goldfield'.[25]

Pastor Ernst Faber, German missionary-botanist

The Protestant missionary and renowned Chinese scholar **Pastor Ernst Faber** arrived in Yichang in April 1887 on his way to Emei Shan, a mountain in Sichuan sacred to Buddhists. Pastor Faber was one of the foremost interpreters of Chinese philosophy and beliefs for Western audiences and he naturally wanted to visit such an important Chinese spiritual site. He was also an enthusiastic botanist and had already made extensive collections in the southern province of Guangdong. Emei Shan, which had never before been explored botanically, presented tempting new territory.

Ernst Faber was born in April 1839 in Koberg in northern Germany and although he first trained as a plumber like his father, his intellectual gifts were soon apparent and his studies and strong faith led him to enter the seminary at Barmen at nineteen, before attending universities in Basel and Tübingen. His interest in natural history was already keen, and he took various courses at the Zoological Museum in Berlin and studied at the Geographical Institute in Gotha, before joining the Rhenish Missionary Society and leaving for China in September 1864. Pastor Faber was first based at Fumen in western Guangdong, which was a relatively large mission, with a chapel and a school, where his botanical knowledge led him to expand the dispensary and where his interest in medicine and natural aptitude led him to develop considerable medical skill, especially in the performance of simple eye operations. However, the Protestant missionaries were not immune from the hostility directed towards the Catholic priests, and Pastor Faber experienced something of the violence that had erupted in the attack on Père Delavay in the same province in 1867, as the chapel and the dispensary at Fumen were twice destroyed. He was, though, luckier than some of his fellows, who were driven out and their missions razed. Despite these difficulties, he did not neglect his scholarly pursuits and developed a strong interest in traditional Chinese philosophy, which led him to translate several important Chinese philosophical works. He also published his sermons in pamphlet form for use by other missionaries and prepared bible commentaries in Chinese.

He became engaged in 1870 to a lady whom he had first known in Germany but she became ill and had to return home, where she died. The death of his fiancée was a personal tragedy for Pastor Faber, but it was the possibility of marriage and a sociable family life that made the lot of Protestant missionaries in China so very different from that of the celibate Catholic priests. Although the Chinese were never reconciled to the Catholic priests living in their midst, they recognised and admired the selflessness of their lives, as asceticism was a facet of their own spiritual tradition. For their part, the Prot-

estant missionary organisations, although deploring what they saw as the priests' theological errors, recognised and admired the dedication and fervour the fathers brought to their work of evangelisation, and encouraged their own missionaries to display similar zeal.

Pastor Faber was an enthusiastic street preacher, but he eventually began to suffer from such chronic catarrh that he had to give up preaching altogether. Unlike their Catholic counterparts, the Protestant missionaries were eligible for furlough, usually after seven or ten years abroad, and in 1876 Pastor Faber went home to Germany. While there, he was offered various academic posts from organisations eager to employ someone with his growing reputation as a Chinese scholar; but he refused them all and returned to Guangzhou in 1878. Differences with the Home Board of the Rhenish Missions caused him and several of his colleagues to resign in 1880, and he went back to Germany in May 1881. He returned only four months later, having raised enough money from friends and supporters to continue working in China as an independent mission-

BELOW

Acer fabri at Westonbirt National Arboretum, Gloucestershire.

ary. He settled in Hong Kong in 1883 and, now that he had more time for botanising, he joined Charles Ford, the Superintendent of the Hong Kong Botanic Garden, on a collecting expedition to the Luofu Shan in Guangdong province in September 1883. There he discovered one of the most attractive of the plants named in his honour, a small and usually evergreen tree, *Acer fabri*, which has slender glossy green leaves that are vivid red and slightly crinkled when young. The star-shaped white and red flowers open from red buds and are followed by red winged seeds. The bark on young branches is greenish and slightly speckled. In spite of its undoubted ornamental qualities, *A. fabri* is not often seen; perhaps because, as a native of southern China, it needs a mild climate and is only hardy in sheltered gardens.[26]

A. fabri had been described by Henry Hance, whom Pastor Faber had got to know during his time in Guangzhou and who had previously received most of Pastor Faber's specimens.

Hance died in 1886 and, when Pastor Faber arrived in Yichang in April 1887, Augustine Henry suggested that he send his specimens directly to Kew. Pastor Faber was happy to agree and, even though he only spent two weeks on **Emei Shan**, his collections were so extensive that when he returned to Yichang in September, it took him and Henry six days to sort through the specimens and select a set for Kew.[27] When the collection was examined at Kew, it was apparent that, even in such a short time, Pastor Faber had found seventy new species, including fifteen ferns.

Prominent among his discoveries was a fine blue fir, *Abies fabri*, which was long considered to be a variety of *A. delavayi* but is now recognised as a separate species. It is found in conifer forests throughout western Sichuan and was introduced by E. H. Wilson.[28] Pastor Faber also found a rose that was similar to Himalayan *Rosa sericea* in that its white flowers only had four petals, but which also had more leaflets and

BELOW

Abies fabri growing on the summit of Emei Shan where Pastor Faber first found it.

BOTTOM

Rosa omeiensis in spring at 3,400 m (11,540 ft), Zhongdian, Yunnan.

BELOW

Rhodendron faberi in late spring.

Photo Credit: Seamus O'Brien

Photo Credit: Ken Cox

thick yellow fruit stalks, and is now known as *R. omeiensis*. As we have seen, Père Delavay collected a form of this rose with pronounced winged thorns (see chapter 8, p.116). Henry also found Pastor Faber's rose in Hubei and it was introduced by Wilson, who collected it in both locations.[29]

On the very summit of Emei Shan, Pastor Faber found *Rhododendron faberi*, another of the many plants named after him. Its white flowers are borne in trusses and the undersides of its leaves are covered with a dense felt, but *R. faberi* is scarce both in the wild and in cultivation.[30] As well as new species, he also collected plants such as *Schizophragma integrifolium*, which Père David had already found at Baoxing, and this provided further information about plant distribution in the region.[31] When Hemsley and his colleagues at Kew studied Pastor Faber's collections from Emei Shan, they realised that he had discovered one of the richest floras in western China and the mountain has subsequently become a magnet for plant-hunters. Pastor Faber continued to collect plants when he returned to the east coast but those native to lower altitudes like *Styrax faberi* from Zhejiang have not proved particularly hardy.[32] As he later lost his own herbarium in a fire in September 1892, it was as well that he had sent specimens to Hance and then to Kew.

In 1886 Pastor Faber had moved to Shanghai to work for the Book and Tract Society of China and thereafter he wrote indefatigably, publishing bible commentaries and translations, as well as several books in both German and English that were aimed at increasing Western understanding of Chinese society. He was awarded an honorary doctorate from Jena University in 1888 and in 1893 attended the Parliament of Religions in Chicago, where he spoke on Confucianism. He died of dysentery in 1899 and is buried where he had lived, at Qingdao in Shandong.[33]

Augustine Henry

During his various collecting forays, Augustine Henry had trained a few local inhabitants to collect specimens and seeds, and they scouted out much of the surrounding region for him. He obtained six months leave of absence in April 1888, which meant that he had time to explore the areas from which his assistants had sent the most interesting specimens. He was accompanied for part of his journey by **Antwerp E. Pratt**, a naturalist of independent means who had come to China at the behest of entomologists at the British Museum to collect insects and other zoological specimens. Pratt was primarily an ornithologist and had not considered making botanical collections but, as he intended to travel much farther west and visit some of the wholly unexplored mountain ranges towards Tibet, Henry recognised that his journey presented an unequalled opportunity for botany and persuaded Pratt to take along one of his own trained Chinese botanical collectors.[34]

LEFT
Antwerp E. Pratt in Chinese travelling dress.

Photo Credit: Pratt, A.E. (1892) To the Snows of Tibet through China

BELOW
Davidia involucrata, *Cercis chinensis* 'Avondale' and *Rhododendron yunnanense* in spring at the Savill Garden, Windsor.

The dove tree

Henry's first expedition was to the Patung (Badong) area of eastern Hubei some 80 km (50 miles) west of Yichang and it was while riding up a river valley on 17 May that he saw 'one of the strangest sights… a solitary tree of davidia in full blow… waving its innumerable ghost handkerchiefs'.[35] This was *Davidia involucrata*, the dove tree that Père David had discovered in Baoxing, although Henry's find later turned out to be a new variety now known as var. *vilmoriniana* (see chapter 11, p.163). It was the only *Davidia* Henry saw during the whole time that he spent botanising around Yichang. He collected specimens from the tree and carefully noted its position so that he was able to send his collectors to gather fruits in the autumn. When these arrived at Kew in 1889 no one thought to sow the seeds in the fruits, but when the senior botanist Daniel Oliver described the *Davidia* from Henry's specimens, he recognised its importance and commented that '*Davidia* is a tree almost deserving a special mission to Western China with a view to its introduction to European gardens'.[36]

In February 1889 Henry was transferred to Hainan Island in the far south, but by then he had sent several thousand dried specimens from Hubei to Kew and discovered some 500 new species. Many of his discoveries featured in the second part of the *Index Florae Sinensis* published in 1889.[37] Henry spent a miserable few months in Hainan, which ended in August when he fell ill with malaria and went home to convalesce. He later visited the herbarium at Kew, working on his specimens and continuing the correspondence he had begun with other collectors in China, including Père Delavay who sent him some *Pedicularis* specimens in November 1889.[38] While living in London, he married and returned to China in 1891 with his new wife Caroline. He had now given up his medical duties and worked as a full time customs officer in Shanghai, before being posted to Formosa (Taiwan) the following year. His wife, though, had tuberculosis and her health gradually deteriorated until she died in September 1894.

Henry's arrival in Yunnan

Henry went home on compassionate leave. On his return to China in July 1896 he was posted to Mengzi in southern Yunnan, where Père Delavay had discovered *Rhododendron vialii*. He continued with his plant collecting excursions whenever possible, but Mengzi was an important customs post close to the border with French Indochina (Vietnam) and his official duties as a customs officer had begun to take up so much of his time that he did not have much leisure for botanising. In 1897 he was posted farther south-west to the newly-opened customs post of Simao. By now his fame as a collector had spread widely in botanical circles and his mail contained regular requests for seeds from directors of large institutions, such as Sir William

Thistleton-Dyer at Kew and Charles Sargent at the Arnold Arboretum, as well as from plant enthusiasts like the Liverpool cotton broker Arthur K. Bulley, whose keen appetite for alpine plants had been whetted by the descriptions of the myriad new species found in the mountains of Yunnan by Père Delavay. Henry sympathised but had to explain to all of them why he could not help: apart from the fact that the flora of southern Yunnan was generally sub-tropical so that any plants he managed to collect were unlikely to be hardy in northern temperate climates, he was just too busy. As he told Sargent, 'seed collecting [is] almost out of the question, as my time is so limited…'[39]

Although Henry could not undertake systematic seed collections, he did collect seeds when he could and living plants of Père David's *Sophora davidii* were raised at Kew in 1897 from seeds Henry had sent back the previous autumn.[40] This was exceptional, though. Henry was fully aware that the most promising sources of hardy plants lay farther north in northern Yunnan, Sichuan and Hubei, and he gave the problem of their acquisition some thought. In 1897 he outlined his conclusions in a letter to Thistleton-Dyer:

> In regard to seed collecting it is not a question of money, but of finding some one with the time on hand and the requisite intelligence and energy, and this is very difficult to find indeed. I would suggest, so great is the variety and beauty of the Chinese flora and so fit are the plants for European climate, that an effort ought to be made to send out a small expedition – the funds… being provided by a syndicate… I would recommend that a man be selected, who has just finished his botanical studies… I mean don't send a collector but a gentleman, a student, and an enthusiast.[41]

Professional plant-hunter: E. H. Wilson

Thistleton-Dyer took this advice to heart and proposed just such an expedition to the nurseryman Harry Veitch, who responded enthusiastically. Veitch had been keen to acquire new Chinese plants for some time, and he had already encouraged Charles Maries, his plant-hunter in Japan, to go to China. However, Maries lacked the determination that marks the great plant-hunters and, after he fell ill during his first visit to China in 1878 and was robbed during the second in 1879, he gave up and returned to England. Maries did bring back some new plants, including Père David's *Primula obconica*, but concluded quite mistakenly that 'almost every worthwhile plant in China has now been introduced into Europe'.[42] Maries lacked staying power: clearly a collector of quite a different calibre was required, and Thistleton-Dyer had just such an individual in mind when he recommended twenty-three year-old **Ernest Wilson**, a well-qualified student of botany who had worked at the

ABOVE

Portrait of Harry Veitch.

ABOVE

E. H. Wilson.

ABOVE

Arthur K. Bulley.

Birmingham Botanical Gardens and at Kew. When Veitch met the young plantsman, he was impressed enough to offer him the job and Wilson jumped at the chance. He left for China in April 1899 and went by way of America, where he visited the Arnold Arboretum and met Charles Sargent, who gave him valuable advice on seed preparation and packaging.[43]

Wilson proved himself a skilled and perceptive collector and fully justified Thistleton-Dyer's confidence. He made two expeditions for Veitch and the seeds and bulbs he sent back to the nursery included many plants first discovered by Père David and Père Delavay, as well as new species he found himself. His success impressed Charles Sargent, who had also received Henry's advice about sending a seed collecting expedition to China, but Sargent had let the idea drop when he was unsuccessful in his attempts to persuade Henry to lead such an expedition himself.[44] Wilson's achievements revived his interest and, in 1906 when Wilson returned from his second Veitch expedition, Sargent asked him to go out to China once more, but this time to collect for the Arnold Arboretum. Wilson accepted and arrived in China in 1907, on the first of the two Chinese collecting expeditions he made for Sargent.

Fashionable alpines

Interestingly, Henry also suggested that Sargent contact some of the American missionaries based around Yichang, but **Arthur K. Bulley** had already tried asking missionaries to collect seeds and commented bitterly that he had ended up with the 'best international collection of dandelions to be seen anywhere'.[45]

Indiscriminate seed collection by well-meaning amateurs was a waste of time, especially when what enthusiasts like Bulley really wanted were alpines, and collecting the seeds of high altitude plants requires a degree of knowledge and a level of determination beyond most amateurs. When Bulley started his own nursery, A. Bee & Co. in 1903, he was determined it would feature the mountain plants of Yunnan, partly because he loved alpines himself, and partly because he knew that alpines were the height of horticultural fashion and that any nursery offering newly-introduced mountain plants from China would be commercially successful.

The interest in growing alpine plants had been started by William Robinson, the garden writer who is best known as the first champion of natural gardens. He had written the pioneering work *Alpine Flowers for English Gardens* in 1870, which was soon followed by books in a similar vein by other alpine enthusiasts such as Henri Correvon, a Geneva-based nurseryman, whose first book *Les Plantes des Alpes*, which appeared in 1884, sold out in fourteen months. Rock gardens at Kew and other botanic gardens fuelled interest in mountain plants, and specialist nurseries such as that of James Backhouse in York expanded to meet the demand for alpines and rock gardens from the increasing number of horticulturalists who wanted to take up this new form of gardening. The fashion for growing alpines was not just confined to Britain and Europe, and the first significant rock garden was built near Boston, Massachusetts in the early 1880s. Alpines appealed to American gardeners, especially those in colder areas, for the same reasons they appealed elsewhere: the plants were generally diminutive and did not take up a great deal of space in small gardens; they did not need to be cosseted in expensively heated glass houses; and they could be looked after by the owner. The demand for alpines increased throughout the period covered by this narrative and Henri Correvon was delighted to observe by 1930 a general and widespread enthusiasm for alpine plants. It was linked to the developing movement in favour of 'natural gardening', which he thought especially strong in North America, where he was greatly impressed by 'The wonderful American women… some of them are the best gardeners I have ever seen.'[46]

Professional plant-hunter: George Forrest

The appetite of the growing ranks of alpine enthusiasts was whetted by reports of the discoveries of dozens of new mountain plants made by Père Delavay and his confrères, and it was this demand that encouraged Arthur Bulley to focus his efforts on acquiring these discoveries for his new nursery. His previ-

BELOW

George Forrest and Lao Chao, his head collector from 1906-1932.

ous lack of success with amateurs convinced him that he would need to employ a professional collector to procure seeds. He wrote to the director of the Royal Botanic Garden in Edinburgh asking if he could recommend anyone and the director suggested **George Forrest**, who was working in the Herbarium at the time. Bulley offered Forrest the job and when he accepted, arrangements were made for his departure in May 1904.[47]

In all, George Forrest made seven lengthy trips to China between 1904 and 1932, usually basing himself at the newly-opened customs post of Tengyueh (Tengchong) on the Yunnan–Burma border. He was methodical and determined, and first concentrated on the mountains of Yunnan, revisiting Père Delavay's territory and, as we have seen, introducing many of the plants discovered by the great missionary-botanist, as well as finding numerous new species. He later ventured farther west into the great river gorges of the Yangtze, Lancang Jiang and Nu Jiang. Forrest was extraordinarily successful, collecting more than 30,000 specimens in all and introducing over 1,000 plants to cultivation. Père Delavay would have recognised him not only as one of the first but also as the most successful of all his botanical 'nephews'.

American plant-hunter: Joseph Rock

One of the botanists with whom Henry had corresponded before he left China was David Fairchild of the US Department of Agriculture. Fairchild was in charge of Foreign Seed and Plant Introductions, and he wrote to Henry when he visited China in 1900 asking how best to procure seeds. Henry's advice was summed up in the terse sentence: 'Don't waste money on postage – send a man.' Fairchild later wrote that 'this word of wisdom made a deep impression on me and had a great influence on my policy when I returned to the United States. Largely because of this advice I inaugurated an exploration of that vast country'. One of the longest-serving of the plant-hunters Fairchild sent to China was **Joseph Rock** (1884–1962) who lived at the foot of the Jade Dragon Snow Mountain in Lijiang and made extensive plant collections throughout the region.[48]

Joseph F. Rock in Tibetan dress on horseback, ca. 1924-1927.

Horticultural developments

Epimediums

One of the *Epimedium leptorrhizum* plants raised from seeds collected by Mikinori Ogisu had long magenta sepals that contrast very attractively with the pale flowers, and this form was propagated and given the name 'Mariko'. Like other evergreen epimediums, *E. leptorrhizum* does best in damp shady spots. The spate of *Epimedium* introductions during recent decades has inspired plant breeders to develop new hybrids and cultivars and many of these are now widely available. In 1990, Elizabeth Strangman crossed *E. dolichostemon*, another Chinese species, with *E. leptorrhizum* to produce the attractive and floriferous pink-flowered hybrid 'Enchantress', while Robin White has developed several good new forms at his nursery in Hampshire, including 'Fire Dragon', a splendid hybrid between *E. davidii* and *E. leptorrhizum* with large yellow flowers set off by pink sepals.[49]

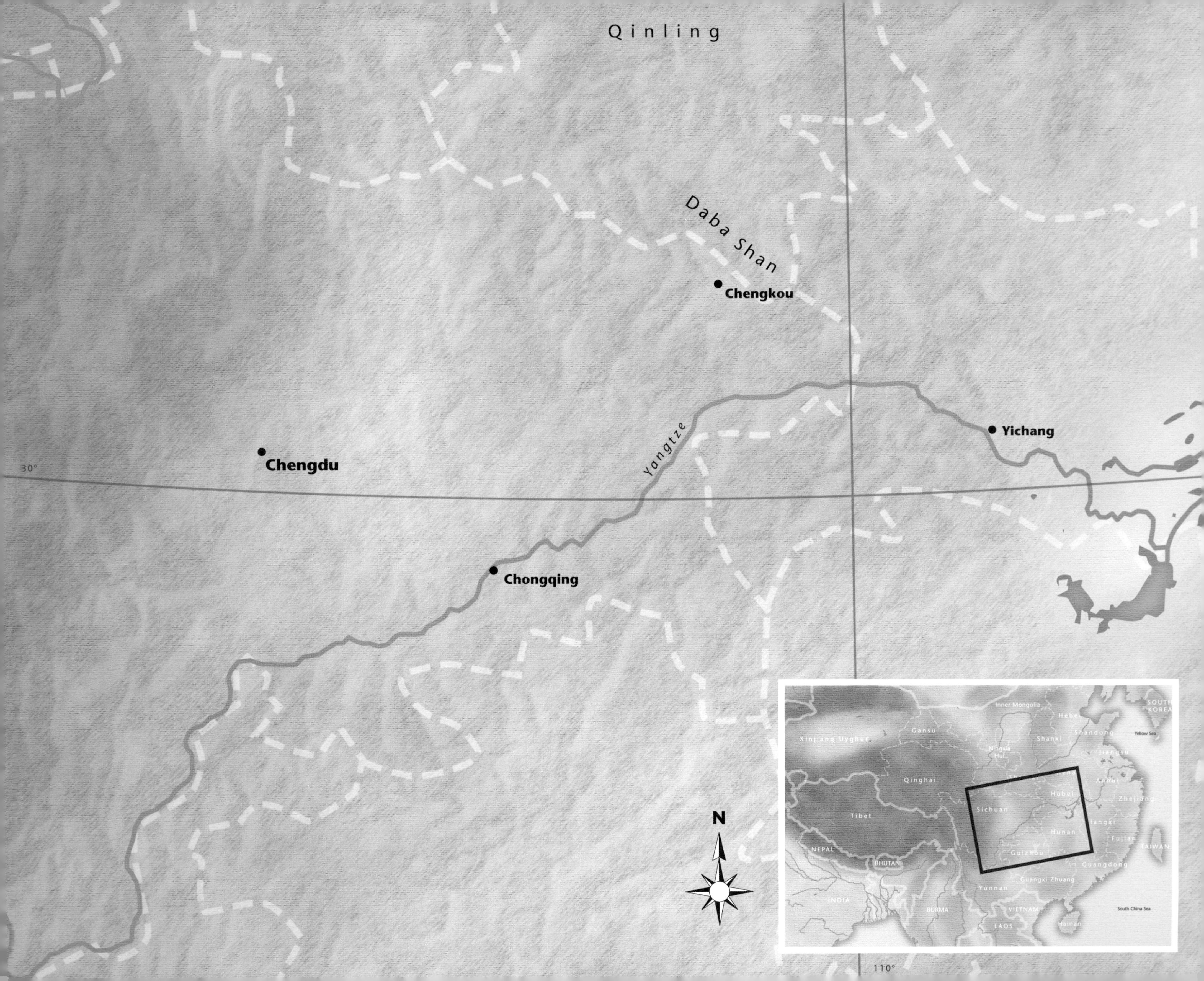

Qinling
Daba Shan
Chengkou
Yichang
Chengdu
Yangtze
Chongqing
30°
110°
N
Inner Mongolia
Hebei
Shandong
Yellow Sea
South Korea
Xinjiang Uyghur
Gansu
Shanxi
Ningxia Hui
Jiangsu
Qinghai
Hubei
Anhui
Zhejiang
Tibet
Sichuan
Hunan
Fujian
Taiwan
Nepal
Bhutan
Guizhou
Guangdong
Guangxi Zhuang
Yunnan
India
Burma
Vietnam
Laos
Hainan
South China Sea

CHAPTER 11

Interesting Specimens[1]

—

God knew the limitations of humanity… and provided the works of creation as a

means by which the Maker might be known…

ATHANASIUS[2]

—

Once descriptions of Henry's Hubei specimens began appearing in Kew's botanical publications and in the *Index Florae Sinensis*, Adrien Franchet realised that if the Paris Muséum was to keep up with the new discoveries he needed an experienced collector in this central area. He made inquiries and learned that **Père Paul Guillaume Farges** of the Missions Étrangères, who was based in a corner of north-east Sichuan, was a keen botanist. Franchet contacted him in 1891 and Père Farges readily agreed to collect for the Muséum for both 'scientific and patriotic reasons', although he pointed out that making plant collections in the area was hard, as the rain was almost continuous in spring, sometimes lasting for months at a time, and the area was heavily populated, so that any land that was not cultivated had usually been stripped of its vegetation for fodder or timber. Collecting specimens was therefore difficult as native plants were frequently only to be found in the most inaccessible spots, which meant that in many places he would have to find men brave enough to climb up and collect for him among the precipices. He went on to say that, as he had only received Franchet's letter in May, he had missed the early-flowering plants but would immediately begin collecting all those that flowered later in the season at higher altitudes.[3] To have found a missionary in this unexplored region who was obviously a competent and experienced plant collector was a stroke of luck for the Muséum and over the next decade Père Farges

provided Franchet and his colleagues with thousands of specimens, including many new species.

Père Paul Farges was from Montclar-de-Quercy near Montauban, Tarn-et-Garonne in south-west France. He had arrived in China in 1867 aged twenty-three, and had lived in the remote Chengkou district of north-east Sichuan since 1871. Even by the standards of the Missions Étrangères, the Chengkou mission was exceptionally isolated, and Père Farges' nearest confrères lived many days' journey to the south. The degree of isolation in which so many of the European priests lived is best brought home to us from the glimpses of their lives provided by Western travellers who visited their missions in the course of various journeys across China. Alicia Little gives a

Photo Credit: Fonds Iconographique des Missions Étrangères de Paris

LEFT

Père Paul Guillaume Farges.

poignant description of one French priest she and her husband met in western Sichuan in 1892:

> The priest himself, a hardy young mountaineer from Central France, showed with some pride the few panes of glass he had just had inserted in the window by his writing desk, thus enabling him to continue working when a Chinese by the darkness of his paper windows is compelled to inaction. Other luxury in his spacious sitting-room there was none, unless we count a book-case of the simplest nature to contain the few books he had brought with him from France. There was no table, three chairs, nothing more! He wore Chinese clothes, with the large fanciful straw hat of the district. He had no wine but that supplied for the Mass... he was one man alone, not a family nor a pair of friends as is so usual in [Protestant] missions. There was no European nearer than a very long day's journey across the mountains, and then not another for days and days. No seven or ten years will entitle him to a trip home to those French mountains, a tiny pictured guide to which he showed us, but which we noticed he did not venture to look at while we there... home scenes sometimes awaken too vivid memories. He received no newspapers, and it seemed few letters. We asked him how he spent his lonely evenings in winter. He said earnestly that was the great trial of the first year, but that after that one got over it.[4]

This priest had been in his mission for eleven years and Alicia Little remarked that he could recount every visit he had ever received as if each one had occurred yesterday, instead of years apart. There were tears in his eyes when they left.

Today, it is hard to contemplate the magnitude of such self-sacrifice. The missionary priests' theological training at the seminary in Paris would have prepared them to some extent for the extended periods of solitude and hardship to which they were committed but, despite their belief that they were sharing the sufferings of Christ, coping with the everyday realities of such loneliness must have been difficult. However, as one historian has commented, even though 'the life of a Catholic missionary was at best physically comfortless and tended to be mentally and spiritually depressing' many found in their faith 'resources which enabled them to grow into men who impressed travellers with the radiant strength and beauty of their lives.'[5] The energy and enthusiasm displayed by the priests for their work of evangelisation did not flag, and each year they reported numbers of baptisms and new conversions. Père Farges was particularly dedicated as his district was very poor and he lived on the same sparse fare as his neighbours, never eating bread or drinking wine. The soil was relatively infertile and agriculture was difficult, which resulted in frequent harvest failures and consequent famines; but Père Farges' community benefited from his interest in botany, which led him to introduce various types of potato and rye that were superior to traditional varieties. These helped increase food production, with a consequent improvement in the lives of the local subsistence farmers.[6]

Photo Credit: Chris Reynolds

BELOW

The Daba Shan, north-east Yunnan.

RIGHT

Valley in the Daba Shan.

Photo Credit: Chris Reynolds

One can imagine that Franchet's request gave Père Farges a welcome intellectual pursuit and, in spite of his frugal diet, he proved an indefatigable walker and for the next ten years he scoured the district in search of plants for the Muséum. Chengkou lay just south of the Daba Shan, a relatively low mountain range – the peak overlooking Chengkou, for example, did not exceed 2,500 m (8,202 ft) – dividing north-east Sichuan from Shaanxi province. The region was not as rich botanically as Yunnan but the flora was diverse, with a large number of woody plants, and Père Farges was able to send Franchet a number of large annual collections. He despatched his first consignment at the beginning of February 1892, having spent the winter – as Père Delavay did – drying his specimens and writing out labels, in which he included the plant's Chinese name and the altitude where it had been found. He had worked hard and, despite not starting to collect until late May, he was able to send back over 800 plants.[7]

New genus of bamboo

Once Franchet began to examine this first collection, he realised that among the specimens was a completely new type of bamboo, for which he had to create a new genus that he called *Fargesia* in honour of Père Farges. *F. spathacea*, the species

found by Père Farges, might have been the first *Fargesia* to be described but it was not the first species of *Fargesia* to be discovered, as Henry had already collected a fruiting specimen of *F. nitida*, the fountain bamboo, in Hubei. Grigory Potanin, the Russian explorer, had collected seed of *F. nitida* in northern Sichuan in 1889, and plants were later raised at Kew. In 1907, Wilson discovered another species of *Fargesia* in western Hubei, which was called *F. murielae* after his daughter Muriel.[8]

Père Farges' first consignment contained specimens of a shrub that he had collected when it was in flower in June 1891 – as soon as he had received Franchet's initial request – and then again when it was in fruit later in the year. Franchet recognised that the new shrub was closely related to the Himalayan *Decaisnea insignis* and he named it *D. fargesii*. In fact, *D. fargesii* had been discovered a few years earlier in neighbouring

ABOVE

Fargesia nitida at Westonbirt National Arboretum, Gloucestershire.

BELOW

Decaisnea fargesii in autumn at Westonbirt National Arboretum, Gloucestershire.

Hubei province by Augustine Henry, but Henry's specimens had only been seen by botanists with access to the herbarium at Kew, whereas Franchet brought the discovery of the new species to the attention of the wider botanical world when he published a description based on Père Farges' specimens the year after they arrived in Paris. Thus, it is Père Farges' specimens of *D. fargesii* that have become the type or reference for the species, even though Henry's were collected first. This pattern was frequently repeated, as many of the plants collected by Père Farges and described by Franchet had already been discovered by Augustine Henry. Some of the missionary-botanists might have felt that their collections were not described quickly enough but, in reality, Franchet and his fellow taxonomists in Paris served their collectors extraordinarily well.[9]

D. fargesii produces creamy-yellow flowers on bare branches early in the year, which are followed by flattened cobalt blue fruit pods that are particularly striking. The fruits were described in the *Gardener's Chronicle* in 1923 as resembling 'big fat blue caterpillars', and children today often refer to the shrub as the 'blue slug plant' (or, with gruesome relish, 'dead man's fingers'). The handsome leaves are large and composed of several leaflets, and *D. fargesii* makes an imposing specimen shrub, especially when hung about with its striking blue fruits.[10]

Maurice de Vilmorin wrote to Franchet in June 1892 for Père Farges' address so that he could ask him for seeds, and the first consignment he received from Chengkou contained seeds of over 300 different plants, including those of *Decaisnea fargesii*. Most of the seeds were sown as soon as they arrived at Les Barres in March 1895, and several of the *D. fargesii* seeds had

BELOW
Decaisnea fargesii seed pods in autumn.

germinated by June. The young plants did well outside, despite a bitterly cold winter in 1895 and had reached 2 m (6.5 ft) by 1899, when they flowered. Vilmorin gave one of the first plants to Kew in 1897 and it flowered in the Temperate House in April 1901, although it did not fruit until 1915.[11]

Père Farges' first collection also included a lily, which Franchet described as a new species called *Lilium sutchuenense*. Vilmorin received seeds in 1894 or 1895 and raised plants that flowered in 1897, which he sent to Kew where they flowered in 1899. The new lily proved easy to grow and propagate and was soon widely cultivated. It gradually became apparent that it was not a new species as had at first been thought, but actually a variety of one of Père David's discoveries, *L. davidii* var. *davidii* (see chapter 4, p.45). It is not surprising that Franchet did not recognise it when he examined Père Farges' specimen because it is hard to produce good herbarium specimens of lilies as the flowers are difficult to dry and the plant's structure is easily lost.[12] Père Farges also sent Vilmorin seeds of floriferous *L. davidii* var. *willmottiae*, which Wilson later introduced from Hubei.[13] Père Farges went on to collect another lily, *L. fargesii*, which brought the number of new species of lily found in Sichuan and Yunnan by Père David, Père Delavay and Père Farges to fourteen.[14] Previously, only ten species had been known from the rest of the world, and Franchet concluded that *Lilium* was yet another genus with its centre in the mountain ranges of western China.[15]

Père Farges' first collection had more than lived up to expectations and Franchet wrote to Père Delavay, then at the sanatorium in Montbeton, to express his satisfaction at having acquired a skilled collector in such a promising area.[16]

Père Farges botanised in other areas in 1892 and, as well as collecting new plants, also collected duplicates of those found the previous year so that Franchet would have plenty of material for comparison purposes. This collection was despatched in March 1893 in four packages containing about 250 different species, and three packs of seeds. Père Farges had to send off his collections by special courier in one large annual consignment as Chengkou was so remote that it lay some twenty days' journey from the nearest point for the ordinary post, which meant that he could not make the frequent regular despatches of small packets of seeds and specimens that Père Delavay had found so useful. Hiring a courier was expensive and Père Farges relied on the Muséum to send him the funds to cover the cost, distributing any surplus among the most indigent members of his community.[17]

Acer griseum

Père Farges despatched his third collection of around 125 previously uncollected species with seeds at the beginning of January 1894, as well as more duplicates. In August, Franchet pre-

M.S.del.J.N.Fitch lith.   Vincent Brooks,Day & Son Ltd Imp

ABOVE

Lilium davidii var. willmottiae as L. sutchuense. Bot.Mag. (1900) No.7715

BELOW

Acer griseum at Edinburgh Botanic Garden.

sented a lengthy paper containing details of some of the first discoveries.[18] One of the most important of these was a maple, which Franchet originally identified as a variety of the Nikko maple but, by 1902, botanists had realised that it was actually a new species and we now know it as *Acer griseum*, the paper-bark maple. This is one of the most beautiful of all trees. Its delicate foliage turns red in autumn and its old bark peels off to reveal a glossy cinnamon underbark that shines as if polished. E. H. Wilson introduced it in 1901 for Veitch's nursery and its ornamental qualities became apparent as soon as the first cultivated specimens began to mature. It is now widely planted in parks and gardens across the temperate world, which is as well as the species is now under considerable threat in the wild.[19] Other maples collected by Père Farges included *A. fabri*, which Pastor Faber had discovered in Guangdong (see chapter 10, p.139); *A. longipes*, and *A. sutchuenense*, a rare species that was reintroduced from Hubei by the 1980 Sino–American Botanic Expedition.[20]

Conifers

The preponderance of woody species in the flora of this region was reflected in the many trees collected by Père Farges, including some fine conifers. One of them was *Abies fargesii*, the common silver fir of the region, which Franchet named after Père Farges, although it had been discovered by Henry. *A. fargesii* once covered the hillsides of north-east Sichuan, western Hubei and adjacent areas, but Wilson saw just a few scattered remnants of these rich forests, and now they are found only on the tallest mountains of the region.[21] Wilson, who introduced *A. fargesii* for Veitch and for the Arnold Arboretum, thought it exceptionally handsome and believed that 'no finer silver fir exists in all the Far East'. This opinion was shared by Joseph Rock who collected the variety *sutchenensis* and sent seeds back to the American Department of Agriculture in 1925. In spite of its early introduction and its undoubted ornamental qualities, the species is still rare in cultivation: perhaps because it is at its best between 2,590–3,040 m (8,500–10,000 ft) and often looks rather miserable at lower altitudes.[22] Another of Henry's discoveries that Franchet described from Père Farges' specimens was *Picea brachytyla*, an exceptionally handsome spruce with silvery needles and an elegant weeping habit that is quite distinctive.[23] Père Delavay had found a very similar conifer in August 1889, which has now been named *P. brachytyla* var. *complanata*. Wilson collected *P. brachytyla* shortly after his arrival in Hubei in June 1900 and he collected a similar spruce in 1910, which he introduced to the Arnold Arboretum. This introduction was originally called *P. sargentiana* in honour of Charles Sargent, but it was later realised that it was not a new species at all but the typical *P. brachytyla*. Although *P. sargentiana* may no longer be a valid species name, Charles Sargent, who did so much to further the Western discovery of

BELOW

Abies fargesii at Shennongjia, Hubei.

RIGHT

Picea brachytyla var. *complanata* at Westonbirt National Arboretum, Gloucestershire.

Photo Credit: Chris Reynolds

the Chinese flora, is still commemorated in the popular name 'Sargent's weeping spruce', which is commonly applied to *P. brachytyla* and its varieties.[24] These spruces are large trees and certainly require space but they are so handsome that it is a great pity that they are not more widely available, especially as the remaining stands are under considerable pressure from timber felling in the wild.

One of Père Farges' conifer discoveries appeared to be closely related to *Juniperus squamata*, a widespread species originally known from the Himalayan massif, which Père Delavay had found in Yunnan and Pastor Faber on Emei Shan. Vilmorin's nursery had acquired seeds (probably from Père Delavay), from which plants had been raised by 1893 and Vilmorin was eventually able to send seeds to the Arnold Arboretum. Wilson and his colleague, the botanist Alfred Rehder, identified Père Farges' juniper as a variety of *J. squamata*, which they called var. *fargesii* as they thought its upright tree-like habit distinguished it sufficiently from the low-growing typical form. *J. squamata* is now known to be such a variable species that most botanists

Tsuga chinensis in the Daba Shan.

no longer maintain separate varieties: the exception is the *Flora of China* which recognises four varieties, including var. *fargesii*, although these are based not on growth habits but on differences between the leaves.[25]

Another of Père Farges' conifer finds was one that Père David had seen in Baoxing and Shaanxi, which he called by its Chinese name of *thié-sha* (*tiesha* in Pinyin), which means iron fir. When Père Farges' specimen was examined it was realised that the *thié-sha* was not a silver fir as Père David had thought but was actually a hemlock, which was given the name *Tsuga chinensis*. The Chinese hemlock is widely distributed across central and western China and Wilson collected it on several occasions.[26]

Franchet described a new nutmeg yew, *Torreya fargesii*, based on Père Farges' specimens, although the species had been discovered by Augustine Henry in 1888. *Torreya* is yet another genus that is represented in China and North America, but nowhere else.[27] In 1895, Père Farges sent back seeds of *Pinus armandii*, which Père David had discovered in the Qinling in 1873, and Henry sent seeds of the same species to Kew in 1897.[28]

New trees

On 6 June 1894, Père Farges came across a small spreading tree in flower and when Franchet examined it, he found that it belong to a new genus that he called *Carrierea*. He named Père Farges' discovery *C. calycina* and a second species has subsequently been found. *C. calycina*, the goat horn tree, produces terminal panicles of pale waxy cup-shaped flowers that E. H. Wilson thought very beautiful. These are followed by conspicuous curved grey seed-pods that resemble horns and give rise to its common name. These pods are set off by remarkably fine foliage and it is the large, pendulous, very glossy leaves hanging from scarlet leaf stalks which make *C. calycina* such an exceptional ornamental tree. Père Farges sent seeds to Vilmorin in 1896 and Wilson introduced it to the Arnold Arboretum in 1908; unfortunately most of the trees raised from Wilson's seeds died during the twentieth century and surviving trees are very rare. In 1994, a century after its first discovery, the late Peter Wharton found the goat horn tree in Datou Shan in northern Guizhou near the Sichuan border and distributed seeds around the world. As a result, there are now many young trees of *C. calycina* in cultivation.[29]

Another of Père Farges' ornamental finds was a handsome catalpa or bean tree, which Professor Bureau called *Catalpa fargesii* as he based his description on Père Farges' specimens, although it had actually first been collected by Henry. *Catalpa* is another genus with representatives in eastern North America and in eastern Asia, especially China where *C. fargesii* is one of four indigenous species. *C. fargesii* is a handsome ornamental

Carriera calycina in early summer.   *Catalpa fargesii* f. *duclouxii* in summer.

species, reaching 18 m (60 ft) in warm sheltered sites, and producing tapering heart-shaped leaves and clusters of pink flowers with yellow-brown throat markings in late summer. The foliage, as with so many Chinese plants, turns a good clear yellow in autumn. Père Ducloux collected a form known as C. *fargesii* forma *duclouxii*, which was also found by Père Delavay and Augustine Henry, and this is the form now seen most often in cultivation. The only difference between f. *duclouxii*

and typical C. *fargesii* is that the shoots and leaves of the typical form have a down-like covering of hairs, while all parts of f. *duclouxii* are perfectly smooth. Wilson introduced C. *fargesii* f. *fargesii* from Hubei in 1901 and again in 1907, when he also introduced f. *duclouxii*. Kew acquired plants from Vilmorin in 1908, but C. *fargesii* first flowered in Britain in the Isle of Wight in 1919.[30]

Père Farges also collected several birches, the best known of

BELOW

RIGHT

Betula utilis subsp. *albosinensis* in spring at Edinburgh Botanic Garden.

Paulownia fargesii flowering in early summer.

which is *Betula utilis* subsp. *albosinensis*, the Chinese red birch, discovered by Henry in 1888. It is exceptionally handsome in its finest forms, with bark that ranges in colour from mahogany-brown through pink to purple-grey and peels off in thin papery sheets. The bright foliage is also attractive. As soon as Wilson introduced *B. utilis* subsp. *albosinensis*, in 1901, it was recognised by horticulturalists as a first-class ornamental plant. It is a variable species that has been collected on several occasions and several cultivars are now available.[31] Much less familiar is *B. insignis*, a handsome hardy birch with very long catkins and good autumn colour that thrives in heavy soil. Although first introduced by Wilson in 1900 and reintroduced from Guizhou in 1985, *B. insignis* is still only found in specialist collections.[32]

Père Farges' fourth collection was despatched at the end of January 1895. It included *Paulownia fargesii*, a handsome fast-growing tree, which Franchet described from Père Farges' specimen although Henry had discovered it in 1888.[33] Other new trees included *Carpinus fargesiana*, a very handsome hornbeam, and *Corylus fargesii*, a recently-introduced hazel with attractive peeling bark.[34] Père Farges had also sent seeds of *Euptelea pleiosperma*, a small tree which flowered at Les Barres in 1900. This species had been discovered in the Himalaya in 1864 and had subsequently been collected in China on several occasions, first by Père David at Baoxing and then by Père Delavay, Pastor Faber, Henry and Wilson: but all these Chinese forms had been considered new species, and it was only when Joseph Hooker examined all the specimens at Kew and Paris in 1905 that he realised that they all belonged to the same widely-distributed species. Botanists were beginning to realise that the range of some Himalayan species extended much farther east than had hitherto been suspected.[35]

A tree raised by Vilmorin from one of Père Farges' first seed consignments proved to be *Eucommia ulmoides*, another of Henry's discoveries. It baffled Kew's botanists when they first studied it as it is anomalous and a new genus had to be created to accommodate it. It is still the only species in the genus. *E. ulmoides* is known as the hardy rubber tree as its leaves and bark contain lengthy elastic gutta-percha fibres, which have long been prized in China as important medicines. The tree is rare in the wild but extensively cultivated for its medical properties. Vilmorin sent a small sapling to Kew in 1897 and French officials had supervised trial plantings in Vietnam by 1901.[36]

Rhododendrons

As well as trees, Père Farges discovered a number of shrubs among the woody flora, but there were few rhododendrons among them. This lack of new rhododendron species, compared to the number found farther west by both Père David and Père Delavay, indicated that Père Farges was collecting in an area that lay beyond the main rhododendron heartland and

ABOVE

Rhododendron oreodoxa var. *fargesii* in spring at Glendoick Gardens, Perthshire.

helped confirm the growing botanical consensus that the centre of the genus lay in the huge mountain ranges and narrow river valleys bordering Tibet, south-west Sichuan and north-west Yunnan. Père Farges did find two new species: tree-like *R. maculiferum* and *R. adenopodum* with pale pink flowers, which Vilmorin raised from seed he received from Père Farges in 1901, although Wilson had already introduced it the year before. Other rhododendrons included: early-flowering, deep pink *R. sutchuenense*, discovered by Henry and suitable only for the largest gardens; a new variety of *R. oreodoxa*, discovered by Père David, now known as var. *fargesii*; and a subspecies of *R. fortunei* called ssp. *discolor*, which E. H. Wilson described as 'the common rhododendron of the woods up to alt. 2,300 m [7,545 ft] in western Hubei'.[37] Père Farges also sent Vilmorin seed of *R. augustinii*, which had originally been discovered by Augustine Henry near Badong. Wilson, who collected *R. augustinii* for Veitch in 1900 and for the Arnold Arboretum in 1907, described it as, 'an exceedingly common species in Hubei, delighting in rocky situations fully exposed to the sun'. The flower colour varies from violet-blue to rosy-purple, and many of the bluest forms have been hybridised or selected to create a number of fine named cultivars.[38]

New rhododendrons might have been relatively scarce but Père Farges collected a number of other ornamental shrubs including a willow and a holly previously discovered by Henry, although Franchet named them after Père Farges. *Salix fargesii* is a handsome willow that is much more frequently seen in cultivation than Père David's *S. moupinensis*, to which it is closely related although separated geographically. *S. fargesii* is an altogether larger plant and strikingly ornamental, with glossy mahogany stems, dark strongly-veined leaves that are among the largest of any willow, and brilliant red leaf buds that shine like flares in the bare winter garden. It was introduced by Wilson for the Arnold Arboretum in 1910.[39] *Ilex fargesii* is a fine holly with long narrow spineless leaves and clusters of red berries on female plants. It is widely distributed across south-west China and is very variable, with several different forms having been collected over the years, the best of which make fine ornamental evergreens for the garden.[40] Another of Père Farges' woody finds was *Enkianthus chinensis*, a shrub with pendulous flowers that are usually pink, which Franchet described in 1895 from Père Farges' specimens, as well as from those collected on Cang Shan by Père Delavay. Wilson introduced *E. chinensis* in 1900 and Forrest also collected seeds.

Distribution patterns

As Franchet had hoped, having a collector closer to the central area was proving immensely useful. Père Farges' specimens were providing much new information about plant distribution within China, as he had found species in north-east Sichuan that had previously only been known from farther west in the Himalaya, or from collections made in Baoxing by Père David and north-west Yunnan by Père Delavay. Augustine Henry's botanical explorations in Hubei had begun this process of clarification. As more collectors ventured into the central provinces over the next few years and as more plants were described, the way that species were distributed across China became clearer and botanists were able to see how these distribution patterns related to those of the wider region.[41]

Père Farges' 1895 collection included a new *Clethra*, which Franchet called *C. fargesii*. This elegant shrub is remarkable for

the long pendulous racemes of small scented white flowers that are produced in mid to late summer. It is closely related to *C. delavayi* and was introduced by Wilson in 1901. New forms of *Clethra* were also discovered by Père Bodinier and Père Cavalerie in Guizhou, and by Pastor Faber in Guangdong.[42] Another exceptional ornamental shrub collected by Père Farges was a smooth-leaved form of *Clerodendrum trichotomum*, which horticulturalists call *C. trichotomum* var. *fargesii* or the harlequin glory bower. The young leaves are purple and in late summer clusters of sweetly-scented white flowers appear, followed in autumn by bright blue berries surrounded by starlike carmine-pink calyces. Père Farges sent seed to Vilmorin in 1898 and Wilson also collected it for Veitch in 1900.[43]

Deutzias

Vilmorin shared some of the *Deutzia* seeds from the 1895 consignment with colleagues and Georges Boucher, the Paris nurseryman who also raised some of Père Delavay's plants, reported in April 1896 that his young deutzias had produced clusters of starry white flowers. He exhibited a flowering branch on 8 April the following year. Both Boucher and Lemoine had stock of this latest *Deutzia* introduction for sale that autumn, although its exact identity was much debated. Victor Lemoine, his son Émile and Maurice de Vilmorin discussed the possibility that it was the same as a *Deutzia* from north China called *D. parviflora*, which had originally been discovered and described by Alexander von Bunge and later collected at Chengde in 1864 by Père David;[44] but eventually they agreed with Franchet that Père Farges' new *Deutzia* was a separate species and it was given the name *D. corymbiflora*. Botanists now believe that it is actually a variety of *D. setchuenensis*, also discovered by Père Farges, and it has been named *D. setchuenensis* var. *corymbiflora*.[45] It is one of the most beautiful of all the white-flowering deutzias.

The Lemoines, with their customary acumen, had recognised the breeding potential of both these deutzias and began to use them to develop new forms. One of the first they raised was *D.* x *myriantha*, the result of a cross between *D. setchuenensis* and *D. parviflora* (obtained by the Lemoines from the Arnold Arboretum, which had received seed from St Petersburg), but the great plantsman Sir Harold Hillier was unimpressed and remarked that *D.* x *myriantha* was 'in no way superior to *D. setchuenensis*'. Like many horticulturalists, Sir Harold thought that these deutzias were already so good that they could hardly be improved; and he praised *D. setchuenensis* var. *corymbiflora*, with its wide long lasting flower clusters made up of over 300 small individual flowers, as 'one of the best of all summer-flowering shrubs, and perhaps the best for chalky soil'.[46]

Vilmorin received seeds of another *Deutzia* from Père Farges in 1897 that was at first thought to be a new species called *D. vilmoriniae*, and then to belong to Père David's discovery

D. longifolia, before it was finally agreed that it was actually *D. discolor*, a closely-related species originally found by Augustine Henry and described in the *Index Florae Sinensis* in 1887. Lemoine went on to cross *D. discolor* with *D. scabra* to create the excellent *D.* x *magnifica* group of hybrids, still among the best white-flowered deutzias. Wilson also collected *D. discolor* for Veitch, but his introduction was originally called

ABOVE

Deutzia setchuenensis var. *corymbiflora* at midsummer.

BELOW

Deutzia discolor (also known as *D. longifolia* 'Veitchii') flowering in summer.

Photo Credit: Raymond Evison

LEFT

Clematis potaninii var. fargesii.

Père Farges' 1895 consignment of seeds also included two *Philadelphus*: *P. subcanus* var. *magdalenae* (syn. *P. magdalenae*), a new discovery, and *P. sericanthus*, which had been discovered by Henry. Both these shrubs were quickly introduced to commerce by Vilmorin, and Wilson also sent back seeds but, even though *P. subcanus* var. *magdalenae* is scented and flowers early, it is not much planted today. *P. sericanthus* is even less garden-worthy and was dismissed by one expert as 'a scentless also-ran'. They were taken up at the time as interesting novelties, but neither are first class and other *Philadelphus* including *P. delavayi* and its forms, to which both these species are closely related, are now preferred by horticulturalists.[49]

Climbers

Climbing plants were also represented and two were described in 1894. The first was a vigorous clematis with clusters of pure white flowers in June that Franchet called *Clematis fargesii*. It is very similar to *C. potaninii*, which was discovered in Sichuan by the Russian explorer Grigory Potanin, and many botanists now consider them the same, although horticulturalists still recognise plants with three or more flowers in each cluster as *C. potaninii* var. *fargesii*. Wilson introduced it to the Arnold Arboretum in 1910.[50] In 1962, a hybrid between *C. potaninii* var. *fargesii* and *C. vitalba* (Travellers Joy or Old Man's Beard) called *C.* 'Paul Farges' was raised in the Ukraine Botanic Garden. This new clematis has proved very vigorous with masses of creamy white flowers.[51] The second 1894 climber was the black kiwi, *Actinidia melanandra*, a hardy species with reddish fruits that had been discovered by Henry and was later introduced by Wilson. Père Farges sent seed of *A. chinensis*, the Chinese gooseberry or kiwi fruit, to Vilmorin and a living plant was raised at Les Barres in 1899. Wilson also collected seed of *A. chinensis* in Hubei in 1900 and plants were offered by Veitch in 1904.[52]

By 1896, Franchet was able to report that he had received more than 2,000 species from Père Farges' corner of north-east Sichuan. He went on to say that what he found particularly striking about the collections he had seen so far was the number of Père Farges' plants that had affinities with the flora of Europe and North America. He had been examining Chinese plants for fifteen years and he was now sure that the floras of north-east North America and China were disjunct remnants of a common flora that had once stretched continuously around the northern hemisphere. Franchet also remarked that several of the woody plants found by Père Farges, such as *Fargesia spathacea* and *Carrierea calycina*, required the creation of new genera to accommodate them; but even then, their place within

D. veitchii before being correctly identified.[47] The confusion over all these deutzias is hardly surprising as botanists had produced a flurry of descriptions of very similar species in a short time, and plantsmen, who were still completely unfamiliar with any of the new shrubs, were trying to identify them from descriptions based solely on dried herbarium specimens. It was only when the various introductions had been cultivated for several years that botanists and nurserymen in different countries could compare the living plants with each other and with the dried herbarium specimens, and then work out exactly which deutzias had been introduced. It also became apparent that some of the supposed 'new' species were actually variants of species that had already been identified, which meant that some names were dropped and others reinstated. However, not all botanists are convinced by current attributions and continue to identify *D. vilmoriniae* and *D. veitchii* as varieties of *D. longifolia*.[48]

the existing system of classification was not always clear.[53] These additions to the overall system of botanical classification required considerable background study, which took time and slowed his progress in determining Père Farges' collections.

Incarvillea

Among Père Farges' herbaceous finds was a form of *Incarvillea mairei* with particularly large flowers, which horticulturalists often distinguish by the epithet *grandiflora* (see chapter 9, p.126). Père Farges collected seeds in August 1894, which Vilmorin sowed the following April. Half the seeds germinated that year and half in 1896, and the plants flowered in 1897 in mid-May, about a fortnight earlier than *I. delavayi*. The large bright pink trumpet flowers lasted about three weeks and by August Vilmorin had ripe seeds to distribute. He had already sent seed to Kew and their plants flowered in April 1898. Plants from Père Farges' introduction were not widely distributed and did not persist in cultivation but *I. mairei* soon became established in temperate gardens as Forrest and then Frank Ludlow and George Sheriff all collected seeds, and further collections have been made recently. *I. mairei* does well in poor dry sites and is therefore a better choice for gardeners who do not have the rich fertile soil required by *I. delavayi*.[54]

Père Farges collected three *Asarum* species, including *A. caulescens*, which had originally been described from specimens collected in Japan, although finding it in north-east Sichuan led to a considerable extension of its geographical range.

BELOW

Incarvillea mairei var. *grandiflora* in early summer at Edinburgh Botanic Garden.

When Franchet wrote a paper on *Asarum* in 1898, in which he included *A. debile*, a new species found by Père Farges, and *A. chinense* (syn. *A. fargesii*) which Père Farges also collected, he noted that *Asarum* was yet another genus centred on the mountains of eastern Asia, especially western China, which had representatives in Europe and North America. He went on to say that in the case of *Asarum*, the link with Europe was unusually weak as it was limited to a single species, while the connection with North America, which has fifteen indigenous species, was strong. Close study of Père Farges' discoveries, as with those of Père Delavay, was providing Franchet with enough new information to allow him to draw definitive conclusions about the wider distribution patterns of an increasing number of genera.[55]

Epimediums

Among Père Farges' herbaceous discoveries, which Franchet described in 1894, were two epimediums: *Epimedium fargesii* and *E. sutchuenense*. *E. sutchuenense* was not introduced until the 1980s, when it was rediscovered by the American nurseryman Darrell Probst, and it is still not well known in cultivation. Evergreen *E. fargesii* is a variable but most attractive species with velvety, purple-coloured young leaves and delicate purple and white star-like flowers that nod above the foliage on long wiry stems. It is yet another of the ornamental epimediums collected and introduced by the Japanese plant-hunter Mikinori Ogisu during the 1990s, which have so swiftly found favour with horticulturalists. Ogisu also collected a form of *E. fargesii* with purple and pink flowers that he named 'Pink Constellation.' A new species of epimedium that he discovered in 1987 was given the name *E. franchetii*.[56]

Roy Lancaster collected an evergreen epimedium on Emei Shan in 1980 that was later identified as *E. acuminatum*, which had originally been discovered in 1858 in Guizhou by Père Perny. This attractive species has thin arrow-shaped leaves and nodding flowers, which are usually pink or purple and white, although the colour is variable. *E. acuminatum* is decidedly ornamental and has proved relatively easy to grow, so it is now well established in gardens. Several selections are now available and it has also hybridised with *E. fangii* to produce *E. x omeiense*.[57] Recent explorations by modern collectors in western China have resulted in the introduction of a number of new garden-worthy epimediums and our knowledge of the genus has increased dramatically. In 1938, only twenty-one types of epimedium were known and only a single Chinese species was cultivated (and that had been introduced by way of Japan). Today, forty-four types of epimedium have been described and more than thirty new forms of Chinese epimedium are listed by specialist growers, including several species first discovered by missionary-botanists such as Pères David, Perny, Bodinier and Farges. New species continue to be discovered and introduced.[58]

Photo Credit: Roy Lancaster

ABOVE

Epimedium fargesii in the wild.

Another of Père Farges' finds was a meconopsis, which was at first thought to be the same as yellow-flowered *Meconopsis integrifolia*, discovered by Przewalski in Gansu in 1872 and later collected by Père Delavay in Yunnan, although it has recently been identified as a subspecies of *M. pseudointegrifolia*.[59] Other herbaceous collections included *Cypripedium fargesii*, a slipper orchid discovered by Augustine Henry that is closely related to Père Delavay's *C. margaritaceum*.[60]

New Arisaema

A new species of *Arisaema*, *A. fargesii*, was included among Père Farges' herbarium specimens but it was not described until 1911, when Vilmorin raised living plants from tiny tubers he had been sent by Père Marie Louis Thermes (1874–1953), one of his missionary contacts in northern Sichuan. Vilmorin sent a female plant to Kew in 1917, which flowered in 1919.[61] *A. fargesii* is closely related to Père David's *A. franchetianum*, with a similar tall pseudo-stem, trifoliate leaf and hooded flower sheath. It can be very hard to tell them apart just by the leaves,

although there is a difference in the shape of the leaf sections, as those of *A. fargesii* are generally fatter and more evenly-sized that those of *A. franchetianum*, where the lower leaf section is much wider than the two lateral sections. *A. fargesii* also has a larger and more robust inflorescence. The growth of *A. fargesii* is divided into various stages: the leaf emerges in the first year as a single shield shape and only in the second year does it divide into the mature trifoliate form; and young (or weak) plants only produce pollen, as they are not strong enough to set seed and so do not develop female parts. As the plant matures and strengthens, female parts are produced alongside male ones, and if the flowers are fertilised, a cluster of red berries develops. The strongly arched hood and spathe-tube are strikingly marked with thin reddish-brown and white vertical stripes, and these together with the large handsome foliage leaf and striped pseudo-stem make it a most striking ornamental herbaceous plant.

Introduction of the dove tree

In 1893, Franchet asked Père Farges for specimens, and more importantly for seeds, of *Davidia involucrata*, Père David's dove tree. Franchet was particularly anxious to acquire this remarkable tree as it had been discovered and originally described by Frenchmen and he wanted France to have the honour of raising the first living specimen.[62] This ambition was now threatened as *D. involucrata* had most recently been collected in Sichuan by Augustine Henry and Franchet knew that Kew might acquire seeds at any time. If this was to be avoided, it was imperative that Père Farges devote himself to ensuring that seeds of *D. involucrata* reached France first. The species was fairly common around Chengkou and Père Farges promised to do all he could to acquire seeds, but he warned Franchet that they were hard to come by as, even though the trees flowered prolifically, they only produced a few fruits each year, many of which fell when only half-ripe. He found none in 1894 and the following year was very wet so no fruit were produced.[63] He had better luck in 1896 and in October was able to send Franchet several of the walnut-sized fruits, which he thought would be ripe as he had collected them just as the trees were dropping their leaves.

Père Farges also sent thirty-seven fruits to Maurice de Vilmorin who received them in June 1897 and immediately planted the seeds out in three different sites and in different conditions. Yet, in spite of his care, few of them germinated and even those that did were all dead by September 1898. To Vilmorin's delight, a healthy seedling was noted at Les Barres in June 1899, two years after sowing; and in September, when the plant was 15–20 cm (6–7.5 inches) tall, he took a photograph of the foliage to send to Kew, so that the botanists there could confirm that it matched the specimens of *D. involucrata* collected by Augustine Henry. Père Farges had suggested that the species would be hardy to about −15 °C (5 °F) as he had collected the fruits at 1,400 m (4,593 ft), but Vilmorin and his team were taking no chances and the precious little plant was cosseted in a glass case padded out with leaves during the next two winters. In 1901, four cuttings were taken and a branch was layered. Only two of the cuttings survived, one of which was sent to the Jardin des Plantes and one to Kew, and the layered plant was sent to the Arnold Arboretum. By July 1902, the young dove tree at Les Barres had reached 1.60 m (5ft 2 ins) and a picture of the flourishing sapling with its lime-like foliage was printed in the *Revue Horticole*. Vilmorin's persistence in encouraging Farges to seek out seeds of *D. involucrata* had paid off; and when Farges was able to send a further consignment of fruits a year or two after his original despatch of seed, the species' French future was assured.[64]

Franchet was right to be anxious about British plans concerning the dove tree. Nurserymen like Harry Veitch were well aware of its importance and were keen to get hold of it as they knew that it would be a commercial sensation when it was introduced. When Veitch decided to send Ernest Wilson to

China in 1899, he stressed that the overriding objective of the expedition was to acquire seeds of *D. involucrata*: nothing else mattered. Consequently, as soon as Wilson arrived in China in 1899, he made his way to Simao to meet Augustine Henry and find out exactly where in Sichuan he had discovered the tree. Henry gave him as much information as he could and drew 'a sketch of a tract of country about the size of New York State' on half a page torn from a notebook; but when Wilson journeyed to south Wushan in Sichuan in 1900 and found the site Henry had indicated, he was appalled to discover that the original tree had been cut down. Wilson would not be defeated this easily and immediately began his own search, which resulted in the discovery of eleven other *D. involucrata* trees. He made sure they were carefully monitored during the year and in November he was able to collect 'a rich harvest of seeds', which he sent back to Veitch's nursery. He found a further hundred trees

the following year but they were a disappointment, producing only a few dozen seeds between them, and he commented that 'during subsequent visits to China extending over a decade [he] never again saw davidia fruiting in the manner it did in 1900'.[65]

The fruits arrived at Veitch's nursery early in 1901 and were sown straightaway. Those sown outside germinated in April 1902, and later that year Wilson and an assistant potted up over thirteen thousand seedlings. Wilson had also brought back three or four living plants from China and these grew on well. By 1903, Veitch was able to offer young dove tree plants for sale and enthusiasts in Britain and France snapped them up. Wilson was mortified when he discovered that Père Farges had actually been the first to introduce the species, although it was undoubtedly his own collection of seeds in 1900 that resulted in the introduction of *D. involucrata* to widespread cultivation. It was also galling to learn that the French dove tree at Les Barres

BELOW

The fruit of *Davidia involucrata*.

came into flower in May 1906, while the British had to wait until 1911 for the first of Veitch's trees to flower at the Coombe Wood nursery. In 1925, the *Gardener's Chronicle* considered that one of the most handsome and free flowering of all the dove trees in cultivation was one growing at an arboretum in Vendome, France, owned by M. Gerard, a wealthy industrialist who had acquired it from Veitch around 1905. The tree at the Arnold Arboretum grown from Vilmorin's original layer did not flower properly until May 1931.[66]

When all the specimens of *Davidia involucrata* were compared, it became clear that those collected in north-east Sichuan by Père Farges and those collected in eastern Sichuan by Augustine Henry and Wilson all had smooth leaves, and this glabrous variety is now called var. *vilmoriniana* to distinguish it from the original downy-leaved type found farther west in Baoxing by Père David. Both the type and the variety thrive in moist sheltered locations, although var. *vilmoriniana* is said to be hardier and, as it is more vigorous and easier to establish, it is now more common than the type in cultivation, especially in northern Europe. The bracts are undoubtedly the tree's chief glory, but it is ornamental at other seasons as its leaves can colour richly in autumn and older trees develop attractive bark. Wilson thought that '*Davidia involucrata* is at once the most interesting and beautiful of all trees of the north-temperate flora', a verdict that it would be hard for anyone to dispute, especially if they have seen those remarkable white bracts fluttering surreally in the half-light at dawn or dusk.[67]

Franchet could not have been better pleased with the collections Père Farges despatched to the Muséum throughout the 1890s, and he kept him supplied with funds and with the botanical publications that he had learned from Père Delavay were so vital to the morale of lonely isolated collectors. Père Farges continued his careful exploration of the region, investigating new areas and despatching two or three cases of specimens every year but, in October 1900, he reported that new plants were becoming much harder to find and his next three collections consisted of only 300-odd specimens each, instead of the usual 2–3,000 specimens. They were the last Père Farges was to make for the Muséum. He was now nearly sixty and twenty-nine years of frugal living at Chengkou, with meagre rations and little comfort, had taken their toll of his constitution. He was no longer really strong enough to endure days of walking and fatigue. His confrères realised that he needed a more sheltered position and, in 1903, he was made almoner of the hospital at Chongqing. Père Farges remained there until he suffered a stroke in 1909, after which he was transferred to a small Christian community in the nearby countryside, close to another missionary. He died three years later at Chongqing in December 1912, aged sixty-eight. He had served as a missionary in China for forty-five years.[68]

Horticultural and research developments

Fargesia murielae

One of Wilson's living *F. murielae* plants numbered 1462 survived the journey back to the Arnold Arboretum and when an offshoot was sent to Kew in 1913, it was extensively propagated. Consequently, almost all the *F. murielae* plants in cultivation for the next ninety or so years came from Wilson's number 1462. It was only when those clones began to flower, virtually simultaneously, in the 1970s and 1980s that botanists had their first opportunity to study the flowers of the species.[69] It is important to remember that, as bamboos often have flowering cycles that span decades, collectors are rarely able to include flowers with their specimens, which means that botanists usually have to base their descriptions on just leaves and culms (stems). Once all the 1462 *F. murielae* clones had flowered, they died; and it will be many decades before their seedlings are ready to flower. We now know that several species of *Fargesia* such as *F. nitida* and *F. murielae* are important staples in the diet of the Giant Panda, and this periodic mass dying of their essential food plants presents further problems for the species' long-term survival. Several new species of *Fargesia* have been introduced to cultivation within the past couple of decades.[70]

Tsuga chinensis

Hemlocks are beautiful conifers that generally display an elegant weeping habit. In addition to its ornamental qualities, *T. chinensis* appears to possess considerable resistance to attacks by the hemlock woolly adelgid, a destructive insect pest that has ravaged native populations of *T. canadensis* and *T. caroliniana* in eastern North America. As a result of several recent collections, *T. chinensis* is now grown in arboreta throughout North America, where its resistance to hemlock pests can be studied, and perhaps used to strengthen the resistance of native species.[71]

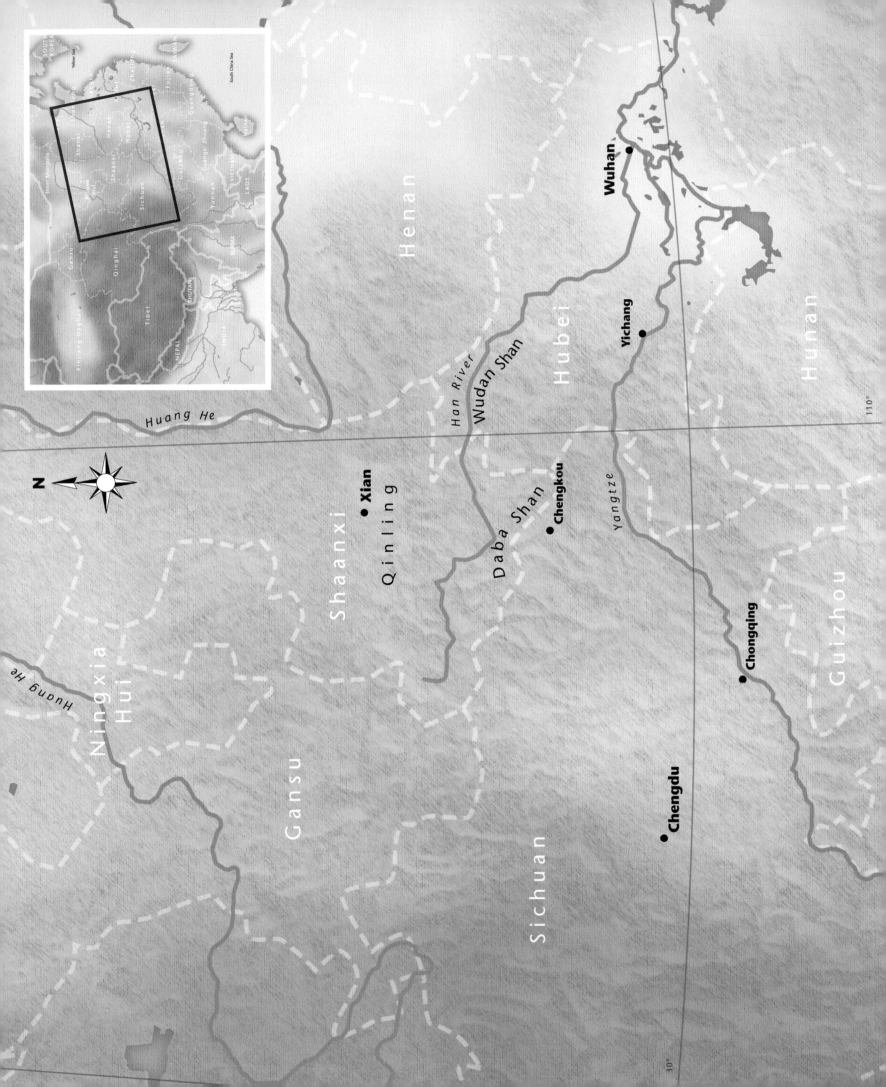

CHAPTER 12

North of the Qinling

—

...for the personal character and the work of many a Roman Catholic missionary,

whom I have met in China, I have conceived a profound respect.

HENRY NORMAN, 1895[1]

—

So far, all the missionary-botanists we have mentioned, apart from Pastor Faber, have been French; but other nationalities were involved in missionary work in China and some of these missionaries were also botany enthusiasts. The most notable was **Padre Giuseppe Giraldi**, an Italian priest based in the **north of Shaanxi province** from 1888–1901, who proved as enthusiastic and dedicated a botanical collector as any of his French colleagues, sending back over 8,000 herbarium specimens, seeds and tubers to the Botanical Institute in Florence.[2]

When the Italian Church was assigned responsibility for arranging missions in Shaanxi province, the Order of Friars Minor, popularly known as Franciscans, which had long been based in neighbouring Hubei province, was one of the Italian organisations appointed to Shaanxi. The Franciscans originally operated in the centre and south of the province around their episcopal headquarters at Xi'an (where Père David stayed in 1872) but, from 1890, they began to establish missions in northern Shaanxi. Padre Giraldi, who had arrived in China at the end of 1888, was one of the first to work in the northern district. He was born in June 1848 at Larciano, near Pistoia in northern Tuscany, not far from Florence, and he decided when still a boy that he wanted to go to China as a missionary. He had to wait until he was twenty-five before laws limiting entry to the seminaries were relaxed and he was able to join the Order of Friars Minor in January 1873. Giuseppe Giraldi spent

his years as a novice at the monastery of Giaccherino, San Vivaldo, where he taught the younger seminarians grammar, while applying himself to his own studies, especially Latin and various aspects of theology. He was kind and enthusiastic, with a simple friendly manner that made him a very popular member of the community; but his devotion to the monastic life never made him forget his missionary vocation and when he took his final vows and became a priest five years later, he hoped to be sent abroad immediately. He was as disappointed as Père David had been when the order he longed for did not come and he

LEFT

Padre Giuseppe Giraldi OFM.

Photo Credit: Museo di Storia Naturale dell'Università di Firenze

was assigned to other duties. In his case, these included preaching in the neighbourhood and teaching at the Church of the Ozzervanza in Siena.

He also worked in the monastery gardens, where he got to know **Antonio Biondi,** a wealthy landowner and dedicated botanist who lived at nearby Castelfalfi. The two men became friends, and Biondi inspired Padre Giraldi with his own love of botany, showing him how to collect and prepare plant specimens to a high standard, and 'transforming him from a skilful gardener into an expert botanist'.[3] Among Padre Giraldi's students at Siena were two Chinese Christians from whom he learned at first-hand about the state of Christianity in China, and this prompted him to challenge his superiors again, saying that he was now nearly forty and it seemed that he would be an old man before ever being sent out to China, even though this was why he had originally entered the Order and become a priest. His vehemence had its effect, but he was first offered missionary posts in South America and the Holy Land before finally being given permission to join the Franciscan mission in northern Shaanxi province in 1888.

Padre Giraldi was so delighted that he left immediately for Naples, without even visiting Rome to collect some of his possessions and, after making a brief pilgrimage to the Holy Land, sailed for China from Jaffa in August 1888. He spent his first few months in northern Shaanxi, living in the Bishop's residence while he learned Chinese, but he began visiting outlying missions as soon as he had picked up enough of the language to act as preacher and confessor. He became a familiar figure as he traversed his district and was known to all as 'Blackbeard'.

Arrival in Northern Shaanxi

When Antonio Biondi heard that Padre Giraldi was bound for unexplored northern Shaanxi, he urged him to collect the plants of his new district and send them back to him at the Botanical Institute in Florence. Northern Shaanxi has cold, relatively dry winters and hot summers, with a predominantly woody flora and from the moment he arrived, Padre Giraldi devoted his free time to botanising. He also took what time he could to explore the Qinling Mountains in the south, which Père David had investigated in 1872–3. He trained his servant and some local youths to help him collect and began sending regular consignments back to Biondi. Padre Giraldi was particularly interested in cryptogams – non-flowering and seedless species such as ferns, fungi and mosses – and two-thirds of his specimens represented these groups. He discovered nearly 200 new species and although none of them are of great interest to horticulturalists, they considerably expanded specialist knowledge of many important genera. The other third of his specimens included some fine ornamental plants that are now widely cultivated in temperate parks and gardens.[4]

Several of Padre Giraldi's trees turned out to be forms of species already known from other areas, which added to the increasing detailed and accurate picture of the East Asian flora that had been emerging since 1863, although the intricacies of the relationships between species in this region still perplex botanists. For example, Padre Giraldi's *Acer grosseri* is so closely related to *A. davidii*, Père David's snakebark maple, that some botanists consider it to be merely a subspecies of *A. davidii* rather than a species in its own right.[5] Similarly, plants of *Ailanthus*, raised from seeds he, Père Farges and Père Ducloux had collected, were originally thought to represent undescribed new taxa and were called *A. giraldii* and *A. giraldii* var. *duclouxii*, although they are so similar to *A. altissima*, the tree of heaven, that several botanists consider them to be the same.[6]

One of the most ornamental of the trees Padre Giraldi came across was the Chinese form of *Cornus kousa*, the Japanese dogwood, which is now known as *C. kousa* subsp. *chinensis*. He found it in 1897, although it had been discovered earlier by Augustine Henry. It was not introduced to cultivation until E. H. Wilson encountered it near Yichang in 1907 while collecting for the Arnold Arboretum, which then sent living material to Kew in 1910. *C. kousa* subsp. *chinensis* is a hardy medium-sized tree with edible red fruits and good autumn colour, but its real glory lies in the long-lasting inflorescences that are produced in summer after most trees and shrubs have finished flowering. The actual flowers are tiny and crowded into a small central cushion, but they are surrounded by four large creamy white bracts that gradually flush pink as they age; it is these bracts that transform a tree in full flower into an unforgettable billow of white. W. J. Bean thought that 'this beautiful flowering tree should be in every garden' and it is hard to disagree with him – if only we all had the space.[7]

The beauty bush

Perhaps the best known of Padre Giraldi's ornamental shrub discoveries is *Kolkwitzia amabilis*, commonly known as the beauty bush, which he first collected in fruit in 1891 in northern Shaanxi, and again in 1895 near Hua Shan in the Qinling some 100 km (63 miles) east of Xi'an. When these specimens were examined, it was apparent that they represented a completely unknown species and a new genus – *Kolkwitzia* – was created to accommodate it, of which *K. amabilis* is still the sole species. A plant raised from seeds collected by Wilson in Hubei in 1901 first produced its small pink and white tubular flowers at Veitch's nursery in 1910; but it is only when *K. amabilis*, a vigorous and extremely floriferous species, reaches its full size that its true ornamental effect can be fully appreciated, as the branches then vanish under a thick smothering of blossom. The beauty bush is popular with gardeners, especially in North America, where it has always been a favourite, and Wilson thought it one of his best introductions for colder parts of the country. Some excellent cultivars are now available, including 'Pink Cloud' raised by the Royal Horticultural Society at their Wisley garden in 1946 from seeds received from the Morton Arboretum in Illinois.[8]

Lilacs

When **Carl Sprenger,** a German plantsman with a large nursery in Naples, learned of Padre Giraldi's plant collecting activities, he contacted him and offered to send him funds in return for seeds. Padre Giraldi responded immediately and Sprenger had a great deal of success raising plants from the seeds he received from Shaanxi. He also distributed some of Padre Giraldi's seeds to fellow plantsmen in Europe, especially Germany, so that it was not long before several of the newly-discovered plants were available commercially. One of the most interesting of Padre Giraldi's packets of seeds contained a lilac he had discovered in March 1891, which proved to be a fine example of *Syringa*

BELOW

Cornus kousa subsp. *chinensis* flowering in summer at Kew Gardens.

RIGHT

Kolkwitzia amabilis 'Pink Cloud' in early summer at Kew Gardens.

oblata (syn. *S. oblata* var. *giraldii*). Sprenger's Naples nursery grew the new lilac successfully and offered it for sale in 1903; but Victor Lemoine had also acquired some of the seed and he distributed the lilacs he raised as *S. giraldii* (this should not be confused with the cultivar 'Giraldii', which has single purple-pink flowers). One of Lemoine's *S. giraldii* plants was sent to Kew where it flowered in the Temperate House in April 1901.[9] It was then apparent that Padre Giraldi's lilac was the same as one that had been collected by Potanin in western Gansu in 1885, but never described. One can only admire the perseverance of botanists as they picked their way through these taxonomic minefields, often dependent on the delayed introduction of living material for confirmation (or otherwise) of original identifications.

Another of Padre Giraldi's ornamental lilacs has long been known to horticulturalists as *Syringa microphylla*, although it has now been recognised as a form of the variable species *S. pubescens* and its correct name is *S. pubescens* subsp. *microphylla*. *S. pubescens* and its various forms are widely distributed across northern China and *S. pubescens* subsp. *microphylla* had been collected by others, including Augustine Henry in 1882, but it was only identified in 1901 when the German botanist Ludwig Diels described two fruiting specimens that Padre Giraldi had collected in October 1896. *S. pubescens* subsp. *microphylla* grows into a large bushy shrub and produces clusters of small scented pale pink flowers in early spring. William Purdom sent seed from Gansu to Veitch in 1910. The new lilac flowered for the first time at the Arnold Arboretum

BELOW
Syringa pubescens subsp. *microphylla* in spring.

in 1915 and Sargent noted that it went on to flower a second time that year. In fact, *S. pubescens* subsp. *microphylla* flowers sporadically throughout the year and Joseph Hers, a Belgian radio engineer who worked in China and sent specimens to the Arnold Arboretum, called it the four-season lilac because of this repeat flowering habit. In 1933, the nurseryman M. Cassegrain introduced the cultivar 'Superba', which has slightly larger, darker pink flowers than subsp. *microphylla*, and this is the form that is most commonly cultivated. It is worth planting close to paths where passers-by can catch its early scent.[10]

Beautyberry

A new species of *Callicarpa* or beautyberry that had been discovered by Augustine Henry in 1886 was introduced by Padre Giraldi when he sent seeds to Ludwig Beissner, supervisor of the Bonn Botanic Garden and an authority on woody plants. Herr Beissner gave some of the seeds to Herr Hesse, a nurseryman at Weener near Hanover, who named the species *Callicarpa giraldii*. The new shrub was first raised at Kew from seeds sent by the Arnold Arboretum in 1908 and the first berries were produced in 1914. Some botanists consider Padre Giraldi's introduction to be merely a floriferous variety of *C. bodinieri* and call it var. *giraldii*. In gardens, it is usually represented by the cultivar 'Profusion'.[11]

Euonymus phellomanus or cork spindle, which Padre Giraldi discovered in 1894, also has very attractive fruit but is much less well known than *Callicarpa* 'Profusion'. It takes its common name from the four corky strips that run along the young green shoots, giving them a squarish appearance. These remain as the wood hardens and are remarkable in themselves. Dan Hinkley reports that he was 'gobsmacked' when he first encountered *E. phellomanus* in the wild as, 'it possessed a woody framework clad by corky ridges of an insane width; on many branches the ridges were three times wider then the stem itself'.[12] Like many spindles, *E. phellomanus* has fine autumn colour but it also has distinctive, rather square, bright pink fruits. It was introduced by Reginald Farrer from Gansu province.[13]

Not all Padre Giraldi's discoveries were sizeable trees or shrubs and one of his smaller finds was a compact deciduous daphne called *Daphne giraldii*. It is hardy and relatively easy to grow, producing a mass of slightly scented yellow flowers in spring. Padre Giraldi found it on various occasions, the first being in northern Shaanxi in 1894 and then in the Qinling in 1897, but *D. giraldii* was not introduced until 1911 when William Purdom collected seeds for Veitch in western Gansu.[14] Most cultivated plants today probably descend from Purdom's original collection.

Daphne giraldii was also collected by **Father Hugh Scallan**, another Franciscan working in Shaanxi. As his name indicates, Hugh Scallan was not Italian but Irish. He was born in Septem-

ABOVE
Callicarpa bodinieri 'Profusion' in autumn at Kew Gardens.

BELOW
Euonymus phellomanus in early autumn.

ber 1851 in Dublin and studied at Douai in France, before joining the Franciscans in Belgium in 1874. He had been christened John, but he was given the religious name Hugh when he took his final vows in 1878. After being ordained priest in 1882, he taught in Manchester until 1886 when he left for China to serve as a missionary in Shaanxi with his Italian brethren. Father Scallan suffered various degrees of ill treatment during his years as a missionary, including imprisonment and robbery. Once, in 1893, he was almost stoned to death when his mission centre was attacked and his students tortured. He was seriously ill after this but recovered quickly and soon returned to his duties.

Rosa hugonis

In 1894, Father Scallan, who was a keen amateur botanist himself, asked Padre Giraldi to send a collection of plants to the botanists at the British Museum and promised to collect some of the specimens himself. This he did, and Padre Giraldi sent a box of plants and a covering letter from Father Scallan to London in October 1895. The following month, he sent the Museum some rosehips that he and others had collected from 'musk roses' in the mountains of Scian-ko (Shangzhou), southeast of Xi'an. Padre Giraldi later wrote offering to make further collections for the Museum but his letter was in Latin and as Father Scallan had begun sending collections back on his own behalf, the Museum did not take up Padre Giraldi's offer. It was obviously easier for the botanists in London to deal with the English-speaking Father Scallan rather than corresponding in Latin with Padre Giraldi.[15] The first collection was received in London in June 1896 and the consignment with the rosehips the following month. Padre Giraldi also sent some of these rosehips to his contacts in Germany, where the species was successfully raised. A further collection from Father Scallan was received at the Museum in 1899 and a rosehip from one of his collections reached Kew the same year. When the rose grown from its seeds flowered in 1905, Hemsley called it *Rosa hugonis* to commemorate Father Hugh. It was introduced to commerce by Veitch in 1908.

Rosa hugonis grows into a large vigorous shrub, reaching over 4 m (13 ft) in height and thriving in poor dry soil. It is one of the earliest roses to flower, with clear yellow cup-shaped flowers followed by small dark red fruits. Although Padre Giraldi had referred to 'musk roses', it was soon apparent that *R. hugonis* was not a musk rose at all but a briar rose and one

that was very closely related to *R. xanthina*. The similarities are such that some botanists consider it merely a form of *R. xanthina*, but there are differences between the two roses, including the fact that the natural habitat of *R. hugonis* lies farther south and west than that of *R. xanthina* f. *spontanea*, which Père David found in Inner Mongolia.[16]

Father Scallan served in Shaanxi for the rest of his life, dying there in 1928 after forty-two years in the province. He proved a diligent collector, sending over 4,000 specimens to the Museum between 1896 and 1904.[17] Among them was a deciduous shrub with terminal panicles of white flowers in summer, which had first been found by Père David at Baoxing and had then been identified as a new species of *Meliosma* and called *M. cuneifolia*. It was also collected by Pastor Faber and Augustine Henry. It was apparent from the first that this *Meliosma* was closely allied to *M. dilleniifolia* from northern India and some botanists now classify it merely as a subspecies, but the *Flora of China* maintains *M. cuneifolia* as a distinct species.[18] It is certainly hardier. As was becoming clear, distribution patterns in western and central China involve a tangled mix of species, including some which range across thousands of miles and others which inhabit just a single narrow valley.

By 1896, Padre Giraldi's missionary duties were increasing and the heat and poor food were beginning to weaken his once robust constitution, so he asked one or two of his confrères to collect plants and he also trained a local man to collect independently for him. Nevertheless, the size of the collections he sent back to Italy decreased. All botanical activities in the province were further curtailed during the Boxer Rebellion in 1899–1900, with its attendant anti-foreign and anti-Christian riots. The violence in Shaanxi culminated in July 1900 when the Italian bishop and his companions were taken prisoner by the rebels and murdered. To add to the tragedy, Shaanxi was suffering the worst drought for sixty years, which led to a dreadful famine and mass starvation. Padre Giraldi shared everything he had with his starving followers, going hungry himself. The following year a typhoid epidemic ravaged the area and Padre Giraldi caught the fever while attending a sick member of his community. He died in May 1901, a dozen years after arriving in China. He was buried in a small local cemetery where, twenty years later, Antonio Biondi and his colleagues in Florence arranged for the installation of an ornate headstone commemorating his botanical achievements.

RIGHT

Daphne giraldii. (*Bot.Mag.* (1917) No.8732)

M.S.del.J.N.Fitch lith.

Vincent Brooks,Day&Son Ltᵈimp.

L.Reeve&Cº London.

M. Smith del.

Meliosma cuneifolia, Franch.

Meliosma cuneifolia. (Bot.Mag. (1911) No.8357)

Descriptions of Padre Giraldi's discoveries

Antonio Biondi worked closely with the botanists at the Institute in Florence and once Padre Giraldi's collections began arriving in the early 1890s, he and his colleagues published descriptions for many of the new plants. Biondi also distributed specimens to Paris, Berlin and Kew, so that accounts of Padre Giraldi's discoveries in Shaanxi appeared in various European publications. Many of them can be recognised by names that include specific epithets such as *giraldii* or *giraldiana*. By the turn of the century, botanists in Berlin decided that enough was now known about the plants of central China to warrant the preparation of a checklist: and in 1900–1901, Ludwig Diels published *Die Flora von Central-China*.[19] Diels' *Flora* included Père Delavay's discoveries in north-east Yunnan and those made by Père Bodinier and his confrères in Guizhou, as well as the large numbers of new plants collected by Augustine Henry and Père Farges. It also included the published descriptions of Padre Giraldi's Shaanxi discoveries but, as Biondi had sent more than 900 of Padre Giraldi's specimens to Berlin, Diels and his colleagues were able to examine these plants for themselves and the *Flora* consequently includes some species that had not previously been described, such as *Kolkwitzia amabilis*, *Euonymus phellomanus* and *Syringa pubescens* subsp. *microphylla*.

Three years after Padre Giraldi's death, **Padre Cipriano Silvestri**, another Italian Franciscan with botanical leanings, arrived in China. Cipriano Silvestri was born in May 1872 in Bardalo near Pistoia, not far from Padre Giraldi's home. He joined the Franciscans in 1887 and, after taking his final vows in 1895, was ordained priest in 1896. Padre Silvestri also studied botany under Antonio Biondi and when he went out to China as a missionary in 1904 it was natural for him to want to emulate Padre Giraldi and collect the plants of his own district for Biondi and his colleagues.

Padre Silvestri was based in Siang-yang (Shiyan) on the northern flank of Wudang Shan in **north-west Hubei province,** a region that was still botanically unexplored. He began to investigate the area himself and, like Padre Giraldi, also trained a local man to collect for him. This arrangement, though, was not without its difficulties. It seems that the local people, while prepared to tolerate the inexplicable activities of a foreigner, were suspicious of one of their own behaving in the same incomprehensible fashion. Their misgivings often centred on the man's collecting bag, which he wore strapped to his back. Occasionally these doubts resulted in direct action: once, the collector was pursued as a chicken thief and on another occasion he was arrested and Padre Silvestri had to appear in court to obtain his release.

In spite of these complications, Padre Silvestri was able to send numerous specimens back to Florence, and as his corner of Hubei lay north of the areas investigated by Augustine Henry in the 1880s and by E. H. Wilson between 1900–1910, his collections helped extend contemporary botanical knowledge of the region. He did not find many new species and few of those he did find were of horticultural value, but his specimens often revealed new forms of species that had already been described and added further detail to the complex jigsaw of plant distribution patterns across central China that botanists were gradually assembling.[20]

Padre Silvestri was recalled to Italy in 1921 so that he could set up a teaching and training programme at St Antony's College in Rome for those of his confrères who were going out to China as missionaries. He died in 1955.

Padre Cipriano Silvestri.

Photo Credit: Museo di Storia Naturale dell'Università di Firenze

Horticultural developments

Syringa oblata

Victor Lemoine had already crossed *S. oblata* with *S. vulgaris* to produce double-flowered lilacs (*S.* x *hyacinthiflora*) and, when he recognised that Padre Giraldi's early-flowering example of *S. oblata* was even more attractive than the *S. oblata* in general cultivation, he began to use it. This cross was extremely successful and resulted in the creation of a series of outstanding lilacs known as the Early Hybrids, which were first listed in Lemoine's catalogue in 1911.[21]

This illustrates the continuing importance of seed collection (always provided, of course, that all collections are authorised and sustainable) as not only might we discover more garden-worthy strains of familiar ornamental plants, as happened with Padre Giraldi's *S. oblata*, but, more importantly, fresh seed collection widens the cultivated gene pool and helps give the species in cultivation some of the diversity and resilience found in a wild population. These qualities increase disease-resistance, which fosters the species' chances of survival in cultivation – this is particularly important if the species is under threat in the wild. Botanists also then have an opportunity to study living plants to discover if there are any small variations between those collected in different areas, or at different heights, thus adding to the knowledge of an individual particular species.

Rosa hugonis

In the 1920s, a chance rose seedling was discovered in Cambridge Botanic Garden, which was eventually identified as a natural hybrid between *R. hugonis* and *R. sericea* that is now called *R.* x *pteragonis* 'Cantabrigiensis'. This new rose was first shown in 1931 and was immediately recognised as a very fine garden plant. Its creamy scented flowers are born in profusion along tall wand-like stems and, as *R.* 'Cantabrigiensis' is not as large as either of its parents and can be kept in further check if its travelling rootstock is carefully watched, it is a better shrub for smaller gardens. Perhaps it is a better choice altogether: Graham Stuart Thomas, a great rose expert, considered *R.* 'Cantabrigiensis' as beautiful as *R. hugonis* and, even, in some ways superior as it was 'a more compact plant with even more lovely flowers'.[22]

LEFT

Rosa hugonis. (Willmott, Ellen (1914) *The Genus Rosa* Plate 95)

BELOW

Rosa x *cantabrigiensis* in the author's garden in early summer.

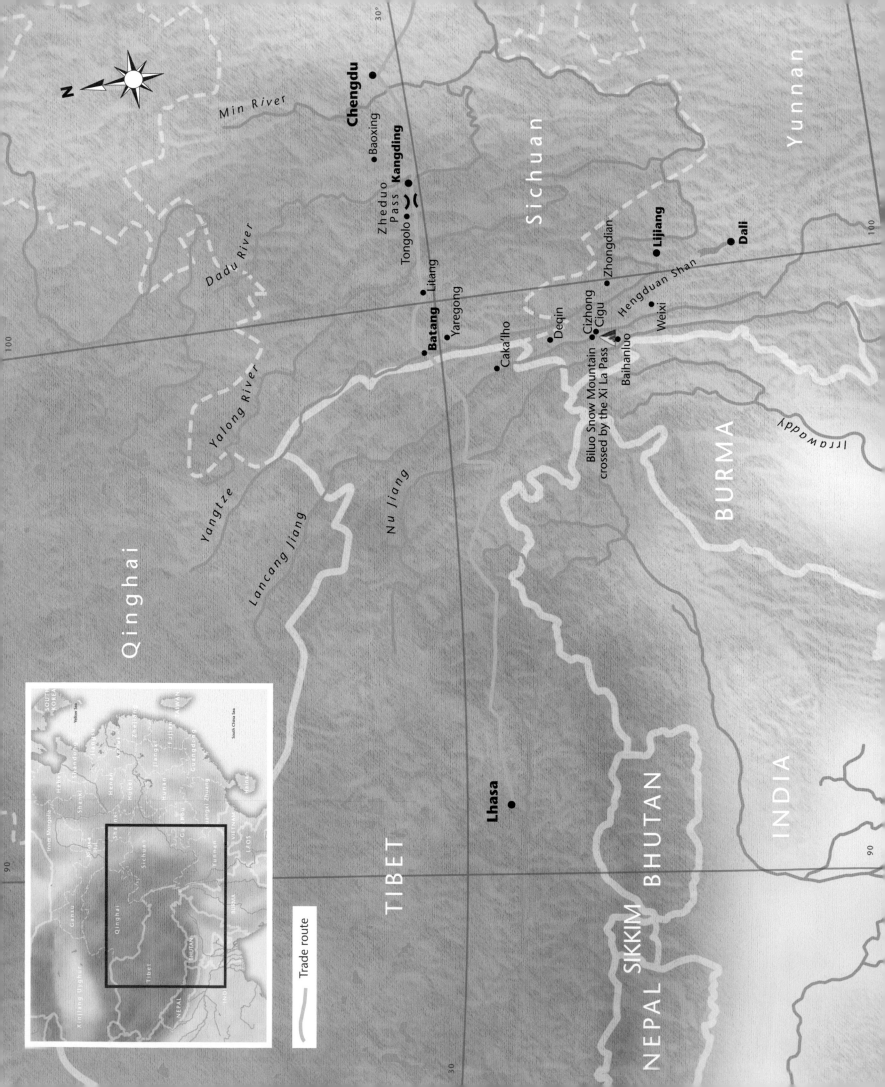

CHAPTER 13

The Tibetan Borders

—

The devotion of the French missionaries in general to the cause of their religion deserves notice. No work

is too hard for them, no living too poor. They are not deterred by epidemics of sickness, or by threatened

massacre. They have simply devoted their lives to the propagation of their religion, and nothing can turn

them from their purpose.

A. E. PRATT, 1892[1]

—

By THE 1880s, ROME COULD BE PLEASED with the progress that was being made by the various Catholic orders and missionary societies operating in China. The support of France for their work had been invaluable and the French ministers at Beijing had championed the cause of the missionaries so effectively that, in spite of considerable unwillingness on the part of the Imperial authorities, the freedoms granted to Christians in successive treaties had been upheld. The knowledge that each infringement of missionary rights would be vigorously challenged and that the French would insist that all attacks on priests and their communities were punished had served to deter the missionaries' opponents. Missionary organisations, fuelled by the apparently inexhaustible supply of young men eager to spend their lives in the service of Chinese evangelisation, had therefore been able to spread across China and expand into the remotest corners of each province.

The exception to this relatively successful picture was in the Tibetan borderlands: the mountainous area dividing the lowlands of central China from the high Tibetan plateau. Today, the region forms the border counties of western Sichuan and of north-west Yunnan but, in the nineteenth century, the succession of soaring ranges and deep narrow river valleys, interspersed with high rolling grasslands, created a landscape where travel and communication were so difficult that the exercise of effective control – whether from Lhasa or Beijing, or from closer Chinese provincial centres – was problematic. In theory, Chinese suzerainty extended throughout the Tibetan borderlands, but, in practice, the Imperial authorities controlled little more than the lands adjacent to the section of the main Chengdu–Lhasa trade route that stretched for some 362 km (225 miles) between the Chinese frontier town of Tatsienlu (Kangding) and the Tibetan town of Batang in the west.

BELOW

The view west from Baima Shan, 4,120 m (13,517 ft), towards the Yangtze-Lancang Jiang divide.

Otherwise the region between the Dadu river and the upper Salween (Nu Jiang) was chaotic and lawless, with various tribes, warlords and lamaseries jockeying for power, and marauding brigands preying on any travellers or merchants bold enough to brave the winding tracks and pack routes that linked the isolated hamlets and villages.

It was in this wild inaccessible country, where the Imperial writ ran only intermittently, that the Missions Étrangères established its Tibetan missions and where, in the teeth of opposition from the Buddhist lamas, the priests clung stubbornly to their posts. The lamas hated the missionaries, seeing them as intruders who wanted to seduce the local inhabitants from their traditional beliefs and traditional loyalties, and they only tolerated the priests because they were forced to do so by their Chinese overlords. Their anger and resentment led them to make repeated raids on the missionaries and their communities and all that kept them from killing the interlopers outright was their

fear of Chinese reprisals. The missionaries in their remote stations could not rely on distant Chinese officials for any immediate help when they were attacked, and they and their followers had to try and fight off their assailants on their own.[2]

The hostility of the lamas towards the missionaries was important because they were involved in virtually every aspect of Tibetan life and were able to exercise considerable influence over the local population, as almost every family had a son who was a lama. The great lamaseries were usually the principal landowners in their districts, with the locals either working directly for them or farming the land as their tenants, which meant that very few people were in a position to ignore the lamas' edicts against the missionaries. Taxes were very high so most of the inhabitants lived in grinding poverty and some were further indebted to the lamaseries as the lamas also acted as moneylenders, taking the land of those who could not pay and sometimes enslaving them as well. As the century wore on, the lamas' hostility to foreigners was fuelled by the apparently inexorable spread of British power in India northwards into the Himalayan massif and up to the very borders of Tibet

BELOW

High peaks west of Kangding.

ABOVE RIGHT

Lamas at Batang.

itself. Several European travellers were intercepted as they tried to reach Lhasa. Although these individuals protested that they were merely traders or explorers, they were regarded as spies and invariably turned away. The Tibetans were well aware that the British rulers of India had first arrived there as traders, and they were consequently suspicious of any foreigner trying to enter Tibet, no matter how apparently innocuous the reason. The rising number of foreigners they intercepted increased the Tibetans' feeling of being under siege and they vented their suspicions and growing anger on the missionaries, who were not only foreigners living in their midst but ones openly trying to convert the local people to an alien religion.

Pères Gabet and Huc, Père David's Lazarist confrères, had managed to reach Lhasa and open a chapel in 1846 but were swiftly expelled. It was not until 1852 that the Missions Étrangères de Paris, which had been given responsibility by Rome for the conversion of Tibet, first attempted to establish missions in the region. By 1865, missions had been founded at Yerkalo (Caka'lho) and Tsekou (Cigu) on the upper Mekong (Lancang Jiang), and over the next twenty years, in spite of repeated attacks and constant harassment and ill treatment, missions were set up at Batang and at Atuntse (Deqin). The Missions Étrangères never succeeded in penetrating the Tibetan uplands, although it remained their ultimate goal.

The missionaries were forced out of Batang in 1873 and Père Félix Biet and his confrère were evicted from Caka'lho, but on this occasion the protests of the French representative at Beijing were sufficient to secure their return. Five years later, Père Biet was made Bishop of the region and moved to the bishop's residence at Kangding. Kangding was an important trading centre on the main route to Tibet, and was always crowded with a shifting throng of tribesmen and traders. Although the majority of the inhabitants were Tibetan and the King of the old Tibetan province of Chiala lived there, it was firmly under Chinese control, which made it the safest place for the Missions Étrangères

in Tibet to have its headquarters. However, Kangding lay some eighteen days east of Batang, and was even farther from the beleaguered missions at Deqin, and Caka'lho and Cigu on the Lancang Jiang.

In 1868, an English merchant named Thomas Cooper made a determined effort to discover the direct trade route from Sichuan to northern India. He managed to reach Batang but the authorities there, who had no intention of letting foreigners investigate the lucrative trade route across the Tibetan highlands, prevented him from travelling further west. He reported bitterly that the lamas '(taught by the Chinese) looked upon my coming as the forerunner of the annexation of their country by the Palin, or white Conquerors of India, and met me everywhere with scowls of hatred, and muttered curses.'

He turned south towards Yunnan where he hoped eventually to reach Dali. His progress was slow and hazardous and he was frequently glad of his pistol, which he found 'commanded respect.' He visited Père Jules Dubernard at Cigu and then struggled south to Weixi, where the local tribal chief refused to let him continue and the Chinese authorities imprisoned him for five weeks. He had no choice but to return to Sichuan the way he had come. Cooper was lucky to survive his journey and as the narrative he published later made clear, the difficulties and dangers faced by any travellers other than those sanctioned by the authorities, particularly the lamas, rendered travel in the region almost impossible for outsiders. Cooper was very grateful for the help he had received from Italian and French missionaries along the Yangtze and at Kangding; but at Batang, when he was trying to get permission to travel through to Assam, he knew that he had to distance himself from them if he was to have any hope of succeeding with his plans. His experience of the dangers of missionary life in this remote region – where the unprotected priests lived in isolated settlements, far from their own compatriots, surrounded by hostile forces and constantly facing the threat of attack, expulsion and death – led him to comment: 'The history of the Tsekou [Cigu] mission may, from the date of its establishment be traced in the blood of numbers of brave and nobleminded missionaries who have fallen by poison and the knife in the cause of their religion.' Cooper wrote this in 1871, and although it was true generally of the Tibetan mission, Cigu had, until then, been one of the most successful missions in these remote mountains. Its tragedy lay ahead.

In spite of determined opposition, the priests living at Batang, Deqin, Caka'lho and Cigu persisted with their work, building churches, founding schools and orphanages, and establishing pharmacies. They also began to vaccinate the inhabitants against smallpox, a local scourge, and the success of this programme brought herdsmen, nomads and distant tribesmen to the missions for vaccination. The missionaries then began to tackle other endemic fevers and locally prevalent conditions such as goitre. These activities won them many friends among the inhabitants. A few families became Christians but the majority, in the face of death threats issued by the lamas, dared not convert, even if they had wished to. The handful of converts in each mission were usually drawn from those who grew up in orphanages established by the priests, or from those whose debts they had paid, or whom they had redeemed from slavery. These limited successes were enough to provoke further violence from the lamas and, in 1881, a missionary was murdered at Batang. Bishop Biet invoked the help of the French minister at Beijing and, after the Chinese viceroy in Sichuan had intervened, a simmering calm was restored. Such Chinese cooperation was not always forthcoming because, as Cooper had recognised, the Chinese authorities in the area often sided with the lamas and were generally unwilling to support the missionaries, as they did not want to encourage an influx of foreigners into a region so close to Tibet. The Chinese might not have been able to protect the priests in the isolated missions to the south, but they could regulate the situation in Batang if they chose, as they had a number of officials there and maintained control. Even though Chinese support for the missionaries was vital to their continued existence in the region, when it was given, it further antagonised the lamas, as it emphasised the priests' association with the detested Imperial overlords.

Arrival of Père Soulié

Remarkably, amidst all the other cares that pressed upon him, Bishop Biet recognised the importance of making accurate information about this unexplored and wholly unknown region available to the West, and he encouraged the Tibetan missionaries to make scientific observations and collect zoological and botanical specimens from their areas. His enthusiastic support for such scientific pursuits resulted in the despatch of thousands of specimens of birds, animals and insects to the Muséum, many of which had been secured by Père Dubernard; but it was only after the arrival of Père Soulié in 1886 that the flora of the region was collected in any systematic fashion.

Jean André Soulié was born in 1858 at Saint-Juéry near Rodez in the Aveyron, in the Midi-Pyrénées region of southern France.[4] He was a serious, rather reserved boy, studious and hardworking, and he attended the *petit seminaire* at Belmont. When he was twenty-six he entered the seminary at Rodez. During his training for the priesthood he decided that he wished to work as a missionary in the East. He wrote to the Missions Étrangères in Paris in June 1884 but did not tell his family of his decision, as he knew how much it would distress them. Nevertheless, only a few hours after he had left them to travel to Paris, his parents learned of his plan and followed him, desperately trying to persuade him to change his mind. They knew, as did all parents with a son in the service of foreign missions, that once he had left France it was extremely unlikely that they would ever see him again. One can only imagine the anguish of families where several sons became missionaries – Bishop Biet was the third of four brothers to join the Missions Étrangères.[5] Even in an age of faith, it must have been very hard to lose more than one beloved son to foreign missions. However, Père Soulié was determined to follow his vocation and he left for China in October 1885, three months after he had been ordained. The news that he was destined for the Tibetan mission must have been devastating for his family, as the dangers faced by priests in the Tibetan borderlands were well known.

The boat in which Père Soulié travelled up the Yangtze was joined at Yichang by Père Bodinier, who was on his way back to Guizhou. The two missionaries became friends and, having discovered a mutual interest in botany, they made several botanical expeditions along the banks of the river while the boat was laboriously hauled up through the gorges.[6] Père Soulié's interest in botany was of long standing and one of his contemporaries at Belmont and then at Rodez was Hippolyte Coste (1858–1924), a passionate botanist who went on to pub-

RIGHT
Street leading to the south gate, Kangding.

FAR RIGHT
Kangding from the south.

LEFT
Père Jean Soulié.

lish the first illustrated work on the French flora. One can imagine the two young seminarians botanising together, although the conscientious Jean Soulié probably did not skip his studies to explore the countryside, as Hippolyte Coste frequently did.

Père Soulié was destined for the Batang mission but, when he and **Père Annet Genestier**, with whom he had travelled out from France, reached Kangding, Chinese officials in the town tried to prevent them from travelling further west. While he waited for some resolution to the difficulty, Père Soulié would no doubt have explored the town, described by one traveller as 'a meanly built and filthy city', with very narrow uneven streets and low wooden houses on shale foundations. The buildings were crammed into the narrowest of valleys, and the Tuo River divided the town, filling the air with the sound of its rushing waters as it tumbled under the bridges. There were eight lamaseries in the vicinity and the lamas 'swaggered though the streets with an insolent mien'.[7] Père Soulié had his first sight of Tibetan monks dressed in red, their heads closely shaved, twisting their prayer cylinders, and muttering at the same time the universal prayer, 'Om Mani Pemi Hom'. It must have made him uneasy, knowing as he did how much his confrères had suffered

at their hands, and recognising that they would become his own most determined enemies.[8]

In the end, the two young priests were only delayed at Kangding for a short time and they arrived at Batang on Christmas Day 1886. Père Genestier then continued on to Deqin while Père Soulié settled in at Batang alongside Père Pierre Giraudeau, who had lived there since being driven from the mission he had established at nearby Yaregong the year before. Père Giraudeau noted how neatly and carefully Père Soulié unpacked and arranged the few possessions that he had brought with him from France – it was these precise orderly characteristics that made him such a good botanist. Père Soulié, who had already begun to learn Chinese, now began to study Tibetan, and was much amused by two children who acted as extra tutors and made him laugh by the imaginative mimicry with which they explained things to him.

Attack at Batang

This tranquil period lasted until 21 May, when the two priests were woken in the night by a hail of enormous stones. At first they thought it was an earthquake. When they realised that

Photo Credit: Pratt, A. (1892) *To the Snows of Tibet through China*

they were actually being attacked by a mob acting at the instigation of the lamas, Père Giraudeau managed to climb on the roof and frighten off the attackers by firing a gun. The Chinese authorities would not provide help and a nearby Christian farm was burned. For the next four weeks, the two missionaries took turns keeping watch, as any more such fierce attacks would wreck the house and lead to their capture. Despite this, an attack came on 20 July that was so violent they had to retreat, although Père Soulié only consented to go after Père Giraudeau convinced him that their deaths, which seemed inevitable if they stayed, would not advance their cause. Chinese soldiers had taken up station either side of the house, ostensibly to guard the missionaries, but as they had no wish to provoke a confrontation with the lamas they did not intervene when the door was forced, or when the chapel and several houses belonging to converts were burned. The priests had taken refuge in the grain store of a neighbouring house, where they were kept hidden for a couple of days by a local Tibetan chief; but they would have been discovered and killed if the lamas had not called off the attacks just in time. The lamas did this only because they did not wish to bring the wrath of the Imperial authorities down upon themselves, which they knew France would insist upon if two of her missionaries were murdered. Two local chiefs, who did not want the priests massacred in their districts, hid them for ten days in a 'noisome hole' and then arranged for their escape. As it was, they had to flee for their lives and the lamas only abandoned pursuit when they saw that the priests had faster horses. It was not until 22 August that they arrived safely back at Kangding. There they learned that the missions at Caka'lho and Deqin had also been destroyed, and that Père Dubernard had been driven from Cigu, although the church he had built was still standing.[9]

Père Soulié was hardly six months into his first posting but he already had first-hand experience of the dangers that beset any priest who ventured into these restive borderlands. Bishop Biet sent to Beijing for help and the French diplomats duly demanded of the Imperial authorities that full compensation be paid and the missionaries allowed to return to Batang. In the meantime Père Soulié, who had resumed his language studies, visited Chapa, a small mission to the south-east, where he gave his first sermon. When he heard the little girls from the orphanage laughing at some of his expressions, he rebuked them soundly – for him and his confrères, the Word of God, no matter how poorly expressed, was never a subject for levity. In his free time he botanised, as he wanted to assist Bishop Biet's efforts to supply French specialists with natural history specimens by making a comprehensive collection of the flora of the area.

Visitors

Père Soulié was at Kangding with Bishop Biet, his secretary Père Dejean and Père Giraudeau in 1889, when the American explorer **William Woodville Rockhill** arrived. Rockhill had been the American chargé d'affaires in Beijing, where he had begun to study Tibetan and had then decided to resign his diplomatic post and try to reach Lhasa. He set out at the end of 1888 and, with the help of his Tibetan language tutor who had accompanied him from Beijing, stayed at a lamasery in Xining for a month before joining a pilgrim caravan bound for Lhasa. Once in Tibet, he learned that the number of foreigners trying to reach Lhasa had disturbed the Tibetan authorities to such an extent that all the approaches to the city were now closely guarded. He realised it would be too dangerous to continue and so turned east to return to China, arriving in Kangding at the end of June.[10] Rockhill had been educated in France and his arrival must have been extraordinarily welcome to the small isolated group of priests, longing for news of Europe and home, and starved of fresh faces and fresh conversation.

RIGHT

The French missionaries at Kangding in 1892. Bishop Biet is seated on the left, next to Père Giradeau. Père Soulié is standing on the right, next to Père Dejean, the Bishop's secretary.

Then **Antwerp Pratt**, the naturalist who had accompanied Augustine Henry on collecting forays in 1888 (see chapter 10, p.141), arrived from Hubei on 4 July. Rockhill and Pratt were the first Westerners, apart from the missionaries themselves, to reach Kangding since 1877, when the British consular official Edward Colborne Baber, Captain William Gill and William Mesny[11] a Major General in the Chinese army, had visited the area. Previously, the only other Westerner to penetrate these inhospitable borderlands, apart from Pères Huc and Gabet, had been Thomas Cooper. Père Gabet's book about his journey, published in 1850, had sparked Western interest in Tibet, and the accounts of travellers like Cooper, Baber and Gill, which contained much information about the Tibetan peoples and their cultures, as well as the missionaries' own writings, fuelled

Fig. 51.—Père de Deken, Prince Henri d'Orleans, Gabriel Bonvalot, photographed by Pratt in Tatsien-lu on the completion of their magnificent journey from Siberia, through Chinese Turkestan and the great plateau of Northern Tibet.

PLANT COLLECTORS IN CHINA
Fig. 52.—Père Dejean, Bishop Biet, Père Jean André Soulié, and William Woodville Rockhill, photographed by Pratt in Tatsien-lu. Rockhill was the famous American explorer of Tibet.

Photo Credit: RHSJnl. (1947) Fig.52 p.100

Western fascination with this mysterious and inaccessible region to such an extent that increasing numbers of Westerners began to brave the difficult roads up to Kangding. Like Rockhill and Pratt, these travellers wrote about their journeys.[12] News of these frequent arrivals only increased Tibetan anxieties. The missionaries' geographical observations and their natural history discoveries, together with the collections made by Pratt and other travellers, gave specialists in Paris and London much new material to describe; and among the Europeans and Americans who made their way west from Sichuan were naturalists, including plant-hunters like E. H. Wilson, who first visited Kangding in 1903 and then again in 1904 and 1908.

Pratt was very appreciative of the help he received from the various missionaries he met during his journey westwards and he found much to admire:

> All the Roman Catholic missionaries have a very hard life, and I think that people at home have very little idea of the sacrifices they make for the sake of their religion. Beyond having cleaner and perhaps, in a trifling way, better houses than the natives, there is no difference in their mode of life… their food is coarse and often scanty, and their lives frequently in danger.[13]

He knew how unpopular the priests were at Kangding and realised that, as one of their associates, he would have to operate cautiously if he was to be able to explore without further antagonising the lamas. He soon found, though, that none of the local Chinese or Tibetans would work for him and he had to take some of the missionaries' followers with him as servants and collectors. He was also accompanied by Augustine Henry's botanical collector.

Above the town of Kangding, immured in its narrow valley, were plateaux and forested slopes where conifers, rhododendrons, willows, honeysuckle and roses abounded. Higher up, above the snowline, were meadows and scree slopes that sheltered a variety of low-growing alpines including *Meconopsis*, *Primula*, saxifrages and *Pedicularis*. Even though Pratt was not a botanist he commented on the plants he came across on his travels, such as the extensive swathes of *Sophora davidii* that he had passed on his way up to Kangding. He was impressed with the richness of the flora in the area, especially the wealth of plants growing just below the snowline.[14] After making some exploratory excursions, he set up camp in a Tibetan tent in conifer woods a little to the north of the town, where he was joined by Père Soulié who had been given permission by Bishop Biet to spend time botanising on the higher slopes.

LEFT

Photographs taken by Antwerp Pratt at Kangding reproduced in the RHS Journal 1947.

BELOW

Slopes above Kangding.

Photo Credit: David Rankin

Collecting expeditions

Pratt went back to Yichang for the winter, but returned to Kangding in the spring of 1890 to make further collections. At the beginning of May, Père Soulié went with him on an expedition south-east along the Dadu River, and they travelled through the extensive rhododendron forests that petered out around 3,657 m (12,000 ft). Pratt was particularly impressed by some immense trees with trunks over a foot in diameter. He found a good place to camp, but noted that his men were using logs from the surrounding rhododendrons to fuel the fire. Since then, the growing population of Kangding has put such pressure on the surrounding forests that little remains today of the old-growth rhododendron woods that Pratt and Père Soulié admired. A fire was certainly necessary, as although the

BELOW

High treeless plateau above Kangding.

Photo Credit: David Rankin

Photo Credit above and above right: Pratt, A. (1892) *To the Snows of Tibet through China*
Photo Credit right: Bonvalot, G. (1892) *De Paris au Tonkin à travers le Tibet Inconnu*

ABOVE

Antwerp Pratt and Père Soulié at the camp in the conifer woods above
Kangding.

ABOVE

Père Soulié holding a butterfly net at
the camp on the site of the log hut.

RIGHT

Gabriel Bonvalot and Prince Henri
d'Orléans.

days were hot, the nights were intensely cold, with severe frosts
and sometimes snow. They hoped to reach the new mission at
Mosimien (Moxi), founded that year by Père Giraudeau, which
lay farther south, but they did not want to use the usual tracks
which were much frequented by groups of lamas, so they tried
carving a way through the virgin conifer forests. When these
proved impenetrable, they had to turn back and Pratt decided
to base himself at the campsite, where he had a small log cabin
built as the tent was too damp to prepare specimens properly.

Pratt now sent Adolf Kricheldorff, a German naturalist who
had joined him as an assistant, north-east to Baoxing with six
local collectors. The visit was unsuccessful as officials at Baoxing prevented Kricheldorff from visiting the higher slopes. Pratt
himself hoped to spend the summer at the log cabin, but the
weather was unseasonable and after heavy snowfall on 2 June
followed by a rapid thaw, the inhabitants began to blame the
bad weather on Pratt's presence in the mountains. The local
Chinese official then refused to allow him to continue living
outside the town and Pratt was forced to return to the Bishop's
residence, where he found his movements so hampered by the
authorities that he decided to leave for Shanghai. It was at this
juncture that a small group of remarkable French explorers
arrived at Kangding.

Their leader was Gabriel Bonvalot (1853–1933) and he was
accompanied by **Prince Henri d'Orléans**, the twenty-three year-
old son of the Duc de Chartres and great grandson of Louis
Philip, the last King of France. Henri d'Orléans had originally

wanted to join the army but
French legislation prevented
this and he had turned
instead to travel and adven-
ture: hunting tigers in
Nepal and visiting Japan
in 1887. In 1889, when Bon-
valot was planning to travel overland from Russia to Tongking
across Tibet, the Duc de Chartres agreed to finance the expedition if Prince Henri could accompany Bonvalot as photogra-
pher and naturalist. The projected expedition was ambitious,
even for an inveterate and experienced explorer like Bonvalot
who had already made three long journeys through Asia, and
it presented the young prince with a remarkable opportunity.
The party started in July 1890 from Moscow and was joined
at Kuldja (Yining) in Chinese Turkestan by Père Constant de
Deken, a Belgian missionary who spoke Chinese and could act
as interpreter when necessary. They had no guide across the
Tibetan plateau, which they reached in December in the bitter-
est weather, so they followed the old pilgrim route, marked by
the skeletons of fallen animals, during which the temperature
fell as low as −40 °C and the mercury froze in the thermometer.
The territory they crossed was so remote that they travelled for
over ten weeks without meeting another soul. Eventually, the
Tibetan authorities stopped them when they were within 150
km (93 miles) of Lhasa and forced them to turn eastwards to
Batang and China.

Bonvalot and Henri d'Orléans reached Batang in June, where they found barley growing among the ruins of the mission. As they continued towards Kangding, they were very conscious that they were travelling through an area where much French blood had been spilt. They reached the Bishop's residence on 24 June and one can imagine the missionaries' delight at the arrival of fellow countrymen from such an unexpected direction. Prince Henri had collected some zoological specimens along the way and had botanised a little at Batang and then at Litang, but it was when he arrived at Kangding and realised the richness of the flora that he began to collect botanical specimens in earnest. He worked hard and by the time he and Bonvalot left a month later, he had put together a collection of around 500 specimens. As Antwerp Pratt was also about to leave, he took the Prince's collections back with him for despatch to Paris. Bonvalot and Henri d'Orléans then made their way south to Dali, where they stayed with Père Leguilcher, and continued through southern Yunnan to Tongking, which they reached at the end of September, having accomplished a quite extraordinary journey of more than 3,750 km (2,330 miles).[15]

New plants from Kangding

When Pratt arrived back in London in December 1890, he brought with him a fine collection of over 700 botanical specimens, which considerably impressed Augustine Henry when he saw them while on leave that winter. He noted that Pratt expected to receive a handsome price for them and they were later acquired by the British Museum.[16] The plants had been collected by the Chinese collector Henry had persuaded Pratt to take with him. Henry later translated the man's Chinese labels and sorted the specimens with William Botting Hemsley, who published descriptions of some of them in 1892. As Franchet and Professeur Bureau had described Henri d'Orléans' collection a year earlier, Prince Henri is often listed as the discoverer of species such as *Syringa tomentella* and *Neillia thibetica*, even though Pratt actually collected them first.[17] This is similar to the situation relating to many of Augustine Henry's discoveries, which are credited to Père Farges as they were first described by Franchet from the missionary's specimens. The work of the taxonomist in the herbarium is critical to plant exploration.

One of Pratt's discoveries was *Rubus cockburnianus*, a bramble related to Père David's *R. thibetanus* although much more vigorous and only suitable for places where it has room to spread. Its maroon arching stems are overlaid with a thick white bloom that give it a ghostlike presence in the winter garden, which is when it is seen to best effect. *R. cockburnianus* was also collected by Padre Giraldi in Shaanxi and named *R. giraldianus* in 1901, but Pratt's specimen – *R. cockburnianus* – had been named first in 1892 so this name has precedence. Wilson introduced the species to the Arnold Arboretum

in 1908.[18] A rhododendron brought back by Pratt, which was later collected by Père Soulié, was described by Franchet in 1895 as *R. prattii*. The species has large white flowers spotted with red and is closely related to *R. faberi*, discovered by Pastor Faber on Emei Shan (see p.141). It has recently been reintroduced.[19] One of the most striking and immediately recognisable of Pratt's finds was *Cypripedium tibeticum*, a slipper orchid with a large deep maroon-purple pouch first collected by Père Delavay, which can be grown outside in cool areas so long as it never dries out.[20] Of course, although Pratt is listed as the discoverer of all these species, the specimens were collected and prepared by the unsung Chinese collector trained by Augustine Henry, while Pratt concentrated on the birds and insects that were his real interest.

Prince Henri's collections included *Cathcartia* (formerly *Meconopsis*) *chelidonifolia*, which had first been collected on Emei Shan in 1887 by Pastor Faber. Yellow-flowered *C. chelidonifolia*, unlike the blue poppies of the open meadow to which it was long thought to be related, is a species for damp woodlands.[21]

Other herbaceous plants collected by the Prince included *Corydalis elata*, a hardier and later flowering relative of Père David's *C. flexuosa*, which has also recently been introduced; and *Lilium lophophorum*, a dwarf lily with narrow greeny-yellow petals that join at the tip to form nodding balloon-shaped flowers.[22] Père Delavay first discovered *L. lophophorum* high on the Lijiang slopes in 1884, and he collected it again above Menhuoying, but Franchet originally described these specimens as a fritillary and it was only after examining the plants collected by Henri d'Orléans that he realised he was actually dealing with a species of lily. Père Delavay's plant is now considered to be the typical *L. lophophorum* var. *lophophorum*. It was later collected by Père Soulié.[23] *L. lophophorum* and its varieties are native to high alpine meadows and have not proved easy in cul-

FAR LEFT
Rubus cockburnianus in
winter at RHS Garden,
Wisley.

Photo Credit: Harry Jans

LEFT

Cypripedium tibeticum
flowering at 3,650 m
(11,975 ft), Jiulong,
Sichuan.

Photo Credit: Harry Jans

ABOVE

Lilium lophophorum flowering at 4,415 m (14,484 ft), Hong Shan, Yunnan.

ABOVE RIGHT

Incarvillea compacta flowering at 4,200 m (13,779 ft) in rubble by the road, Baima Shan, Yunnan.

BELOW

Incarvillea compacta at 2,300 m (7,545 ft), Deqin, Yunnan.

tivation, which is a pity as they are beautiful. An *Incarvillea* collected by the Prince was named *I. bonvalotii* in honour of his companion, but Przewalskii had already discovered this species and the seeds he collected in 1880 germinated in St Petersburg, where the resulting plants were described by Maximowicz as *I. compacta*, which is the earlier name and has priority. *I. compacta* is low growing and closely related to *I. mairei*, which was collected in the region by Père Soulié. It has proved to be yet another lovely species of these high mountains that is depressingly short-lived in cultivation.[24]

From a horticultural point of view, one of the best plants that Prince Henri found was a vigorous white-flowered *Clematis*, which he collected at Batang at the beginning of June. When Franchet came to examine it, he realised that it was the same as a *Clematis* discovered by Père David in Baoxing and later collected by Potanin in Gansu, which is now known as *C. spooneri*. E. H. Wilson introduced *C. spooneri* to the Arnold Arboretum from Baoxing in 1908, and it has been grown in gardens ever since: although it is probably more familiar under some of its older names, particularly *C. montana* var. *sericea* and *C. chrysocoma* var. *sericea*.[25]

Père Soulié's first collections

The missionaries left at Kangding in August 1890 must have felt very flat after their summer visitors had departed; but Père Soulié applied himself to the task of botanical exploration with fresh energy. He had already begun to collect the plants of the surrounding area, but he was now further inspired by Henri d'Orléans' patriotic fervour. Both men wanted to see French botanists remain at the forefront of discoveries in the Tibetan borderlands, which they felt was only right as Frenchmen had been the first to explore the region. Having seen Pratt's collector at work, they realised that there was now a danger that English botanists might steal a march on them. Père Soulié was determined that '*la gloire*' of providing the first overview of the Tibetan flora should belong to French botanists and therefore began to botanise in earnest around Kangding.[26]

Distribution patterns

When Père Soulié's first collection arrived at the Muséum in May 1891, Franchet examined the specimens immediately and was ready to give a paper outlining his first impressions the following month.[27] It was quickly apparent to him that there were strong links between the flora that Père Soulié was investigating in Kangding and its surroundings, and the flora that Père David had discovered in Baoxing about 88 km (55 miles) to the northeast, and these affinities helped consolidate his ideas about the distribution of species in this mountainous region. Franchet went on to comment that until a few years earlier, no one had suspected the existence of such a rich indigenous flora in west-

ern China, and that as he had been studying the flora of the mountains of western Sichuan and Yunnan for the past decade, he was now in a position to draw some considered conclusions. He believed that each parallel mountain chain appeared to have its own particular flora, including species found nowhere else. He thought this was particularly the case in Yunnan, which – based on the collections received to date – appeared to have a greater number of endemic species than either Sichuan or the Himalaya. Franchet noted that although the flora of the western Chinese mountains shared similarities with the plants of the Himalayan massif, these were a product of similar altitudes and climactic conditions, rather then mere physical and geographical continuity. He also concluded that as genera such as *Primula*, *Gentiana* and *Pedicularis*, which appeared to have their origins in this region of Central Asia, were also represented in the mountains of Europe, the flora of the two areas must have been connected at some period in the geological past, even though they were separated by vast distances.

BELOW

Clematis spooneri.

Photo Credit: Roy Lancaster

CHAPTER 14

A Devoted and Indefatigable Collector[1]

—

In hardly any instance has a traveller penetrated in this region to a point

where he has not found a member of these Roman Catholic missions to have

been before him.

CAPTAIN WILLIAM GILL, 1880[1]

—

BISHOP BIET, WHO WAS NOW SERIOUSLY ILL, returned to France in 1891. Père Giraudeau became acting Bishop, while Père Soulié went to occupy the mission at Tongolo, about

three days' journey west of Kangding on the main route to Batang. Apart from short sojourns at Chapa, this was the first time since he had arrived in 1886 that Père Soulié had lived on his own for any length of time; but he was luckier than his isolated confrères, as he could return periodically to Kangding in pursuit of his botanical operations. Nevertheless, the adjustment to a predominantly solitary existence, after living with his confrères at Kangding for so long, must have been hard.

LEFT

Rosa soulieana at Cambridge University Botanic Garden.

BELOW

Upland plateau west of Kangding.

Photo Credit: Edward He

The missionary priests might live among their communities and in the midst of the hubbub of communal Chinese life, but they were effectively alone, far from their compatriots: and the intellectual isolation and lack of companionship resulting from their solitary postings were the hardest things they had to bear. Westerners who travelled alone in China found that the worst thing about their journeys was the length of time they had to spend apart from fellow countrymen and they all comment on the strain of this cultural isolation. A. J. Little wrote that 'the great delight of meeting the Fathers... is in the opportunity of once more being able to hold converse with well-informed men of education',[2] while William Gill comments on 'the delights of hearing a familiar tongue'.[3] When Henri d'Orléans arrived at Kangding, he said that one of the best things was being able to hold long conversations in French, after so many months of hearing only harsh unfamiliar tongues, and he remarked that he and Bonvalot had 'revived' as soon as they breathed 'the air of France'.[4]

William Rockhill visited Tibet again in 1892 and, when he arrived at Tongolo after being on his own for nine months, he was, he said, 'wild for a talk'. He goes on to describe how he found his:

> old friend... Père Soulié living in a little room in a Tibetan house just outside the village... We sat and chatted for a couple of hours and I drank a bottle of wine, which the good fellow insisted on sharing with me, though he had but the one to use in case of sickness. He looked aged and worn, but was the same cheerful, pleasant companion I had found him in former years.[5]

It appears that a year on his own at Tongolo had already left its mark on Père Soulié.

For Père Soulié, though, as for his like-minded confrères, botanical pursuits provided a mentally-stimulating antidote to an existence that might otherwise have proved depressingly monotonous, and he duly set about exploring his new sur-

roundings. Tongolo was situated above the tree limit at about 3,672 m/-12,050 ft in an area of fine tableland with good pastures, which meant that the flora was predominantly alpine in nature, with *Pedicularis* flourishing in the high meadows – E. H. Wilson noted that '*Pedicularis* is the genus par excellence of these alpine meadows' with flowers of every colour except blue. There were also species of *Primula* including *Primula poissonii*, *Lilium duchartrei* and various *Corydalis*, gentians, anemones, dwarf delphiniums, rhododendrons and viburnums, as well as bushes of *Lonicera*, *Spiraea* and *Berberis* in sheltered corners.[6]

Père Soulié approached his task with such energy that he was able to send off a consignment in October containing 368 specimens, most of which he had collected at Tongolo since July. One of the last plants he collected that year was *Delphinium souliei*, a new species with deep blue long-spurred flowers that he found near Kangding.[7] He explained to Franchet that he had not just focused on 'rarities', as he wanted to present a complete picture of the local flora. Franchet asked him to look particularly for rhododendrons, primulas and gentians, but the rhododendrons were over by the time Père Soulié had moved, settled in, and found time to botanise in 1891. He promised to concentrate on them the following year.[8]

He returned to Kangding for a few days in April 1892, and sent off a few more specimens; and he explored one of the high plateaux above the town on a return visit in June. He asked Franchet for some sort of box to put his specimens in while he was collecting, as otherwise he just had to hold them in his hands. The achievements of missionary-botanists like Père Soulié are even more impressive when one considers that, at least at the beginning of their operations, they lacked even the most basic botanical equipment, and they always lacked reference books, depending on Franchet for copies of the Muséum's publications and the latest journals. Père Soulié often deplores the delays that prevented him receiving Franchet's parcels of books, and several of the packages that Franchet sent him seem to have gone astray. Their non-arrival prevented him from catching up with current botanical ideas, but the real problems he faced were much more prosaic. The chief of these was caused by the damp climate, which meant that specimens always took an age to dry out properly, especially in wet weather, and the papers in which they were pressed had to be changed several times. Père Soulié never had enough paper and it was extremely hard to acquire fresh stocks: as he lamented to Franchet, Tibet was not like France, where anything could be obtained. Another problem was cost, and Père Soulié apologised that the plants were not always as well dried as he could have wished, but paper – even when he had it – was too expensive for him to use

Photo Credit: Harry Jans

ABOVE

The Zheduo Pass at 4,298 m (14,101 ft) which Père Soulié used on his way to and from Kangding.

as liberally as was really necessary.[9] As Père David had found at Baoxing, the damp climate meant that great care had to be taken if plants were not to spoil. Père Soulié told Franchet that at one point when he had had to leave a collection drying at Kangding while he went back to Tongolo, the man he had left in charge, although well meaning enough, had let some of the flowers go mouldy.[10]

On receipt of Père Soulié's first collection in 1891, Franchet had mentioned that some of the specimens were too big and had given Père Soulié more appropriate measurements for herbarium collections. He had also sent him an example of the best type of specimen label so that he could see exactly what information was required.[11] Père Soulié was happy to receive this guidance and, once he really began to concentrate on botanical exploration, he realised that the wealth of plants was so great that he would need help if he was to do justice to the flora that surrounded him both at Kangding and at Tongolo. He therefore trained two of the local inhabitants to botanise on their own: one at Tongolo, to help with his own collecting expeditions, and one at Kangding to botanise independently in his absence; although he commented that what was really needed was not just three men, but seven or eight.

G. N. Potanin

What made Père Soulié particularly keen to press on with his botanical explorations, was the arrival at Kangding in April 1893 of the Russian explorer Grigory Potanin and his wife Alexandra. They made extensive botanical collections while

LEFT

Scree slopes above Kangding.

ABOVE

Grasslands west of Kangding.

at Kangding, but had to leave just three months later when Alexandra fell seriously ill; sadly, she died before they reached the Yangtze. Carl Maximowicz, the Russian botanist who had done such sterling work in describing the collections Potanin had made during his first Chinese journey in 1884–86, had died in 1891, which meant that there was now no one waiting in St Petersburg to determine the vast collection of some ten thousand botanical specimens, which Potanin had amassed on the current journey through Shaanxi and across the Qinling and Sichuan, and so they remained unexamined. The fate of Potanin's collection emphasises the importance of the work of the taxonomist in the herbarium, and the debt the missionary-botanists and his colleagues owed to the indefatigable Adrien Franchet in Paris, as he continued with the work of identifying and classifying the various Chinese collections mounting up at the Muséum. His dedication can be seen in the fact that Henri d'Orléans' specimens were described as soon as they arrived at the Muséum whereas Pratt's at Kew had to wait until Hemsley found time to work on them; thus, as botanical honours go to those whose collections are described first, Pratt often plays second fiddle – taxonomically speaking – to the Prince, even in cases where he was actually the first to collect the plant.[12]

Potanin had been accompanied by an assistant, V. A. Kachkarov, whom he had sent to Batang to make collections, but Kachkarov only spent a few days there, as the lamas blamed him for the lack of rain and turned the people against him. Still, this was more than any of the missionaries had been able to do since their expulsion in 1887: in spite of repeated requests from Bishop Biet, permission had still not been granted for their return and they had not been able to venture any farther west than Tongolo. Potanin was very wealthy, which meant that he was able to fund his expeditions on a lavish scale, and Père

Soulié commented that for a mere quarter of the sum Potanin had spent, the Muséum could acquire every plant from Kangding and the whole of the surrounding region.[13] It was not just paper that Père Soulié found expensive but also hiring horses and paying servants and collectors, who also had to be recompensed for leaving their fields, and arranging for collections to be packed up and despatched on their long journey to Shanghai and onwards to France.[14]

Père Soulié and his assistant at Tongolo worked hard during 1892 and 1893 – especially during the spring and summer when they explored the whole area – and in August, before the

BELOW

Rhododendron flavidum.

onset of the rainy season, he was able to send off a large collection. He went to Kangding at the end of October to supervise the despatch of three more cases, which he wrapped in skins to provide further protection from the damp.[15] As he had promised, these consignments contained almost 2,000 plants including several *Rhododendron* specimens, although Père Soulié said that they were not his favourites and admitted that he would not have collected them if Franchet had not specifically asked for them. He also commented that they were hard to dry as the leaves had a tendency to roll up.[16] It was fortunate that he managed to overcome these difficulties, as he discovered some fine new species, both at Tongolo and at Kangding.

Rhododendrons

One of the new rhododendrons found at Tongolo was low growing *Rhododendron flavidum*, with small flowers that are always various shades of yellow in the wild, although a white

form which may be a hybrid has emerged in cultivation. Another discovery was *R. intricatum* with violet-blue flowers, a dwarf species that is very common throughout these high moorlands and occupies the same alpine areas as Père Delavay's dwarf *R. fastigiatum* does around Dali and Lijiang, and as heather does in the uplands of Europe.[17] Père Soulié also discovered *R. vernicosum*, which is much bigger than any of his other Tongolo rhododendrons as it grows into a large shrub or small tree. It is closely related to several species discovered by Père David at Baoxing, notably *R. oreodoxa* and *R. decorum*. *R. vernicosum* has the same glossy foliage – the leaves have a waxy coating that is shiny when rubbed, hence the epithet *vernicosum* or 'varnished' – and similar clusters of bell-shaped flowers. It is a

BELOW

Rhododendron vernicosum festooned with the epiphyte *Usnea*, flowering in spring at 3,500 m (11,483 ft), Zhongdian, Yunnan.

ABOVE

Rhododendron souliei flowering in midsummer at 3,640 m (11,942 ft) on the Zheduo Pass.

variable and easily-grown species with flowers that are usually pink. It was introduced by Wilson in 1904.[18] Franchet named one of Père Soulié's Kangding rhododendrons *Rhododendron souliei* in his honour, and it is a beautiful species, with glaucous young foliage and small clusters of open saucer-shaped white or pink flowers in early summer. *R. souliei* is much prized by those who garden in the cool well-drained sites where it thrives.[19]

Roses

Another fine plant to bear Père Soulié's name is *Rosa soulieana*, a vigorous shrub rose that he found near Kangding, bearing large clusters of wide flat white flowers, which open in midsummer from ivory-yellow buds and are followed by small orange hips. The foliage has a distinctly glaucous tint, which admirably sets off both flowers and fruits. Each year it produces long flexible stems covered with large prickles, and the vigour of these young growths betrays *R. soulieana*'s close relationship with

some of the great climbing rose species. It does not climb itself, but makes a big sprawling shrub that is perhaps better suited to wilder plantings. Maurice de Vilmorin, in his quest for seeds, contacted Père Soulié who began sending him regular packets, among which were seeds of *R. soulieana*. These germinated in 1896 and when the resulting plants flowered in 1899, Vilmorin gave one to Kew. Wilson first collected *R. soulieana* in the Yalong river valley in 1904, where it was very common, and he frequently found it in the warm dry valleys of the border region up to about 3,000 m (9,842 ft).[20] The scented rambler 'Wickwar', which appeared around 1960 in Keith Steadman's garden in Wickwar village in Gloucestershire, seems to be a hybrid between *R. soulieana* and Himalayan *R. brunonii*, both of which grew there. 'Wickwar' is an example of the happy conjunctions that can still occur in our parks and gardens between the missionaries' discoveries and plants from other regions.

Père Soulié also collected a handful of new primulas, including *Primula polyneura*, an attractive woodland species with magenta flowers on tall stems and hairy crinkled leaves, which, as one would expect, requires shade and rich leafy soil. He found *P. souliei* near Kangding, but its habitat is quite different

as it grows in rocky crevices at around 4,000 m (13,120 ft), and despite their best efforts, growers have never been able to keep it happy for long in cultivation.[21] Other alpines have proved more accommodating, including *Leontopodium souliei*, a relative of the European edelweiss, and *Aster souliei*, a low-growing clump-forming aster that produces long-lasting daisy flowers with narrow violet-blue petals in early summer. *A. souliei* is a variable species that has since proved a good ornamental plant for sunny well-drained sites. Several other species of *Aster* were collected in the region, including *A. tongolensis*, a long-flowering mat-forming species and another one that Franchet named *A. bietii* to commemorate Bishop Biet, 'an indefatigable promoter of natural history research in western China'.[22]

Buddleja davidii

As well as specimens, Père Soulié also collected seeds for the Muséum and he found seeds of Père David's *Buddleja davidii* near Kangding in the spring of 1893, which he promptly despatched to Franchet (see chapter 4, p.50). Vilmorin also received seeds, some of which he distributed to other nurserymen, and when their plants and those raised at the Jardin des

Plantes flowered, he noted that the flower colour varied from purple and violet through to a reddish lilac. The foliage is also variable and when Hemsley came to describe the species in 1889 from specimens collected by Augustine Henry and Pastor Faber, he did not realise that they were the same species as the one discovered by Père David and named them *B. variabilis*. Later in 1893, Vilmorin also received seeds from Père Farges, and was able to present young plants to the Paris School of Forestry by 1896. *B. davidii* had first been grown at the Botanic Garden in St Petersburg but that introduction was an inferior form, and it was only when seeds were received from Pères Soulié and Farges that *B. davidii's* potential as a garden plant could properly be appreciated. In 1898, only five years after its introduction to France, Vilmorin was in a position to provide a detailed description of the new *Buddleja*, together with cultivation notes and advice on how best to prune it to maximise its flowering. E. H. Wilson collected seeds on several occasions in Hubei and Sichuan between 1903 and 1908, and his introductions ensured that *B. davidii* became firmly established in British and American parks and gardens.

Left to its own devices, *B. davidii* can look very scruffy but well grown specimens are attractive, and many fine selections have been made over the years, so that cultivars with white, pink, purple-red or purple-black flowers are now available. Nurserymen are still improving *B. davidii* and are breeding neater smaller varieties that are more suited to the restricted size of modern gardens. *B. davidii* cultivars are popular with gardeners, especially in hotter drier areas, as they are reliable late-flowering shrubs that will flower for weeks if regularly deadheaded. The popular name for *B. davidii* is the butterfly

Photo Credit: Chris Reynolds

TOP

Reddish purple form of *Buddleja davidii* in high summer.

ABOVE

Buddleja davidii growing wild in the Daba Shan, north-east Sichuan.

bush, as the long flower panicles produce copious quantities of nectar that readily attract butterflies: another attribute that makes it and its cultivars garden favourites.[23]

The species has since become one of the most familiar of all Père David's discoveries, as the tiny winged seeds soon blew far beyond the boundaries of the nurseries and gardens where the plants first grew, and buddlejas began to appear wherever they landed. As *B. davidii* is vigorous and positively relishes poor stony ground, it soon took root along roads and railway embankments and spread to every other conceivable site, including cracks in neglected roofs and fissures in cliff faces. Its long racemes of lilac flowers are now a common sight throughout the temperate world, including Père David's hometown of Espelette;

although there are places where *B. davidii*'s very vigour makes it an unwelcome intruder and in some areas of the United States, for example, it is classified as an invasive alien.[24]

Money difficulties

Père Soulié's enthusiasm for his task had been fuelled by the sight of *Plantae Davidiana*, Franchet's two-volume work describing Père David's collections, which Potanin had lent him. Père Soulié only had the books for a day – and one can imagine his frustration at having to give back such a treasury of information after such a short time – but they proved a revelation: and when he mentioned the work to Franchet he wrote that he hoped Franchet would have enough material in a few years for a similar work on the flora of Tibet.[25] He again requested copies of Franchet's articles on Tibetan plants, which he knew Père Bodinier in Hong Kong had already received, although nothing had arrived for him. It was not until April 1894 that he received the publications, which Père Bodinier forwarded to him.

Père Soulié now intended to collect around Moxi to the south as he thought that he had more or less exhausted the floras of Kangding and Tongolo; but he found it hard to carry out this plan as he was very short of funds. His letters to Franchet are now almost wholly concerned with his financial difficulties.[26] It appears that whatever money had been entrusted to the bursar of the Missions Étrangères to recompense Père Soulié for his collections had gone astray, leaving Père Soulié penniless. Père Soulié interpreted this as evidence that the Muséum did not really care about him or his work, and he felt considerably aggrieved by this supposed neglect. He told Franchet that he had used up all his own money and incurred debts, and he did not feel able to continue with his collecting until he had received some of the monies he felt were owing to him. He remarked bitterly that as no funds had been forthcoming, he could only conclude that the plants of the area were of no interest to the Muséum and he regretted that '*la France*' would suffer if he stopped sending his collections to Paris.[27]

Franchet understood that Père Soulié was really very upset and wrote back immediately, praising his efforts and assuring him that he was much valued by the botanists at the Muséum: but Père Soulié was not to be mollified by mere 'verbal' encouragements and insisted that he must have money in hand before he could continue his botanical operations. Franchet also confirmed that he had sent Père Soulié funds via the bursar, as had Maurice de Vilmorin, but the money had not yet arrived in Kangding and Père Soulié did not feel that he could resume his collecting operations trusting to promises alone. Franchet wrote again listing all the sums paid, but still nothing arrived. Père Soulié really needed money in hand and he felt very badly used. His frustration and disappointment were made worse by the fact that Bishop Biet, to whom he had applied, did not feel able to let him use any of the money the Muséum had already sent to Kangding, as these funds were specifically for zoological collections.[28] Bishop Biet, though, could see that something had to be done to remedy the situation, and he instructed Franchet to sell any botanical duplicates not needed by the Muséum and to use any money from the sales to fund Père Soulié. Bishop Biet was then able to instruct his confrères in Kangding to advance Père Soulié money to cover his operations for the year, although as Père Soulié noted bitterly, the amount would barely cover the expenses he had incurred in 1893. His immediate money worries might now have been relieved, but he still resented the way the Muséum had taken advantage of him – as he saw it – and he emphasised that, in spite of all the lavish promises that had been made, he had still effectively funded three years' collecting himself.[29]

Changes along the Lancang Jiang

In September, in the midst of these continuing financial anxieties, messengers arrived informing the missionaries at Kangding that Père Benigne Courroux had died at Caka'lho in August. Père Courroux and Père Pierre Bourdonnec had been at Caka'lho when the mission had been completely destroyed in 1887, and they had been driven out and forced to seek refuge farther south. Père Courroux had made an unauthorised return to Caka'lho in 1890 and he and Père Bourdonnec had managed to hang on there ever since, in spite of fierce opposition from the local lamas. As soon as the news of his death reached Cigu, some ten days journey to the south, Père Genestier who had been at Cigu with Père Dubernard, went to join Père Bourdonnec at Caka'lho, which was considered too remote and difficult a mission for one man to cope with on his own. As soon as Bishop Giraudeau learned of these developments, he instructed Père Soulié to go to Cigu to assist Père Dubernard.

Père Soulié's immediate problem was how to get to Cigu, as the lamas continued to resist the missionaries' return to Batang after their expulsion in 1887 and refused to let them use the main road west to the Yangtze and the Lancang Jiang. The Imperial authorities, under pressure from the French minister at Beijing, had issued the priests with all the right passports and permissions but, as the Chinese were not prepared to back up these documents with force, the lamas were effectively able to prevent the missionaries travelling any farther along the road than Tongolo. Père Soulié would have to find another route west.[30]

Père Soulié's journey

Père Soulié knew that he would have to adopt some sort of disguise or the lamas would quickly learn of his whereabouts, whatever route he took. He therefore shaved off his beard and dressed himself in local clothes: wearing a felt hat half covered by a red turban; black and red leather boots secured at the knee; and the long loose Tibetan coat, worn hitched up over a belt to form a capacious pouch, which he found extremely useful for carrying his compass, notebook and barometer. The Tibetan chieftain at Kangding had given Père Soulié a pass for the interior south of the main highway, but he preferred not to use it, as his disguise was good enough for him to get by as a Mongolian trader.

Père Soulié's journey eventually involved travelling over 322 km (200 miles) across the high rolling plateaux of the ancient Tibetan kingdom of Kham, and crossing the Yalong, Litang and Yangtze rivers. On Major Davies' 1908 map of the region, the route is shown as a single red line crossing a vast blank expanse, marking all that was then known in the West about several hundred miles of otherwise unsurveyed territory. The two men who had brought the news of Père Courroux's death to Kangding served as Père Soulié's guides and the little group set out from Tongolo on 11 October, heading south-west towards Deqin. They crossed the Yalong River three days later and found themselves in a high region of wooded slopes and grassy plateaux, where they came across hamlets and shepherds' huts where they could buy tsampa, and occasionally they shot pheasant for the pot.

There was some snow at night but, as Père Soulié later commented, although it sounds very poetic to sleep under the stars on virgin snow, it is hardly practical and travelling became more difficult as the weather deteriorated. Night fell on 24 October before the party could find shelter and in the wind and thick driving snow they could not strike a light to make a fire and so could not warm up or have any hot food. They tethered the horses to the packs and hunkered down as best as they could until dawn. Then they moved off, before the sunlight on the snow could render them snow blind. They struggled on, exhausted and breathless until they found a shepherd's hut where they were able to make a fire. They rested awhile and then continued, only to find themselves in a forest where fallen tree trunks and branches blocked the route. When some local inhabitants found them persevering through the forest, they were much incensed, as the track was forbidden and had been blocked on purpose. It appeared that Père Soulié had, quite inadvertently, walked into the middle of a war, as the whole district had taken sides in a dispute between the local chieftain and the nearby lamasery over the amount of tribute due to the lamas. Père Soulié was able to settle his transgression over the forbidden route by paying a small fine, but the lamas would not give him permission to continue his journey south-west across their territory. He was forced to accept that he would have to make a long detour around the war zone, and he and his companions retraced their steps, turning north towards Litang and the main road, knowing that they would have to find an alternative route west to the Yangtze before they reached the highway, which was still firmly under the sway of the Batang lamas.

The immediate problem caused by the unexpected detour was that Père Soulié's companions now found themselves in unfamiliar territory and had to keep asking the way, which brought them into contact with more of the inhabitants; but there was little fear of Père Soulié himself being recognised as, after almost a month on the road, he blended in so well that he was taken for a Buddhist pilgrim. At night, they could hear

dogs barking in the distance and, much closer, wolves howling. On 4 November, when they were about three days south of Litang, they found someone who was prepared to accompany them as a guide, which meant that they could turn west again and, when the guide left them four days later, they were southwest of Yaregong and well on their way to the Yangtze.

Père Soulié was now faced with a new problem and one that had been worrying him ever since leaving Tongolo: how was he going to cross the river without the lamas at Batang finding out? In effect, it proved quite easy, as one of his companions was friendly with a rich family that was able to arrange the crossing. On 12 November, some of the family fired shots to attract the attention of raft-owners on the opposite bank and

BELOW

Père Soulié's journey is shown as a red line across unsurveyed territory on Major Davies' 1908 map.

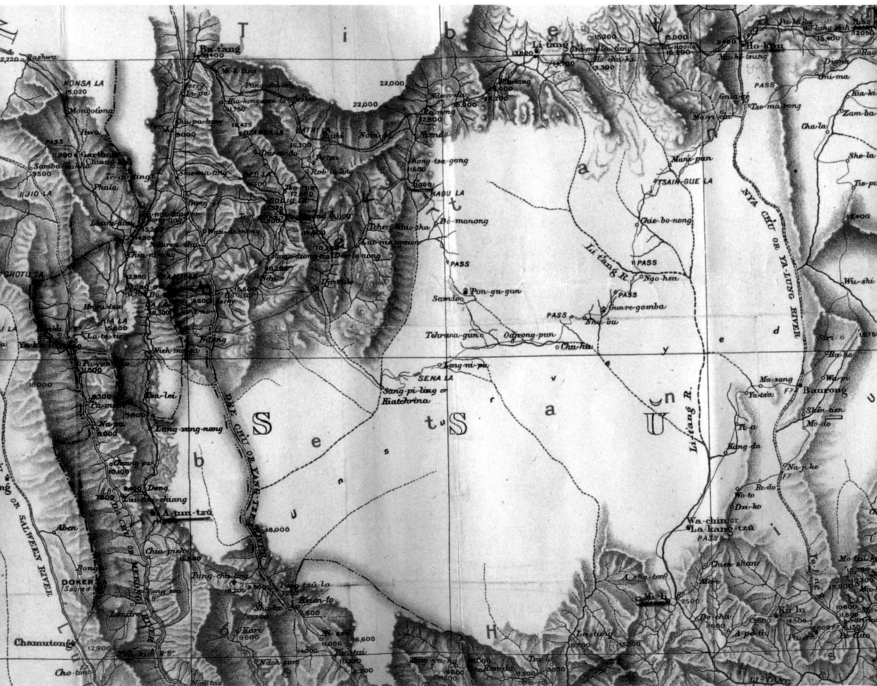

when the men arrived, tricked them into believing that Père Soulié had been issued with the necessary passes at Batang. The raft consisted of about ten pieces of wood 'more or less' roped together and it took three trips to get everything across. Naturally enough, the animals were extremely nervous but were 'encouraged' with cries. Once across, the banks were bare and steep with only a few stunted bushes, so they had to press on, as there was very little grazing and they had to find food for the animals.

Arrival at Caka'lho

By 15 November, they had reached an area where Europeans were known and Père Soulié felt that he had to smarten himself up, so that he did not lower the inhabitants' opinion of the foreigners. On 17 November, they were met by Pères Bourdonnec and Genestier who had ridden out from Caka'lho to meet them. Reunions between the priests were always joyous occasions, especially in these dangerous borderlands, and the three men embraced 'effusively, almost tearfully'. Père Soulié's comment is deeply felt: 'The missionary, after leaving his country, after the heart breaking farewells to his family and to all those he loves, who, inconsolable himself, tries to console them by saying 'we will meet in heaven', has always at the bottom of his heart a hurt that neither time nor suffering can eliminate'. Père Soulié had not seen Père Genestier since they had parted in Batang eight years earlier, and he was delighted to meet his companion again. He had never met Père Bourdonnec, a 'headstrong Bréton', but he knew of the difficulties and dangers that he and Père Courroux had overcome to continue living at Caka'lho, and he was glad to meet at last a priest of whom he had heard so much. His confrères escorted Père Soulié to Caka'lho, which they reached at midday, where he was warmly welcomed by the inhabitants who had assembled to greet him. He now learned that, a few days after Père Courroux had been buried, his coffin had been dug up and opened by adherents of the lamas, and the sacred objects buried with the body stolen. Père Bourdonnec had reburied the remains of his old friend beneath the chapel, in the hope of keeping the grave safe.

Père Soulié spent a week or so at Caka'lho, sharing the small house that Pères Courroux and Bourdonnec had built at their own expense amidst the ruins of their old residence, which had been destroyed when the mission was pillaged in 1887. While Père Soulié rested and recuperated after his hard jour-

Photo Credit: Hedley, J. (1910) *Tramps in Dark Mongolia*

ABOVE

A train of pack mules.

ney, he exchanged information with his confrères, giving them news of Kangding and Bishop Biet, and learning about the current situation at Batang and along the Lancang Jiang. One can imagine the pleasure the three priests found in each other's company: new faces and new conversation were rarities to be enjoyed to the full. William Gill admired the way in which the French priests, in the midst of the noise and poverty in which they lived, never forgot their culture or their manners, and he said that when he met them he could imagine himself back on the Champs Élysées. Perhaps, for the priests, such gatherings reminded them of their days as enthusiastic novices in the seminary in Paris, when the hardships of their lives in China still lay ahead of them.

Towards the end of the month, Père Soulié set out on the 115 km (72 miles) trek to Cigu, following the tortuous pack road that wound southwards along the east bank of the Lancang Jiang; although he commented that, after the adventures of his journey from Tongolo and the warmth of his welcome at Caka'lho, this last leg of his journey was a 'bagatelle'. He arrived at Cigu on 7 December, in good time to celebrate Christmas with Père Dubernard.

Photo Credit: Harry Jans

CHAPTER 15

The Flower Garden of the World

—

…several species of meconopsis all of them surpassingly lovely, miles of

rhododendrons, and acres of primulas of which I counted over a dozen species

in flower, many of which I had never seen before… Those mountains have, rightly

in my opinion, been called the flower garden of the world.

GEORGE FORREST, 1905[1]

—

PÈRE ALEXANDRE BIET, BISHOP BIET'S BROTHER, had founded the little village at Cigu after the original mission at Bonga had been sacked in 1865. He had approached the headman of Cizhong, a village on the west bank of the Lancang Jiang to ask if there was anywhere in the vicinity where he and some members of the Bonga community could settle and the headman had offered to sell him an uninhabited mountain bordering the Lancang Jiang a little way south of Cizhong, at the foot of which was a chasm called the Devil's Hole. In spite of this unappealing name, Père Biet realised immediately that the riverside site was ideal for a settlement, as the slopes were not too steep for agriculture and it was bisected by a rapid mountain stream, which would provide fresh water. He bought it on the spot, although he had to sell some of his vestments and borrow from his brother and other confrères to raise the money. He and his followers began to build houses and clear the land, and gradually other converts joined them, and a small farming village grew up, which became known as Cigu.

Père Jules Étienne Dubernard (1840–1905) arrived at Cigu in 1867 and furthered Père Biet's work, engaging Bai workmen from a Yangtze village to help build a church in the centre of the village, which, when it was completed three years later, stood 19.8 m (65 ft) high, and had triple gables, lattice woodwork and was ornamented with Chinese designs. It was easily the most impressive building in the area.[2] Père Dubernard also founded a school and established a pharmacy and medical practice, where his skills were much needed as the climate was unhealthy, with many cases of malaria and fever. Slowly, he developed a network of friends and allies among the surrounding Tibetan and Lisu villages, as well as with the other tribes and nomads of the region. Whenever possible, Père Dubernard brought back some of the unfortunates enslaved in the course of local disputes, using salt rather than currency to buy their freedom. His careful tenure ensured that Cigu became the most successful of the Tibetan missions, with 300 converts living scattered among the nearby hamlets. This did not protect him or many of his community in 1887 when some of the area's inhabitants, led by the local lamas, took advantage of the violence against the

LEFT

Lilium souliei with *Rhododendron russatum,* at 3,845 m/12,514 ft, Zhongdia, Yunnan.

RIGHT

Père Dubernard.

Photo Credit: d' Orléans, H. (1897) *From Tonkin to India by the Sources of the Irawadi January '95-January '96*

priests which flowed down the Lancang Jiang after the destruction of the mission at Batang, to sack the village and drive Père Dubernard and his congregation out. They wrecked more than fifty houses and ransacked the church, but left it standing.[3]

Missions had been established farther down the Lancang Jiang at Xiao-Weixi and at Weixi in Yunnan, the local Chinese administrative centre some 102 km (64 miles) to the south, and here Père Dubernard found refuge. He was able to return to Cigu seven months later with the support of the Chinese authorities at Weixi. At the beginning of 1893, when his enemies realised that they were not going to be punished for the attacks of 1887 and guessed that they would be likely to get away with further aggression, they destroyed the rope bridge at Cigu, effectively cutting the little

BELOW
Cigu on the west bank of the Lancang Jiang. The old pack trail along the east bank has been replaced with a tarmac road and overhead cables now bring electricity to the village, but apart from these changes, Cigu today would be very familiar to the missionaries who served there over a century ago.

settlement off from the main pack road on the east bank. These rope bridges are a feature of the region, where narrow rivers and steep cliffs abound, and are simple affairs, consisting of a bamboo rope securely anchored to a post above a small platform on each bank, from which a simple sling is suspended by means of a well-greased slider. The traveller is hooked up into the sling and then pushes off from the platform and his weight and gravity do the rest and the slider rushes down the rope to the opposite bank, where he hauls himself hand over hand the last few feet. Everything can be sent across these bridges, even horses and mules. Simple though this arrangement was, it provided an important connection between the village and the rest of the region, and Père Dubernard went in person to Weixi to appeal for its reinstatement: but, although the Chinese officials took his side and ordered the bridge to be repaired, nothing was actually done. This meant that the Cigu inhabitants had to use the bridge at Cizhong, or ones farther south, and these longer journeys exposed the Christian villagers to frequent attacks. The local lamas also repeatedly threatened to destroy the church and twice ransacked the village, and generally made things as difficult as possible for Père Dubernard. Yet, in spite of the hostility and threats, he carried on much as he had always done, treating the smallpox victims and other invalids that were brought to him, teaching in the school, and ministering to those of his community who had remained faithful to his teachings.

First explorations at Cigu

It was into this troubled situation that Père Soulié arrived at the end of 1894, and he must have found it inspiring to join Père Dubernard, a veteran of the earliest days of the Tibetan mission. His new home was a two-storey building, with a roof of Chinese tiles and a terrace, which faced the river and stood in the centre of the village, close to the fine church.[4] Père Soulié's duties included teaching in the little school, but he also had time for botany and his first task was to sort out and clean the seeds he had collected for Maurice de Vilmorin during his journey across the unexplored uplands. The rapid pace at which he had travelled had prevented him from collecting any specimens, but for a plantsman like Père Soulié the journey at the end of the growing season provided an excellent and unlooked for opportunity for seed collecting. In spite of the difficulties he encountered en route, he still found time to amass a fair collection of ripe seed-pods and fruits for Vilmorin.[5] Once these had been prepared, Père Soulié had time to venture beyond the narrow cultivated surroundings of Cigu to investigate the exposed dry cliffs bordering the Lancang Jiang. Even though it was still too early in the season for much plant growth, it was quickly apparent that the vegetation of these exposed crumbling slopes was relatively uninteresting. He then found that all

his attempts to clamber up the streams tumbling into the river, in order to reach the more interesting plants that he was sure must grow farther upstream, were blocked by precipices. There was a track, though, that led west from Cigu over the barrier ridge and once away from the hot dry steep-sided gorge of the Lancang Jiang, the vegetation was very different.

Biluo Snow Mountain and the Xi La

The valleys and gullies beyond Cigu and the river extended westwards to the mountains forming the Lancang Jiang–Nu Jiang divide, of which Biluo Snow Mountain, which reaches 4,400 m (14,350 ft), was the closest. Père Soulié called this mountain Sela (Xi La), but Xi La is actually the name of the pass that crosses the peak at 3,048 m (10,000 ft) and provides a route west to the Nu Jiang. It was here that Père Soulié botanised over the next few months: first exploring the tangled woody scrub of the valleys and the mixed forests that covered the lower slopes of Biluo Snow Mountain; and then, once the Xi La pass became accessible as the snows began to melt at the beginning of June, investigating the pass itself and the farther western slopes of the mountain. Eventually he reached the upper levels, where the rich grassy meadows and high alpine screes are covered with a rich medley of plants from July to September.

Biluo Snow Mountain and the other mountains west of Cigu are part of the Hengduan Mountain system, which lies between the Tibetan–Qinghai Plateau and the Sichuan basin and consists of numerous ranges that run north–south between the great

BELOW
The mission buildings at Cigu from the south.

rivers that drain the Tibetan uplands. One of the most important physical features of this region is the range of heights between the deep steep-sided river gorges and the soaring peaks, which, together with a corresponding range in temperatures and rainfall, has led to the creation of a great variety of different habitats and plant niches, resulting in the emergence of one of the richest floras in the world. These high ranges also affect the timing of the monsoon, so that the rains in June on the Nu Jiang–Lancang Jiang divide produce a profusion of flowers in August, while the midsummer rains on the Lancang Jiang–Yangtze divide just 32 km (20 miles) to the east means the flora peaks in autumn – at a time when snow can already be falling at similar altitudes on the ranges farther west.

Extraordinarily diverse flora

It was quickly apparent to Père Soulié that he had stumbled upon a botanical treasure-trove. He could see that the mixed forests of the lower slopes included maples, birches and alders growing alongside dense thickets of rhododendron, where many plants reached the size of trees at 9 m (30 ft). There were also viburnums, deutzias, currants, honeysuckles, roses, cherries, spiraea and various bamboos, with *Clematis* sprawling through the bushes. Ferns, lilies, *Corydalis* and *Oxalis* were common.[6] He found that rhododendrons grew abundantly all the way up to and beyond the tree line, and he informed Franchet at the beginning of May that he had already found ten or so species of *Rhododendron* that he had not seen around

BELOW
Looking north from Cigu towards the Lancang Jiang–Yangtze divide – landslips are a constant danger along the steep banks.

RIGHT
Biluo Snow Mountain from the north; the Lancang Jiang can just be seen in the depths of the gorge.

Kangding or Tongolo. He thought several of them would be new, if Père Delavay had not already collected them.[7] He was right to think that there would be overlaps with the flora Père Delavay had discovered farther south, but although many of his plants had already been collected by Père Delavay, his specimens did include some new varieties. One example was a *Rhododendron* he collected in July that Franchet named *Rhododendron brevistylum*, but which has turned out to be a variety of *R. heliolepis*, first discovered by Père Delavay near Gualapo.[8] As Père Soulié climbed, he found that the nature of the rhododendrons changed, with larger species giving way to shorter types with smaller leaves that were better suited to conditions at these altitudes, where some of the mountains reach over 5,000 m (16,604 ft) and the peaks never lose their covering of snow. Franchet remarked of *R. saluenense*, one of Père Soulié's discoveries, that it had large flowers relative to its diminutive stature – an adaptation which helps attract insects during the short flowering season.[9] Other adaptations among species native to these high slopes include the production of copious quantities of tiny seeds that ripen quickly and are very light and sometimes winged to facilitate wind dispersal.

Botanical puzzles

Père Soulié soon realised that, among all the new species he was collecting, were several that he recognised from his expeditions around Kangding and he described the flora to Franchet as a real '*macedoine*' or medley.[10] One of the plants familiar from

Kangding was one that had been identified as *Isopyrum vaginatum*, originally discovered in Gansu in 1885 by Grigory Potanin and named by Carl Maximowicz. Père Delavay had also collected it on at least two occasions, but it was only when Franchet received Père Delavay's notes after his death that he was able to link some *Cimicifuga*-like fruits Père Delavay had sent him with this particular species. These fruits and Père Delavay's notes showed Franchet that the plant in question could not actually be an *Isopyrum* as Maximowicz had thought. When he examined Père Soulié's specimens more carefully, he could see that the plant was not a *Cimicifuga* either, although it was obviously closely related to it. To accommodate it, Franchet created a new genus, which he named *Souliea* and the hitherto anomalous *Isopyrum vaginatum* became *Souliea vaginata*. So it remained for a century but DNA studies in the 1990s led to a reorganisation of all these species, and both *Cimicifuga* and *Souliea* were then classified as species of *Actaea*, which meant that the correct name for *S. vaginata* became *Actaea vaginata*.[11] Not all botanists accept this new classification, however, and in the *Flora of China*, *Souliea vaginata* is still accepted.[12]

The plant-hunter **Frank Kingdon-Ward** explored this region in 1911 and his findings bear out Père Soulié's description of the flora as a '*macedoine*', as he saw many plants from farther east growing in these high mountains. In the shade of the shrub belt near Deqin, he found 'two delightful Cypripediums' – *C. tibeticum* discovered by Antwerp Pratt at Kangding and *C. flavum* discovered by Père David at Baoxing – 'and higher

LEFT

Rhododendron saluenense in spring.

BELOW

Souliea vaginata.

ABOVE

Primula souliei.

up at 13,000 feet [3,962 m] the white-flowered *Souliea vaginata* was just blooming, its flowers opening almost before the leaves appeared, as is the case in so many of these shade plants.' In the dampest spots, he found Père Delavay's *Primula sonchifolia* and, whcre it was drier, yellow-flowered bushes of *Paeonia delavayi*.[13] Crossing the Xi La, Kingdon-Ward saw diminutive blue *Primula bella*, discovered by Père Delavay on the summit of Cang Shan in 1884, with 'crimson azaleas, purple columbines and more, forming sheets of colour. Every rock and boulder supported a small garden of saxifrages and tufted alpines, every marsh displayed masses of some rare flower such as *Primula souliei*, with sedges, gentians and sphagnum.' *P. souliei* is perfectly adapted to the climatic conditions in which it has evolved: its small lilac-pink flowers appear in the rainy season in June and hang down to protect the nectar and pollen, while its seed pods are held erect so that the autumn gales can disperse the seeds.[14]

New *Nomocharis*

BELOW

Nomocharis saluenensis in midsummer.

It was while exploring the Xi La pass and the high meadows on Biluo Snow Mountain in late June and July that Père Soulié collected some more specimens that Franchet found perplexing.

The first was a red-flowered *Nomocharis*, which Franchet at first took for a Tibetan form of *N. aperta*, one of Père Delavay's finds, although he later recognised that it was a new species, which he called *N. saluenensis*. The colour of the large saucer-shaped flowers varies from yellow through pinks and reds to deep purple. Père Soulié also found another *Nomocharis* on Xi La, *N. meleagrina*, with heavily-spotted flowers. Both these species are confined to the high mountains in the far west of China, where they thrive, protected by a thick covering of snow through the long winters and springing up during the cool wet growing seasons. These species of *Nomocharis* belong to what is essentially a north temperate flora, even though the latitude is the same as that of Cairo in Egypt. Plants native to these altitudes have been hard to establish in cultivation. One of the reasons is that on peaks such as Biluo Snow Mountain, where thick snow can still be found above 3,900 m (12,795 ft) in the middle of June, the stony alpine slopes run with water from melting snow on hot days. This produces a combination of constant moisture and very sharp drainage that has proved hard to replicate, and high mountain species such as these *Nomocharis* are not easy in cultivation, although they can flourish in the same cool conditions that suit some of the woodland lilies.[15]

Père Delavay's specimen of *Nomocharis aperta* had caused Franchet problems when he first examined it and he initially identified it as a species of *Lilium*, although this was later found to be an error. It is, however, perfectly understandable as lilies are closely related to *Nomocharis*, and they are also easily

confused with fritillaries: in fact, the genus *Nomocharis* has been described as 'a sort of haven of refuge for plants… which will not pass for fritillaries but have too many differences to be included in lilies'.[16] It is easy to see that discriminating between such closely-related genera is hard, particularly as individual plants can vary so much, depending on the age and size of the bulb and the specific growing conditions. This means that separate plants of different ages, even those belonging to the same colony, can exhibit marked variation in height, leaf size and shape, and flower size – very puzzling for the botanist trying to identify dried specimens in the herbarium, particularly when dealing with completely new species.[17] The confusion occurred again when one of Père Soulié's herbaceous discoveries was first described as a *Nomocharis* and then as a *Fritillaria*, before eventually being assigned to *Lilium* in 1950. It is currently known as *Lilium souliei* and has red or purple flowers that are so dark they appear black. Frank Kingdon-Ward saw waist-high colonies of this lily growing farther north at Dokhe La.[18]

New genus

While exploring near Xi La on 15 July, Père Soulié collected a specimen of what turned out to be another species of the anomalous primula-like plants first found by Père Delavay. Franchet had experienced great difficulty classifying Père Soulié's specimen and created a new subgenus of *Primula* named *Omphalogramma* to accommodate it (see chapter 7, p.89). Once he also had Père Soulié's specimen to study, he realised that the strange plants belonged neither to *Primula* nor to *Androsace* – another possibility – and in 1898 he established *Omphalogramma* as a new genus in its own right, so that Père Delavay's finds became *O. delavayi* and *O. vinciflorum* (originally spelled *vincaeflora*) and he named Père Soulié's discovery *O. souliei*. It is now thought that the genus *Omphalogramma* occupies an evolutionary point midway between *Primula* and *Androsace*. Franchet noted in his description of *O. souliei* that it had remarkably large flowers. Heinrich Handel-Mazzetti saw it as he crossed the Xi La pass in 1916:

> Mirrored in the bog pools was the strangest of all primulas, now split off into a genus of its own under the name *Omphalogramma souliei*, with solitary hairy pendent flowers 6 cm [2.3 inches] long in various shades from violet to purple, most of them with six petals, the lowermost projecting obliquely.[19]

One can only admire Franchet's insight and perseverance as he tried to make sense of the relationships between unfamiliar plants, having only dried specimens to work on, without having ever seen living material or even photographs. Professional plant-hunters like Wilson, Forrest and Kingdon-Ward carried cameras and were able to take photographs of plants

BELOW

Omphalogramma souliei in June at 3,342 m (11,000 ft) photographed in 1927 by Frank Kingdon-Ward.

Photo Credit: *Gardener's Chronicle* (1927) Series 3 81 fig 126

Photo Credit: d' Orléans, H. (1897) *From Tonkin to India by the Sources of the Irawadi January '95-January '96*

diately as he was still very upset by the lack of funding from the Muséum for his botanical work.[20] He pointed out to Franchet that plenty of other people wanted his specimens, and he said that Rockhill had asked him for plants for American botanists.[21] It is perhaps not surprising then that Père Soulié did not focus solely on botany but also turned his attention to zoology, which he knew would be funded by the Muséum's grant to the Bishop at Kangding. He organised the local hunters to scour the area, much as Père Dubernard had done some years earlier, and he was able to send back a large number of specimens, including mammals, reptiles and several birds, including two new species, which the ornithologist at the Muséum named after him.[22]

In the middle of August, he learned that his old Kangding acquaintance **Prince Henri d'Orléans** was close by and he went to welcome him.[23] The Prince's current journey of exploration was motivated by his intensely patriotic desire to see France become an effective rival to Britain in the Far East. He thought that one way of achieving this would be for France to expand her commercial interests into the areas north of her colonial possessions in Tongking and Indo-China. To facilitate such an expansion, he was investigating the previously unknown course of the Mekong/Lancang Jiang through Yunnan as far as the Tibetan borders. When he reached Cigu, he planned to swing away west to see if there was a practicable route through to India that could be developed for trade. To this end, he and two companions set out from Tongking in January 1895, joining the Mekong (Lancang Jiang) where the Lagree-Garnier party had left it in 1868. They then followed the river north through Yunnan to Dali, which they reached at the end of May.

Henri d'Orléans and his party left Dali on 14 June and made their way due west to the Nu Jiang, before turning back to the Lancang Jiang. Along the way thieves took their theodolite and various other possessions, but they continued to make scientific observations as accurately as they could. The explorers followed the Lancang Jiang north until, on 19 August, they came in sight of the white mission buildings at Cigu and were directed to the rope bridge at Cizhong, as the one at Cigu had still not been restored. It was here that Père Soulié met them. Père Dubernard was waiting on the west bank – the Prince and his companions were the only Europeans, apart from Thomas Cooper in 1868 and a handful of fellow priests, that Père Dubernard had seen in his twenty-eight years at Cigu.

in situ, which at least provided botanists in the herbarium with information about size and habitat. They also returned home between trips and worked on their own specimens, and could readily answer questions from any other botanists who were identifying their collections. Franchet had none of these advantages: all he had were dried specimens and brief field notes, and sorting out these closely-related species, based on nothing more than dried materials, showed both his skill in interpreting the physical details of the specimens in front of him, and the depth of his knowledge of the relationships linking the genera represented in the Chinese flora.

Père Soulié eventually sent some 700 specimens from Cigu to the Muséum but he did not despatch his collections imme-

Prince Henri was considerably impressed by what he saw of the mission at Cigu:

> Père Dubernard had collected the debris of several mission stations, and has become the rallying-point for those believers whom persecution has driven to the refuge of this agricultural community which he has founded. As I marked his administration of his subjects, his help for the unfortunate, his care of the sick, and saw him supervising the harvest, laying by food for the improvident, and giving instruction to the young; he seemed to me to resemble some beneficent overlord of the Middle Ages… Père Dubernard was venerated throughout the countryside, and looked up to, at once for his wisdom and his care, as the benefactor of the land. His reputation for healing power was widespread…

The Prince admired the missionaries, 'posted like sentinels along the line of the Mekong [Lancang Jiang], await[ing] with unwearying patience, constancy, and alertness, the day when they shall be admitted into Tibet', and he marvelled that in spite of the difficulties and dangers of their lives, their zeal remained as ardent and unquenched as ever.[24]

One of his companions made a brief excursion to Deqin but Henri d'Orléans, who was suffering from fever, stayed with the priests to rest. The Prince's visit must have been a very pleasant interlude for the missionaries, but he and his companions were ready to leave again on 10 September.

During his stay, Père Soulié told him about his journey from Tongolo to Caka'lho across the unexplored plateau and the Prince encouraged him to write an account of his journey for the French Société de Géographie. Over the next two months, Père Soulié used the notes he had made each night during his journey to compile a detailed article describing the country he had passed through and the people he had met. He also drew a map of his route, based on the regular compass bearings he had taken throughout his travels. He sent it off to the Société in December 1895, and it was published in 1897 – it appears on Major Davies' 1908 map as a fine red line crossing a great tract of unsurveyed territory (see chapter 14 p.200).[25] By the time the article appeared, Père Soulié was no longer at Cigu but had been moved north to revive the defunct mission at Yaregong.

Henri d'Orléans' botanical discoveries

Henri d'Orléans travelled fast but he did not neglect his natural history collections and before leaving Cigu he sent some of his mules back to Dali, bearing a large consignment of specimens for the Muséum. When the chests reached Kunming, Père Delavay added 'a little packet' of his own to each horse – specimens from the final collection that he made for Franchet in the summer of 1895.[26] The Prince's botanical specimens included a fine herbaceous plant that Franchet at first thought was an astilbe before deciding it was actually another rodgersia. He named it *Rodgersia henrici* after the Prince, who had found it on 11 July on the banks of the Nu Jiang in a forest of bamboos, larches, magnolias and wild cherries.[27] The new *Rodgersia* has proved to be very closely related to *R. aesculifolia*, which Père David found at Baoxing, but although the foliage is superficially similar, Prince Henri's discovery has narrower more elongated leaflets and the tall pink inflorescences are more robust. These differences are not sufficient, though, for botanists to consider it a separate species and it is now regarded merely as a distinct variety of *R. aesculifolia*, for which the correct name is now *R. aesculifolia* var. *henrici*. *R. aesculifolia* is the most widely distributed of all species of *Rodgersia* and William Purdom collected another variety farther north in Gansu; var. *henrici* would seem to be the common southerly form. Prince Henri's *Rodgersia* has since been recognised as an exceptional ornamental plant for damp shady sites and is now widely cultivated.[28]

On the same day as he collected the *Rodgersia*, Prince Henri had found a new lily, *Lilium henrici*, with flat outfacing white flowers marked with purple blotches at the base, which is only found in the vicinity of the Nu Jiang above 2,800 m (9,200 ft). It was first raised at the Edinburgh Botanic Garden from Forrest's seeds and has been successfully grown elsewhere in Scotland, but it is rarely available.[29]

Geographical results

Henri d'Orléans and his companions reached Sadiya on the Brahmaputra River in December, having travelled over 3,380 km (2,100 miles) in eleven months, of which 2,575 km (1,600 miles) were in unexplored territory. During his pioneering journey, he had investigated the Lancang Jiang as far as the Tibetan borders, verified the routes of various earlier explorers, 'solved many of the mysteries of the complex system of parallel valleys of western Yunnan', and established the sources of both the Nu Jiang and Irrawaddy River.[30] It was a considerable achievement.

CHAPTER 16

Rebellion

—

Be one a believer or not, one has to admit that the missionaries are utterly wholehearted in their endeavours

to persuade the people to throw off the yoke of the lamas and turn to Christianity. …and though under the

protection of the State they all too often forfeit their lives. Yet there are always successors ready to step

joyfully into their shoes.

HEINRICH HANDEL-MAZZETTI[1]

—

YAREGONG LAY 80 KM (50 MILES) SOUTH-EAST of Batang and the mission there had been empty since Bishop Giraudeau, its founder, had been forced out in 1886. The decision a decade later to try to re-establish the mission was the result of developments in the long-running saga concerning the reinstatement of the missionaries at Batang after their expulsion in 1887. No real progress had been made until an energetic new French minister, Auguste Gérard, arrived at Beijing in April 1894. He pressed the missionaries' case and, as he was constantly assured by the Imperial authorities that the priests had permission to return, he encouraged Bishop Giraudeau to put this to the test by trying to reoccupy the mission at Batang. Bishop Giraudeau had already made several attempts to do just this, but local officials always managed to thwart his efforts. Nevertheless, M. Gérard thought they should make one more determined effort and so Bishop Giraudeau instructed Père Soulié to leave Cigu and try to re-occupy the mission at Yaregong. He thought that this would prove a sufficient test of Chinese good faith, while being unlikely to provoke the sort of reprisals from the lamas that would endanger priests sent to Batang or those already at Caka'lho.

LEFT

Cotoneaster franchetii in autumn.

Journey to Yaregong

Père Soulié left Cigu on 21 July 1896, first going north to Caka'lho where he learned that the lamas at Batang were once again planning to attack the mission. He thought that his arrival at Yaregong might deflect them and give them another target, and he hastened his departure. He was joined by five local Christians, including one man who had fled Yaregong in 1886 when the mission there was attacked and who had hidden in the hills for a month with his family, before making his way across the Yangtze to Caka'lho. Shortly before Père Soulié and his party reached Yaregong, they met three horsemen who warned them that the lamas at Batang would not hesitate to run them out once they learned that a European was trying to settle there again. Père Soulié noticed that his companions became markedly less enthusiastic once they heard this; but he insisted they continue and they arrived at Yaregong on 15 August. A local family gave him lodgings, as all the mission buildings had been destroyed.

The following day, Père Soulié asked his companions to go with him to Batang to present his letters to the local chiefs, but they refused point blank and suggested they go instead to Kangding. Père Soulié realised that this desire was prompted as much by the men's fear of the lamas as by the knowledge that the wages he would pay them for the journey would enable them to buy enough grain at Kangding to give them a healthy profit when resold at Caka'lho. He saw that he could not dissuade them and, as he thought that it would be a good idea to give Bishop Giraudeau a report on events to date, he despatched two of the men to Kangding, while sending the others to buy supplies locally. A noisy group of Tibetans armed with swords, guns and sticks arrived at the house on 18 August but Père Soulié's host stood his ground and saw them off. Some of the men then rounded up his mules, which were grazing on the plain beyond the village, and began loading them with his luggage. Père Soulié saw that they intended to argue that as he had nowhere to live, he could not stay and so he quietly moved to a hut attached to the old mission buildings. When the Tibetans invited him to leave with his mules, he was able to respond quite truthfully that he was not going anywhere as he was already at home. At this point, two or three of the men tried to seize him by force, but they were checked by the others and an argument broke out between those who wanted to use violence and those who did not. Père Soulié sat quietly in a corner, waiting for the dispute to run its course. In the end, he was dragged outside and one of his assailants grabbed a large stone. Père Soulié faced him: 'Throw your stone, if you are brave enough'. The man dropped the stone. Another raised a stick and Père Soulié took a few steps towards him: 'Do you take me for a dog?' The man was too embarrassed to continue, and the attack petered out. Nevertheless, Père Soulié now found himself outside, without food or shelter, and had no option but to leave. As he did so he talked to some of his attackers, and they explained that they would have let him stay if the lamas had not given orders for his expulsion.[2]

Second attempt

Père Soulié made his way to Kangding, arriving there on 5 September. He had not seen his confrères for nearly two years, but he could have wished for more propitious circumstances for this reunion. It was now clear that no return to Batang was possible without solid Chinese support. The French decided the stalemate had gone on long enough and their consul at Chongqing threatened to take the missionaries to Batang himself. Such an act would have been an intolerable affront to Chinese prestige and the threat forced the viceroy of Sichuan to send one of his own officials to Batang to bring the lamas under control. Thus, on 15 May 1897, after a decade in exile, Bishop Giraudeau and Père Soulié entered Batang once again. A month later, Père Soulié accompanied by Bishop Giraudeau returned to the deserted mission at Yaregong and, once they had confirmed that there would be no opposition to his re-instatement, the Bishop left him to re-establish the shattered mission.

Père Soulié's first task was to build somewhere to live. He had to do this more or less on his own as the Batang lamas and their tenants had spread dreadful stories about the missionaries in their absence – blaming them for everything from plagues of rats to floods and drought – so none of the inhabitants wanted to help him. In 1898, ten Christian families from Caka'lho joined Père Soulié to form the nucleus of a new community and, as his presence in the area became accepted, he was gradually able to establish good relations with his neighbours. The Red Hat lamas at the nearby lamasery, who were great rivals of the Yellow Hat lamas of Batang, were happy to befriend him. He was eventually able to buy more land and worked hard to make the mission self-supporting and to assist his community, turning his hand to a multitude of tasks, including charcoal burning, pottery, and lime-burning. The area was in the grip of a typhoid epidemic, but Père Soulié had learned a great deal from Père Dubernard and was able to treat some of the sick. He began to vaccinate the local farmers, as well as two living Buddhas, and gradually developed such medical skill that people flocked to him for treatment. So great was his reputation that even lepers and the blind came to him to be cured. By 1902, he was held in such esteem that when he wanted to build a larger house and chapel, the neighbouring Red Hat lamasery provided the wood and, in 1904, sent carpenters and a dozen lamas to help erect the frame. Père Soulié's personal popularity did not result in many conversions – perhaps because the inhabitants of the valleys constituting the district of Yaregong were all tenants of various lamaseries, and so could not afford to antagonise their landlords. By November 1904, this lack of evangelical success had begun to depress him and he wrote to an old friend, lamenting that his neighbours seemed indifferent to his message and admitting that he was considerably discouraged by the failure of all his efforts in the face of the lamas' implacable hostility to his religion.[3]

Explorations

As well as labouring to rebuild the mission, Père Soulié explored the countryside around Yaregong, which was situated at about 3,658 m (12,000 ft), taking compass bearings, learning the network of tracks and marking the principal features of the landscape. After four years he felt he knew enough about the area to draw up a map and write an article for the Société de Géographie, although he apologised for the map's shortcomings, explaining that it was based solely on his compass readings and so could not compare with a proper topographical survey. He thought it might nonetheless be of interest, as Batang seemed destined to fall under French influence and become an important link between Tongking, Yunnan and Russian territory in northern China. It is possible to detect Henri d'Orléans' influence in this nationalist and expansionist agenda, and also to see how advances the British were making in Tibet had turned the thoughts of the missionaries in the region to the compara-

ABOVE

Cotoneaster mairei in autumn.

RIGHT

Cotoneaster bullatus in autumn.

ble acquisition of territory by France. Père Soulié thought that if Tibet fell to the British, France should seize the Tibetan borderlands by force. Such imperialist ideas were fully in tune with the spirit of the times and the Société congratulated Père Soulié on his work, publishing his article in April 1904 and awarding him its Silver Medal.[4]

Père Soulié did not neglect his botanical collections on these exploratory expeditions, although the flora was generally similar to that he had investigated around Kangding and Tongolo, and a collection of 2,000 specimens reached the Muséum in 1907, bringing the total that he had sent back to over 7,000. This collection also included specimens from Caka'lho, which he visited from time to time. His collections were typical of the ground-hugging alpine flora of the high windblown grasslands and screes surrounding Yaregong and included primulas, gentians, *Pedicularis* and *Androsace*.[5]

Death of Adrien Franchet

On 12 February 1900, French botany suffered a grievous blow when Adrien Franchet died suddenly. He was irreplaceable. His prodigious work in the herbarium had ensured that the immensely important Chinese collections of Père David, Père Delavay and Père Farges were properly studied. He had determined the specimens collected by Prince Henri d'Orléans and examined the first collections despatched by Père Soulié. He had also found time to study other collections such as those made in Japan by Père Faurie. During his twenty years at the Muséum he described nearly two thousand new species, established twenty-eight new genera and acquired a supreme understanding of the Chinese flora. His detailed knowledge of spe-

cies' distribution throughout eastern Asia led him to argue that the origin of genera such as *Rhododendron*, *Pedicularis*, *Primula*, *Epimedium* and *Gentiana* could be found in the mountains of western China, particularly in those of north-west Yunnan and south-east Sichuan where they were represented by an extraordinary number of species, and that these diminished in number the farther from their heartlands that one travelled. His conclusions are now generally accepted, and it was a tragedy that his sudden death prevented him from publishing an overall synthesis of his ideas.[6]

Franchet's assistants at the Muséum continued his work as well as they could and over the next few years published numerous articles describing new plants from collections made by Pères Bodinier, Cavalerie, Maire and Ducloux, as well as those from Père Soulié. They also continued to examine specimens collected by Pères Delavay and Farges, and were able to make comparisons between plants collected in several different Chinese provinces.

Seed introductions

Seeds collected by the missionaries were another source of new plants and two cotoneasters were described from plants raised at Les Barres from seeds Père Soulié had sent to Maurice de Vilmorin. The first was produced around 1895 and was named *Cotoneaster franchetii* in Franchet's honour in 1902; although when it was compared to other cotoneasters in the Muséum's herbarium, it was discovered that it was actually the same as one first collected by Père Delavay on Heishanmen in August 1889, which had been wrongly identified as *C. pannosus*. *C. franchetii* makes a large spreading shrub, producing white flowers in May followed by clusters of deep red berries and is now common on the west coast of North America where the winters are not too cold. Vilmorin sent a plant to Kew in 1901; and it has recently been reintroduced to cultivation. A closely-related species, named *C. mairei*, was collected in north-east Yunnan by Père Maire in 1912 and is now found in gardens throughout the temperate world.[7] The second of Père Soulié's cotoneasters was raised from seed he sent to Vilmorin in 1898 and was named *C. bullatus*. At first the new species did not appear particularly ornamental but, once the young plants matured, it became apparent that *C. bullatus* was a very handsome species, with abundant clusters of red fruits from midsummer and bright yellow foliage in autumn. When Veitch exhibited it in October 1912, it was considered the best of the Chinese shrubs in the display and the great plantsman W. J. Bean concluded

that it was 'undoubtedly one of the finest of the [*Cotoneaster*] species introduced in this century'.[8] It is closely related to *C. moupinensis*, a species discovered by Père David and recently reintroduced. *C. bullatus* is now common in cultivation and has escaped into the wild in Europe.[9]

It was apparent by 1905 that a large *Berberis* with narrow serrated leaves and black fruits, which had been raised at Les Barres from seeds collected by Père Farges, was a new species. Vilmorin asked the German botanist Camillo Schneider to dedicate it to Père Soulié. This he duly did, and this handsome species is now known as *Berberis soulieana*. *B. sanguinea* was another Sichuan berberis raised by Vilmorin from seeds sent in all probability by Père Farges. It had been discovered by Père David at Baoxing and was first cultivated at the Jardin des Plantes from seeds he collected.[10]

Vilmorin's contacts with the missionaries and his generous distribution of new species to botanic gardens had given specialists a chance to familiarise themselves with some of the new Chinese discoveries and French nurserymen had made the new plants available to French gardeners; but it was only when nursery owners like Harry Veitch and Arthur Bulley sent their own collectors to China that the new plants became widely available in Britain. Charles Sargent of the Arnold Arboretum and David Fairchild of the US Department of Agriculture also realised the importance of sending their own collectors to China, and from the turn of the century onwards concerted efforts were made to introduce newly-discovered Chinese plants to America.

New arrivals at Cigu

George Forrest, who had been employed by Arthur Bulley to collect seeds for his new nursery, A. Bee & Co., arrived at the customs post at Tengyueh (Tengchong) at the end of July 1904 and then journeyed north. He explored the Zhongdian plateau before visiting Cigu for the first time in November 1904. There he found Père Dubernard, now sixty-four, and still labouring among his converts, in spite of the continuing hostility of the neighbouring lamas: only eighteen months earlier Père Dubernard and his community had foiled an attack on the mission by two hundred armed men.[11] Père Dubernard had been joined earlier that year by Père Théodore Monbeig from Salies-de-Béarn near Bayonne in the Pyrénées-Atlantiques, the same area of south-west France that Père David came from. Père Monbeig was also a keen botanist and revelled, as Père Soulié had done, in the richness of the flora in the valleys and ranges west of Cigu towards the Nu Jiang.

There was now added reason for the missionaries to cross the Xi La pass, as they had managed to establish a new mission at Balhang (Baihanluo) on the east bank of the Nu Jiang in June 1899. It was run by Père Soulié's original travelling companion, Père Annet Genestier.[12] One of the greatest difficulties for this new mission was that although in summer Baihanluo was only two or three days' journey from Cigu, for six or seven months of the year, when snow blocked the Xi La pass, Père Genestier was completely cut off from his confrères at Cigu. This made Baihanluo the loneliest of missions – Frank Kingdon-Ward wrote of the 'sense of paralyzing isolation' he felt in these mountains – and Père Genestier also had to deal with the determined hostility of the lamas at nearby Champutong (Binchunglu). During one of their attacks, the mission was burned to the ground and Père Genestier lost his entire herbarium containing 1,200 specimens of the local flora.[13]

The priests told Forrest enough about the wonders of the

flora to the west of Cigu to excite his interest, although he did not think that they knew much about botany or that the specimens they were preparing for the Muséum were much good. Forrest did not speak much French and the missionaries hardly any English so perhaps he did not realise that Père Monbeig was actually a competent botanist, although he did buy some seeds from him, which he sent to Bulley at Bees Nursery.[14] He decided to return the following year to make an extended visit because, as he wrote to his brother, 'I simply cannot leave those flowers to be discovered by and named after Frenchmen'.[15] There was a strong competitive streak in all those – even the missionary-botanists – who collected plants in China.

Crisis

Forrest had not chosen his time well. The uneasy status quo that had prevailed for so long in the region had been destroyed in August 1904 when Sir Francis Younghusband had entered Tibet at the head of a British force and marched into Lhasa. This invasion unsettled the lamas in the Tibetan borderlands, stirring up anger towards their Chinese overlords who had encouraged Chinese farmers to move in and occupy Tibetan territory, and forced them to tolerate the hated foreign missionaries. The lamas in Batang began openly to defy the Chinese official in charge and Père Soulié was so perturbed by this news, as he knew that the missionaries' safety depended on the lamas' fear of Chinese reprisals, that he went to confer with Père Georges Mussot at the Batang mission on 15 February. He only stayed a few days, and it was understood that Père Mussot would make for Yaregong if he should have to flee, as they did not think that any trouble would spread beyond Batang itself. They both underestimated the danger, even though they each had personal experience of the lamas' hatred: Père Soulié in 1887, when he had been attacked and hounded out of Batang, and Père Mussot in 1901, when he had been and captured and brutally ill-treated at his previous mission near Kangding.[16] Neither he nor Père Soulié properly appreciated the fury that was building up in many of the lamaseries, where the missionaries had come to symbolise everything the lamas hated about foreign domination of their lands.

Murder

The situation in Batang escalated and, at the end of March, the lamas rose up and murdered several Chinese officials. Père Mussot fled to the house of an official stationed at a nearby Yangtze crossing-point but, on 1 or 2 April, the lamas and some of their tenants dragged him outside and took him back to their lamasery at Batang. They kept him chained up for three days and, after further ill treatment, finally shot him. He was fifty-four and had been in Tibet since 1883. Rumours of these disturbances reached the Red Hat lamas at Yaregong the next

Photo Credit: Edward He

day and they warned Père Soulié that trouble had broken out at Batang and that his life was in danger. He did not believe them, as he thought that the hostile lamas would merely ransack the mission again. He decided that he had time to pack up his things and leave them with the Red Hat lamas for safekeeping.

He had wholly misjudged the situation and when armed lamas and about sixty of their tenants, who had been compelled to accompany them, surrounded his house just before sunset on 3 April, Père Soulié was trapped. He immediately created a diversion that gave some of his followers time to escape and then opened the door, demanding to know what the crowd wanted. His defiance was of no avail and he was attacked and beaten. A sword was thrust into his side as his legs were being chained and he received a brutal blow from a rock. His assailants tied him to a tree. The following day, the crowd had trebled in size and lamas from Batang arrived to interrogate him. His possessions were seized and all his religious objects destroyed. Père Soulié was forced back into the house and chained to the staircase, along with his servant Nicolas. He was kept prisoner for ten days, during which he repeatedly asked his tormentors to spare Nicolas. He could not rest because of the pain of his wounds, and by the time he was taken to a valley beyond the village on the morning of 14 April, he was very weak and could hardly walk. He was once again tied to a tree. One of the villagers asked his forgiveness for not being able to protect him. At that moment, he was shot in the head from behind, and then again in the chest.[17]

Père Soulié was forty-five and had been in Tibet for nineteen years. We must hope that his faith and his belief that he was sharing the sufferings of Christ to whom he had devoted his life were of comfort during his final ordeal. After his murder, the lamas released Nicolas, who took the news to Kangding, whence it was telegraphed to France at the end of April.

Cigu

The Batang lamas and their brethren farther south were now in full revolt and they and their followers laid siege to the Chinese garrison at Deqin. Garbled rumours of these events filtered down the Lancang Jiang over the next few weeks, and some of the lamas' followers from Deqin began to prowl around Cigu so that, when Forrest arrived at the village at the end of April, he found the priests and their community living in hourly expectation of an attack and 'practically in a state of siege'.[18] When nothing happened immediately, Forrest got on with organising his collecting expeditions, but the rumours of trouble farther north were persistent and the priests doubled the sentries guarding the bridges over the Lancang Jiang. The

lack of definite or reliable news meant that the missionaries could not know how dangerous their situation had become. Even when they learned in May that Père Soulié and Père Mussot had been murdered, they made no preparations to leave and continued to go about their daily activities, even though they knew that Deqin was surrounded and that if it fell, Cigu would be next. Père Monbeig crossed the Xi La pass in July to visit Baihanluo, and Forrest continued with his collecting.

The tension increased and Cigu learned on 17 July that Deqin was close to collapse: but still the priests hesitated. Père Dubernard was desolate as he contemplated the ruin of his life's work and Père Bourdonnec wildly proposed fighting it out, although the mission was completely indefensible. At 5 p.m. on 19 July they heard that Deqin had fallen and the Chinese garrison slaughtered. Rapid flight to Yeichi, some 48 km (30 miles) to the south, where there was a friendly headman and some Chinese troops, was now the only hope.

By the light of the rising moon, the priests, mounted on mules and accompanied by about eighty Christians, together with Forrest and his collectors, made their way southwards along the east bank. As they passed close to a lamasery some of the party inadvertently made a noise, which alerted the lamas to their flight. The next morning they learned that these lamas were making a forced march to get ahead of them and block the road south. Speed was now vital if they were to escape the cordon, but Forrest could not get the old priests to hurry. By noon, they could see columns of thick smoke rising behind them where the pursuing lamas had put Cigu to the torch: but now when speed was critical, the priests stopped to confer with their followers. Forrest, in despair at their dilatoriness, climbed higher to reconnoitre and saw bands of armed lamas racing full tilt along the road in pursuit. He ran to give the alarm and the group scattered.

Père Bourdonnec rushed off into the forest, where he was later tracked down and shot with poisoned arrows before being cut down with swords. Père Dubernard managed to evade his pursuers for a further day but he was eventually captured, hideously tortured and shot. It was a tragic end for men who had devoted their lives to helping the people among whom they had made their homes.

Trapped

Almost all their followers were murdered and Forrest was only saved because, as the lamas blocking the road in front sighted him and gave chase, he skidded round a bend and tumbled hundreds of feet down a precipitous scrub-covered slope, which hid him from view. The lamas posted sentries with large mastiff dogs along all the ridges, effectively trapping him in the long valley. Bands of lamas began combing the slopes for him and when he realised that they were following his footprints, he took his boots off and buried them in a streambed. At night, he climbed up towards the ridge summits, seeking a way out: but the guards and the dogs were unavoidable. During the day, he tried to find somewhere to hide and rest; but the bands of lamas hunting him were relentless and flushed him from his hiding places. He had several narrow escapes: once his pursuers came close enough to fire their poisoned arrows at him, two of which pierced his hat brim. He still had his gun and was tempted many times to fire on them, but he knew that this would be suicidal, as it would only serve to pinpoint his location. He had no food, although after three days he did find some ears of wheat and dried peas, and he became increasingly weak and light headed as he sought desperately to escape the cordon. By the end of the eighth day he knew he could not last much longer, and the following morning he staggered into one of the Lisu villages at the bottom of the valley, where he collapsed. The villagers fed him and hid him from the lamas for three days, before smuggling him out of the valley and guiding him to Yeichi. The circuitous journey over the highest slopes, where snow still lay on the ground, took almost a week and Forrest suffered dreadfully in the monsoon downpours. To make matters even worse, he stepped on a sharpened bamboo stake and severely wounded one of his bare feet.

Survivors

At Yeichi, the Chinese official gave Forrest food and clean clothes and took him to Xiao-Weixi where he found Père Monbeig, who had been on his way back to Cigu from Baihanluo when a message from Père Bourdonnec had warned him that trouble was imminent. He had got close enough to Cigu to see the church and village burning, before turning southwards and making his way as best he could without food or shelter to the mission at Xiao-Weixi.[19] His state of mind after witnessing the destruction of his church and home, guessing at the fate of his confrères and being in fear for his own life can only be imagined. An armed troop then escorted Père Monbeig and Forrest to Dali. Père Monbeig made his way to Kunming where Père Genestier had sought refuge, while Forrest stayed at the China Inland Mission so that his feet, which were in an awful state after being shredded by broken stones and pierced by the bamboo stake, could be treated and he could recuperate.[20]

George Forrest had been astonishingly lucky to survive. His plight on the mountains south of Cigu was truly desperate, and he later recounted how at one point Père Dubernard had 'saved' him by appearing to him when lamas were close by and silently pointing out the correct direction to take. He affirmed that without this spectral help, the marauding bands of lamas would undoubtedly have caught him. The priest's apparition to Forrest at this perilous juncture seems to have been another of the remarkable instances of what has been called the phenomenon

ABOVE
Rosa moyesii in the late Jim and Jenny Archibald's garden.

RIGHT
Ornamental hips of *Rosa moyesii*.

of 'the third man': that is the comforting presence of a benevolent visual or auditory guide who appears to those in extraordinary danger offering help and support. 'The third man' might be some sort of hallucination brought on by exhaustion, hunger and fear: but several mountaineers and explorers owe their lives to the phenomenon.[21] Forrest recorded his gratitude to the elderly priest by naming a primula *Primula dubernardiana* after him, although in fact this little primula had already been discovered by Père Delavay.[22]

Rosa moyesii

Ernest Wilson's first expedition in 1899 to find *Davidia involucrata* for Veitch's nursery was such a success that Harry Veitch sent him back to China in 1903 to search for yellow *Meconopsis integrifolia* in the mountains of western Sichuan. Wilson visited Kangding for the first time later that year and then spent the following summer investigating the flora of the surrounding area. On one of these long exploratory trips he was accompanied by James Moyes, a Scottish missionary at Kangding, who spoke Tibetan. One of the plants Wilson came across most frequently was a large rose with deep ruby-red flowers, which had first been collected by Antwerp Pratt, although it was not described until Wilson returned and named it *Rosa moyesii* in honour of James Moyes 'to whom [he was] much indebted for hospitality, assistance and companionship on one long and interesting journey in Eastern Tibet.' *R. moyesii* was first exhibited by Veitch in 1908 and Wilson collected it again on several occasions for the Arnold Arboretum. Another form with large pink flowers, which is also common in the wild, was first known as

Photo Credit: Seamus O'Brien

R. fargesii, before being identified as a form of *R. moyesii* that horticulturalists still call var. *fargesii*, although many botanists no longer consider it to be sufficiently distinct from the species to warrant a separate name. As the young plants distributed by Veitch began to mature, it became apparent that *R. moyesii* was an exceptionally fine species and that its intense blood-red flowers were followed by magnificent flagon-shaped orange hips. Graham Stuart Thomas, an expert plantsman, considered it one of the finest of all flowering shrubs.

Rose breeders soon recognised *R. moyesii*'s potential and within thirty years or so of its first introduction, Pedro Dot in Spain had used it to develop white-flowered 'Nevada', which has since been recognised as one of the best shrub roses of modern times, and the Royal Horticultural Society had raised the hybrid *Rosa* 'Geranium' at its garden at Wisley.[23]

James Moyes, commemorated in this superb rose, was a member of the China Inland Mission, one of the largest of the Protestant missionary organisations working in China. He was a Baptist from Lochgelly in Fifeshire who had been a miner and a shopkeeper before going out to China in 1896 aged twenty-four to work as a missionary in Kangding. He went home on leave in 1901 and returned to Kangding in October 1903. In March 1904 he welcomed a small group of American missionaries led by Dr Susie Carson Rijnhart, a remarkable Canadian missionary who, with her husband Peter Rijnhart, had previously established an independent mission near Xining in Qinghai. In May 1898, the Rijnharts had tried to reach Lhasa but their attempt to journey through the heart of Tibet had ended disasterously. Their year-old son had died in August and the following month they were attacked and all their supplies and mules stolen. A few days after that Peter Rijnhart vanished, leaving Susie Rijnhart to struggle on to Kangding alone. When she reached the town on 26 November, half dead from starvation and frostbite, James Moyes was the first to greet her. Peter Rijnhart was never seen again and it was eventually presumed that he had been murdered.

Dr Rijnhart went home but spent the next four years raising funds in America so that she could return to Tibet with some companions to establish another mission. When they arrived in Kangding, James Moyes arranged temporary lodgings for them until they were able to move into their own premises. In October 1905, he and Susie Rijnhart married; but her health was now failing and Moyes resigned from the China Inland Mission so that they could move to Chengdu. They returned to Susie Rijnhart's' home at Chatham in Canada in 1907 but she died in February 1908, two months after giving birth. James Moyes subsequently applied to rejoin the China Inland Mission and return to Tibet, but his request was refused.[24]

CHAPTER 17

Final Years

—

Recognition should be given to the general respect entertained by foreigners

of opposing Christian creeds, for the lifelong devotion to their task, on the

slenderest stipend, of the Roman [Catholic] priesthood.

COLONEL HOWARD VINCENT[1]

—

THE CHINESE MOVED SWIFTLY TO PUT DOWN the rebellion and by the autumn of 1905 they had established complete control of the Tibetan borderlands. In September, Père Grandjean and twenty soldiers from Batang entered Yaregong and recovered Père Soulié's body, which had lain with just a few stones and branches tossed over it before one of the Red Hat lamas had had covered it properly. Père Monbeig returned to Cigu, which must have taken considerable courage, and found only charred ruins. He buried the bodies of Père Dubernard and Père Bourdonnec in the little cemetery and began the laborious task of re-establishing the mission. The old church and mission buildings had stood in the centre of the village but Père Monbeig decided to leave the original site bare, and when Heinrich Handel-Mazzetti visited Cigu in 1915, he saw that 'A ruined wall of white stone, the graves of the missionaries murdered by the lamas, and some Chinese burial mounds with names inscribed above them were all that remained of the mission burnt down in 1905.'[2] The desolate site was eventually planted with some of the vines that the Missions Étrangères introduced to the area after 1905, in the hope that winemaking would provide the villagers with another source of income, and rows of vines now cover the area where Père Dubernard's church once stood. Today, there are vineyards throughout the district and the grapes are used to produce a rough red wine.

In 1908, Père Monbeig was joined by Père Jean Lesgourgues and the following year, when Cigu appeared to be threatened by the landslips that are a feature of the loose riverside slopes, they moved the mission upriver to Cizhong, where a small detachment of Chinese troops were based. The first task was to build a new church and Père Monbeig – perhaps inspired by the memory of the fine church he had first known at Cigu – resolved that the new church at Cizhong would be built of stone

Photo Credit: Fonds Iconographique des Missions Étrangères de Paris

LEFT

Deutzia monbeigii in the author's garden at midsummer.

LEFT

Père Théodore Monbeig.

in the same style as French churches and would have a proper bell-tower. It was an ambitious project and marked the priests' renewed confidence and sense of security, now that the Chinese were in full control of these turbulent borderlands. Remarkably, the impressive new church was completed just two years later. It was consecrated at the beginning of January 1911.

The missionaries' zeal for making natural history collections was similarly undaunted. Père Genestier at Baihanluo began to remake his herbarium. His specimens had suffered considerably from insect attacks, but even so Heinrich Handel-Mazzetti was happy to purchase eleven sheets of spring-flowering plants when he passed through in August 1916, as he would otherwise have missed these early season species altogether. He later named a new primula species *Primula genestieriana*.[3]

Père Monbeig's collections

By 1907, Père Monbeig had also resumed his botanical collecting and he was able to send four consignments of plants

BELOW
The site of Père Dubernard's church and the mission buildings is now occupied by the school and vineyards.

Photo Credit: Fonds Iconographique des Missions Étrangères de Paris

to the Muséum in 1910, and a further collection at the end of 1913.[4] Many of his plants had already been collected by Père Soulié but, as George Forrest's had realised 'these unknown hillsides are a veritable botanist's paradise', and Père Monbeig found several new species. These included a hornbeam, *Carpinus monbeigiana*, which has only recently been introduced but appears to do best in warm sites, and a mountain ash with white-pink fruit and remarkable crimson autumn foliage that was named *Sorbus monbeigii*.[5] It was reintroduced to the West in 1991.

One of Père Monbeig's finest woody discoveries was a white-flowered deutzia that he collected near Cigu in May 1912, which was named *Deutzia monbeigii*. It was introduced by Forrest in 1917 and has subsequently proved one of the most ornamental of the white deutzias, with a spreading habit and peeling chestnut bark. Harold Hillier ranked *D. monbeigii* 'among the best of all [Forrest's] many attractive introductions.'[6]

Père Monbeig also sent specimens to the botanists at the Edinburgh Botanic Garden, among which was a fine new species of *Pedicularis* that he had collected near Cigu. It was named *P. monbeigiana* and is similar in appearance to *P. davidii*, although it is taller with remarkably handsome flower spikes. He found another *Pedicularis* near Weixi that was identified as a variety of *P. rhinanthoides* now known as subsp. *tibetica*. This subspecies has been successfully cultivated from time to time, but it is hard to keep going as almost no seeds are produced in cultivation, because there is nothing to pollinate the very long narrow flowers.

Murder

Père Monbeig was transferred to Batang in November 1913. The following spring, he visited the nearby missions at Caka'lho and Yaregong, before deciding to go on to Litang. However, on 12 June he and his servant were attacked by a band of robbers as they neared the town. The first shots killed Père Monbeig's horse, but although his servant bravely dismounted and offered him his own horse, there was no time to remount and they were quickly overwhelmed. In the ensuing melée, both men were shot and then cut down with swords. Père Monbeig was buried at Kangding the following month. His brother Père Émile Monbeig, also a member of the Missions Étrangères, arrived in the area later in the year to investigate the murder and to visit his brother's grave at Kangding.[7]

Père Monbeig was the last missionary to make significant plant collections in China and the outbreak of war in Europe in the summer of 1914 brought almost all work on existing Chinese collections in the Muséum's herbarium to a halt. Hector Leveillé, one of the China specialists, did try to carry on and even managed to publish works on the flora of Yunnan and Sichuan in 1915 and 1918, before his own death in 1918. The battles raging across northern France made it impossible for

ABOVE

Deutzia monbeigii.

RIGHT

Pedicularis monbeigiana at 3,500 m (11,482 ft) between Moxi and Kangding.

him to compare specimens with those in foreign herbaria or to discuss problematic identifications with colleagues from other countries, which meant that many of Leveillé's names have subsequently proved invalid. Others who had played important roles in the Western discovery of the Chinese flora also died in 1918: Maurice de Vilmorin, who had been the first to grow so many new Chinese plants, died in April and Professor Édouard Bureau in December. It was Bureau who had engaged Franchet in 1880 to work on Père David's collections and determining these specimens had equipped Franchet with the solid knowledge of the Chinese flora, which made possible his subsequent prodigious labours as he examined the collections made by other missionary-botanists. Professor Bureau had also recognised the importance of Franchet's work and found him the paying position at the Muséum that enabled him to continue working on the Chinese collections. Although he had been retired for many years by 1918, Professeur Bureau had done so much to encourage the fruitful collaboration between the French missionary-botanists and Adrien Franchet (and his Muséum colleagues), that his passing can be seen as a symbolic marker signalling the end of the missionary contribution to knowledge of the Chinese flora.

ABOVE

Pedicularis rhinanthoides subsp. *tibetica* in its native habitat near Bita Hai, Zhongdian with *P. longiflora* var. *tubiformis* and *Primula sikkimensis*.

ABOVE

The new church at Cigu stands above the village.

The next stage

From 1914, work on collections from western China moved from Paris, which had been the centre of all such activity for the last forty years, to other institutions in the UK and US: to Kew, to Edinburgh Botanic Garden, where the collections made by George Forrest, Reginald Farrer and Frank Kingdon-Ward were studied; and to Boston where Ernest Wilson and the systematic botanist Alfred Rehder worked on the collections Wilson had made for the Arnold Arboretum. Joseph Rock's specimens were also examined in America, while Heinrich Handel-Mazzetti returned to Austria in 1919 and began classifying his own extensive herbarium. Today, Western botanists work closely with their Chinese colleagues: new plants are still being discovered and the specimens collected by the missionary-botanists are now used as essential reference points by all those studying the Chinese flora.

After the suppression of the 1905 revolt, the missionaries were largely left alone by the lamas and were able to minister to their communities without the continual threat of violence that had rendered the borderlands so dangerous for so long. It must have taken those who remembered the old days – for instance Bishop Giraudeau and Père Annet Genestier, who had

begun his missionary service at the same time as Père Soulié and who had survived the horrors of 1905 – some time to get used to such unfamiliar tranquillity. Yet, by the time they died – Père Genestier in 1931 and Bishop Giraudeau in 1941 – the threat to the missionaries no longer came principally from the lamas, but from the increasingly powerful Chinese Communist Party. In recent decades, however, Christians have once again been allowed to practice their faith and Cigu is now a prosperous and predominantly Christian village, with only two families remaining Buddhist. A new church has been built above the village and a priest comes regularly to take services.

Père Monbeig's fine church still stands at Cizhong: a monument to the tenacity and endurance of the priests of the Missions Étrangères and now a place of pilgrimage for Chinese Christians. Christianity is currently the fastest growing religion in China, with over ten million Catholics. We might celebrate the missionaries who feature in this narrative for their botanical discoveries but botany was only ever a sideline for these isolated dedicated priests: the conversion of the people they had exiled themselves to serve was the real purpose of their lives, and they would see the continuing expansion of the religion for which many of them died as the true measure of their success.

LEFT

A cheerful member of Cigu's Christian community.

Sources and General Bibliography

NOTES ON SOURCES

Letters to Adrien Franchet from Pères Delavay, Farges and Soulié giving an account of their botanical activities are to be found in the Franchet Archive at the Manuscript Department, Muséum d'Histoire Naturelle, Paris.

Biographical details and other records relating to the missionaries can be found in the Archives of the Missions Étrangères de Paris at their headquarters at 128 Rue du Bac Paris 75007 (http://archives.mepasie.org)

ABBREVIATIONS USED FOR WORKS FREQUENTLY CITED IN THE NOTES

TAXONOMY AND FLORAS
Flora of China: FOC
http://flora.huh.harvard.edu/china/index.html

Germplasm Resources Information Network [Online Database]: GRIN
http://www.ars-grin.gov/cgi-bin/npgs/html/taxgenform.pl
USDA, ARS, National Genetic Resources Program
National Germplasm Resources Laboratory, Beltsville, Maryland

Grey-Wilson, C. and Cribb, P. (2011) *Flowers of Western China* Royal Botanic Gardens, Kew Publishing *Flowers of Western China* (2011). An invaluable work with pictures of many of the plants featured in this account.

The Plant List: http://www.theplantlist.org

The Royal Horticultural Society Horticultural Database: RHSDatabase
http://www.rhs.org.uk

PERIODICALS
Alpine Garden Society Bulletin: AGSBulletin

Curtis's Botanical Magazine: Bot.Mag.

Botanische Jahrbücher für Systematik, Pflanzengeschichte und Pflanzengeographie: Bot.Jahrb.Syst.

Bulletin de la Société Botanique de France: Bull.Soc.Bot. France

Bulletin des Nouvelles Archives du Muséum d'Histoire Naturelle: Bull.Nouv.Arch.NHM

Bulletin of Miscellaneous Information Royal Botanic Gardens Kew: KewBulletin

Gardeners' Chronicle: Gard.Chron

Hooker's Icones Plantarum: Icon.Plant

Journal of the Arnold Arboretum: Jnl.ArnoldArb

Journal of the Proceedings of the Linnean Society (Botany): Jnl.Linn.Soc.

Journal of the Royal Horticultural Society: RHSJnl

Notes from the Royal Botanic Gardens, Edinburgh: Notes. RBGE

Nouvelles Archives du Muséum d'Histoire Naturelle: Nouv. Arch.MHN

Repertorium Specierum Novarum Regni Vegetabilis: Repert. Spec.Nov.Regni.Veg.

Revue Horticole Journal d'Horticulture Pratique: Rev.Hort.

The Garden Magazine of the Royal Horticultural Society: RHS The Garden

GENERAL
All works published in London unless otherwise specified.

Bean, W.J. (1970-1980) *Trees and Shrubs Hardy in the British Isles* Eighth Edition 4 vols Bean

Bretschneider, E. (1898) *History of European Botanical Discoveries in China* St Petersburg 2 vols and map: Bretschneider, E. (1898)

Grimshaw, J. and Bayton, R. (2009) *New Trees Recent Introductions to Cultivation* Kew Publishing *New Trees* (2009)

Farrer, R. (1919) *The Rock-Garden* 2 vols: Farrer, R. (1919)

Fournier, P. (1932) *Voyages et Découvertes Scientifiques des Missionaires Naturalistes Français XVe –Xxe Siècles* Paris: Fournier, P. (1932)

Franchet, A. (1888) *Plantae Davidianae* 2 vols Paris: *Pl.Dav*

Franchet, A. (1889-1890) *Plantae Delavayanae* Paris: *Pl.Del.*

Hinkley, D. (2009)*The Explorer's Garden Shrubs and Vines from the Four Corners of the World* Timber Press: Hinkley, D. (2009)

Howgego, R.J. (2008) *Encylopedia of Exploration 1850-1940* Vol.2 *Continental Exploration* Hordern House, Australia: Howgego, R.J. (2008)

Lancaster, R. (2008) *Plantsman's Paradise Travels in China* Antique Collectors' Club Second Edition: Lancaster, R. (2008)

Lauener, L.A. (ed. D. Ferguson) (1996) *The Introduction of Chinese Plants into Europe* SPB Academic Publishing, Amsterdam: Lauener, L.A. (1996)

O'Brien, S. (2011) *In the Footsteps of Augustine Henry and his Chinese Plant Collectors* Garden Art Press: O'Brien, S. (2011)

Sargent, C.E. ed. (1913-1919) *Plantae Wilsonianae An Enumeration of the Woody Plants collected in Western China for the Arnold Arboretum of Harvard University during the years 1907, 1908 and 1910 by E.H. Wilson* 3 vols. Cambridge University Press: *Pl.Wils.*

Vilmorin, M. and Bois, D. (1904) *Fruticetum Vilmorianum* Paris: *Fruticetum Vilmorianum* (1904)

Wilson, E.H. (1986) *A Naturalist in Western China* 1913 Cadogan Books: Wilson, E.H. (1913-1986 ed.)

Notes and References

INTRODUCTION pages 1-3

1. Mazzetti-Handel, H. (1927) *A Botanical Pioneer in South West China* Vienna Translated and published David Winstanley (1996) Essex p.94

CHAPTER ONE pages 4-7

1. Davies, H. (1909) *Yun-nan The Link between India and the Yangtse*, Cambridge University Press, p.217
2. Launay, A. (1894) *Histoire General de la Société des Missions-Étrangères* Paris and (1890) *Atlas des Missions Étrangères*, Lille
3. Guennou, J. (1963) *Les Missions Étrangères* Paris (1869) *Annales de la Propagation de la Foi* Lyons v. 41 pp.270-276
4. The phrase is John Donne's: *Biathanatos*, Sullivan, E.W. ed. (1984) Assoc. Univ. Presses p.29

CHAPTER TWO pages 8-23

1. Wilson, E.H. (1906) *Leaves from my Chinese Notebook*, *Gard.Chron.* 39 p.419
2. Boutan, Emmanuel (1993) *Le Nuage et le Vitrine Une Vie de Monsieur David* Editions Raymond Chabaud Bayonne; *Abbé David's Diary being An Account of the French Naturalist's Journeys and Observations in China in the years 1866-1869* translated and edited Fox, H. (1949) Harvard UP

3. Scott, B. (2004) *Père Jean Pierre Armand David CM, Oceania Vincentian* 5 Sept
4. Bretschneider, E. (1898) *History of European Botanical Discoveries in China* St Petersburg II pp.346-352; *Bot. Mag.* (1846) No 4269
5. Bretschneider, E. (1898) II pp.323-342
6. See Kilpatrick, J. (2007) *Gifts from the Gardens of China* Frances Lincoln for an account of these discoveries
7. Hance, H. (1871) *Notes on some Plants from Northern China, Jnl.Linn.Soc.Bot.* XIII pp.74-95
8. Blanchard, É. (1871) *Les Récentes Explorations des Naturalistes en Chine, Revue de Deux Mondes* 91 Feb.
9. Franchet, A. (1884) *Plantae Davidianae* Paris I p.60: described in 1880 by Maximowicz from a specimen collected by Przewalski in Outer Mongolia
10. Bretschneider, E. (1898) *History of European Botanical Discoveries in China* St Petersburg 2 vols and maps
11. Bretschneider, E. (1898) II pp.837-870; *Bull.Soc.Bot. France* (1904) p.9; *Rev.Hort.* (1902) pp.476-478; Sargent, C. *Trees and Shrubs* (1905) p.121; Spongberg, S. (1990) *A Reunion of Trees Discovery of Exotic Plants and their Introduction into North American and European Landscapes* Harvard UP p.171; Muir, N. (1983-1984) *A Survey of the Genus Tilia, The Plantsman* 5 pp.206-242; FOC 12:248
12. *Bot.Mag.* (1929) No.9284 gives a useful summary of this species' various introductions
13. Kilpatrick, J. (2007) pp.61-65

14. Fiala, J. (revised Vrugtman, F.) (2008) *Lilacs A Gardener's Encyclopaedia* Timber Press p.77; McKelvey, S.D. (1928) *The Lilacs A Monograph* New York

15. *Garden and Forest* (1889) p.309

16. Drake del Castillo, E. (1900) *Notice sur la Vie et les Travaux de A. Franchet, Bull.Soc.Bot.France* pp.158-172 with list of publications; Colombel, Marc *Adrien Franchet* http://www.rhododendron.fr/articles/article 42.pdf

17. Fox, H. (1949); Grey-Wilson, C. (2000) *Clematis The Genus* p.193

18. Verlot, B. (1867) *Clematis davidianae, Rev.Hort.* pp.90-91; Decaisne, J. (1877) *Décade de Plantes Nouvelles ou Peu Connues, Flore des Serres* p.161, also (1881) *Revision des Clematites du Groupe des Tubuleuses cultivées au Muséum, Nouv.Arch.MHN* pp.195-214; *Pl.Dav.*I:13

19. *Pl.Dav.*I:163

20. Henry Hance had identified a *Clematis* specimen collected by one of his correspondents near Beijing as *C. tubulosa Jnl.Linn.Soc.* (1871) 13 p.75; FOC 6:370 *C. heracleifolia: C. h.* var. *davidii, C. h.* var. *davidiana* and *C. tubulosa* itself are all considered synonyms

21. Brown, M. and Toomey, M. (June 2013) *Herbaceous Clematis, The Plantsman* NS 12 (2) pp.106-111

22. First described by de Jussieu (1830) as *Cedrela sinensis* and by M. Roemer in 1846 as *Toona sinensis; Flore des Serres* (1869-1870) pp.124-125; *Rev.Hort.* (1891) pp.573-576; Edmonds, J. and Staniforth, M. *Toona sinensis Curtis'sBot.Mag.* (1998) 15 (199) No.348 pp.186-193; FOC 11:113

23. Boutan, E. (1993) p.41

24. Fox, H. (1949) p.295

25. Op.cit. p.177

26. *Pl.Dav.* I:121; *Gard.Chron.* (1902) 32 p.95; *Rev.Hort.* (1907) pp.39-41

27. FOC considers *A. c.* var. *davidii* synonymous with *A. chinensis* 8:275; The Plant List considers both synonymous with *A. rubra.*

28. *Jnl.Linn.Soc.* (1871) 13 p.89; *Pl.Dav.*I:320-321; Franchet, A. (1897) *Les Carex d'Asie Orientale, Nouv.Arch.MHN* p.195; See Goydar, D. (2005) No.527 *Lathyrus davidii, Curtis' Bot.Mag.* 22:(2) pp.109-113 for discussion of another specimen David sent to Hance.

29. *Pl.Dav.*I:103; *Rev.Hort.* (1872) p.74; Henry, L. (1902) *L'Amandier de David, Rev.Hort.* pp.290-292; FOC as *Amygdalus davidiana* 9:394

30. Bean, W. (1896) *The Genus Prunus, The Garden* 50 p.165; Bretschneider, E. (1898) II p.860

31. *Pl.Dav.*I:266-268; *Pl.Wils.*III:261-262

32. Boutan, E. (1993) p.4: letter to David dd 14.11.1864

33. *Flore des Serres* (1869-1870) Vol.18, No.1899 pp.123-126

34. Fox, H. (1949) p.69; *Rev.Hort.* (1872) pp.291-292

35. *The Garden* (1875) p.524

36. *Revue de Deux Mondes* (1871) 93 p.333

37. Fiala/Vrugtman (2008) p.270

38. Green, P. (1984-1985) *The Chinese 'Common' Lilac, The Plantsman* 6 pp.12-13

39. Henry, L. (1900) *Crossings Made at the Natural History Museum of Paris from 1887-1889, Jnl.RHS* 24 pp.222-225

40. Notably *S. x prestoniae.* Second generation x *henryi* and x *prestoniae* hybrids are sometimes listed as *S.* (Villosae Group)

41. *Gard.Chron.* (1902) 32 p.122

42. Information from Malcolm Pharoah, holder of the National Collection of Astilbe (UK)

43. *New Trees* (2009) pp.880-884

CHAPTER THREE pages 24-31

1. Sources: *Bull.Nouv.Arch.MHN* 3 (1867) pp.18-96, 4 (1868) pp.3-83 contains Père David's Journal of his time in Beijing and his journey to Inner Mongolia. It was translated and abridged by Helen Fox (1949) *Abbé David's Diary being an Account of the French Naturalist's Journeys and Observations in China in the years 1866 to 1869* Harvard UP pp.3-140. The most complete biography is by Emmanuel Boutan (1993) *Le Nuage et le Vitrine Une Vie de Monsieur David* Editions Raymond Chabaud Bayonne. Bishop, G. (1990) *Travels in Imperial China* is not always accurate.

2. Fox, H. (1949) p.29

3. Op.cit. p.3

4. From the poem *God's Grandeur* by Gerard Manley Hopkins

5. Fox, H. (1949) p.9

6. Op.cit. p.8

7. Op.cit. p.60

8. *Flore des Serres* (1869-1870) pp.124-125

9. Loc.sit; *Pl.Dav.*I:117; FOC 9:352 recognises two forms of *R. xanthina* f. *normalis* (syn. f. *spontanea*) - the single form, and f. *xanthina* – the double form; *Bot.Mag.* (1899) No.7666; Phillips, R. and Rix, M. (2004) *Ultimate Guide to Roses* p.30

10. Fox, H. (1949) pp.82-83

11. Op.cit. p.96

12. Op.cit. pp.122-123

13. Colquhoun, A.R. (1883) *Across Chrysé ... a Journey of*

Exploration through the South China Borderlands II p.248.

14. *Rev.Hort.* (1872) p.451; *Pl.Dav.*I:231; *Bot.Mag.* (1928) No.9219; *FOC* 17:43-47

15. Op.cit. p.127

16. Fox, H. (1949) pp.4, 143

17. As *R. xanthina* f. *spontanae*; reputedly raised at Osterley Park: Thomas, G.S. (1994) *The Graham Stuart Thomas Rose Book* Timber Press p.106

18. Simmonds wrote to W.J. Bean in 1943 explaining how he had discovered the new hybrid in his garden: Beharrell, N. and Culham, A. (2006) *The Origin of Caryopteris* x *clandonensis, The Plantsman* (NS) pp.192-197

CHAPTER FOUR pages 32-55

1. The full narrative of Père David's journey to Baoxing in given in his *Rapport à MM les Professeurs-administrateurs du Muséum d' Histoire Naturelle* (*Bull. Nouv.Arch.MHN* (1871) VII pp.75-100) and in his *Journal d'un voyage dans le centre de la Chine et dans le Thibet oriental* (*Bull.Nouv.Arch.MHN* VIII (1872), IX, (1873) X, (1874)) These accounts were translated and abridged by Helen Fox (1949) *Abbé David's Diary being an Account of the French Naturalist's Journeys and Observations in China in the years 1866 to 1869* Harvard UP. Fox, H. (1949) p.168

2. Op.cit. pp.187-188

3. Op.cit. p.287; *Bull.Soc.Bot.France* (1873) p.157; *Pl.Dav.*I:357, II:59

4. Fox, H. (1949) p.194

5. *Pl.Dav.*II:120; *Bot.Jahrb.* (1901) 29 p.278; *Bot.Mag.* (1925) No.9079 *as S. bockii*, now a synonym; Newsome, C. (1992) *Willows The Genus Salix* p.139; *FOC* 4:263

6. Fox, H. (1949) p.253

7. Op.cit p.265

8. *Bull.Nouv.Arch.MHN* (1871) VII p.85; *Nouv.Arch.MHN* (1874) X p.11

9. *Pl.Dav.*II:868; Newsome, C. (1992) p.100 only females are in cultivation; *Pl.Dav.*II:94; Rix, M. (2005) *Curtis'sBot. Mag.* 22:2 No.529 p.119

10. *Pl.Dav.*II:93-96; Farrer, R. (1919) II p.165; Richards, J. (1993) *Primula*

11. *Pl.Dav.*II:865, Fox, H. (1949) p.277

12. *Nouv.Arch.MHN* (1874) X p.12; *Pl.Dav.*II:8 9 syn. *H. chinensis*

13. Fox, H. (1949) pp.278-279

14. Op.cit. p.297

15. Père David called it *R. rotundifolium* Fox, H. (1949) p.282; *Pl.Dav.*II:85; *Pl.Wils.*I:540

16. Fox, H. (1949) p.282

17. *Pl.Dav.*II:88

18. Fox, H. (1949) p.283

19. Boutan, E. (1993) *Le Nuage et le Vitrine Une Vie de Monsieur David* Editions Raymond Chabaud Bayonne p.131

20. Fox, H. (1949) p.278

21. Op.cit. p.285

22. *Pl.Dav.*II:30

23. *Pl.Dav.*II:69; *Pl.Wils.*I:111; *Bot.Mag.* (1923) No.8980

24. *Pl.Dav.*II:2; *Pl.Wils.*I:326; Grey-Wilson, C. (2000) *Clematis The Genus* p.107

25. *Jnl.Bot.Soc.France* (1903) pp.524-526; *Pl.Wils.*I:334-335; *FOC* 6:344 as *C. chrysocoma*

26. Farrer, R. (1919) I p.225

27. Rolfe, R. *Orchid Review* (1903) XI pp.289-292; Cribb, P. and Butterfield, I. (1999) *The Genus Pleione* second ed. RBGKew pp.101-109. See p.109 for complex synonymy

28. Fox, H. (1949) pp.285, 286

29. *Pl.Dav.*II:83-87

30. see *Pl.Wils.*I:503-549

31. Stearn, W.T. (2002) *The Genus Epimedium and other Herbaceous Berberidaceae* RBGKew pp.55-60; Rix, M. (April 2007) *The Second Coming, Gardens Illustrated* 124 pp.64-71; Rice, G. (March 2008) *Raising Rising Stars, RHSTheGarden* pp.160-165

32. Fox, H. (1949) p.288; *Pl.Dav.* p.131

33. Fox, H. (1949) p.289

34. Op.cit. p.291

35. Farrer, R. (1919) II:499 as *C. luteum*; Lancaster, R. (1982) *RHSTheGarden* pp.425-430

36. *Bull.Nouv.Arch.MHN* (1874) X p.48

37. Wilson, E.H. (1916) *Aristocrats of the Garden* Boston p.291

38. Baillon, H. (1871) *Adansonia* 10:115; *Pl.Dav.*II:60

39. Crépin, F. (1874) *Bull.Soc.Roy.Bot.Belgique* 13 pp.253-254; *Pl.Dav.*II:42; *Pl.Wils.*II:322-323

40. *Nouv.Arch.Mus.* (1874) X p.179 as *Stransvaesia davidiana: FOC* 9:120; *Bot.Mag.* (1912) No.8418, (1920) No.8862, (1923) No.9008; *Pl.Wils.*I:192

41. *Fl.desSerres* (1877) XXII p.168; *Rev.Hort.*(1885) pp.136-137; Fryer, J. and Hylmo, B. (2009) *Cotoneasters: A Comprehensive Guide* Timber Press

42. *Pl.Dav.*II:66 as *Panax davidii*; *Brittonia* (2001) 53:117; Hinkley, D. (2009) p.249; *New Trees* (2009) p.778

43. 'belle espèce' *Pl.Dav.*II:39; *Kew Bulletin* (1909) pp.258-259; *Pl.Wils.*I:54

44. *Pl.Dav.*II:81; *Pl.Wils.*I:560-61

45. Hinkley, D. (2009) p.193

46. *Pl.Dav.*II:44; *Icon.Plant.* (1890) XX No.1934; *Pl.Wils.*I:41; *Bot.Mag.* (1913) No.8520 and (1923) No.8991; Nevling, L. (24 June 1964) *Climbing Hydrangeas and Their Relatives, Arnoldia* pp.30-31; various forms are recognised: cf. FOC 8:408, RHSDatabase and GRIN

47. *Bot.Jahrb.Syst.* (1881) 1:487

48. *Bull.Nouv.Arch.MHN* (1871) VII p.54; Fox, H. (1949) p.297

49. *Pl.Dav.*II:105; Birks, J. and H. (2001) *Louseworts in Sichuan and Qinghai, Bull.AGS* 70 pp.296-300

50. Fox, H. (1949) p.297

51. *Pl.Dav.*II:53 as *Astilbe chinensis*; Booker, H. (2005) *Rewarding Rodgersia, Hardy Plant Jnl.HardyPlantSoc.* 29 1 pp.43-47

52. Batalin, A. (1893) *Acta Horti Petropolitani* XIII p.96; *Pl.Dav.*II:46 as *R. podophylla*; Booker, H. (2005)

53. Fox, H. (1949) p.298

54. Loc.sit.

55. Op.cit. pp.299-301

56. David, A. (1871) *Rapport à MM les Professeurs-Administrateurs du Muséum d'Histoire Naturelle, Bull. Nouv.Arch.MHN* VII pp.75-100

57. *Pl.Dav.*I:288; Bretschneider, E. (1898) II p.847; *Pl.Wils.* II:39-40; *New Trees* (2009) pp.426-427; O'Brien, S. (2011) p.149. (Syn. *Abies sacra*)

58. *New Trees* (2009) pp.199-201

59. Rice, G. and Strangman, E. (1993) *Hellebores* Timber Press pp.60-61; Lancaster, R. (1992) *Helleborus thibetanus, RHSTheGarden* 117 4 pp.156-159; Hall, T. and Matthew, B. (1998) *Helleborus thibetanus, Curtis'sBot.Mag.* 15 No.353; Basset, C. (2008) *Sur les Traces du Père David, Hommes et Plantes* 65 pp.4-9

60. Veitch, J. (1904) *Far Eastern Maples, RHS.Jnl.* 29 p.348; *Pl.Wils.*I:96; Harris, J. (2000) *Gardener's Guide to Growing Maples* Timber Press p.61

61. *Bot.Mag.* (1913) No.8493

62. *Pl.Dav.*II:15; Rix, M. (1993) *Corydalis flexuosa from Western China, The Plantsman* pp.129-130; Gardner, D. (1998) *Curtis'sBot.Mag.*15(i) No.332; Hinkley, D. (1999) *The Explorer's Garden* Timber Press pp.131-132

63. *Bot.Mag.* (1900) No.7715 and *Rev.Hort.* (1899) pp.475-476 both as *L. sutchuenense*; Wilson, E.H. (1925) *The Lilies of Eastern Asia* pp.81-82

64. Synge, P. (1980) *Lilies: A Revision of Elwes' Monograph of the Genus Lilium and its supplements* pp.71-73; Jefferson-Brown, M. (2003) *Lilies A Guide to Choosing and Growing Lilies* RHS pp.74-78

65. Syn. *L. farreri Gard.Chron.* (1919) 65 p.76, (1920) 67 p.281, (1924) 76 p.33; Wilson 1926 pp.64-66; Stern, F.C. (1960) *A Chalk Garden* p.124; Synge, P. (1980) p.77

66. *Pl.Dav.*II:43; *Bot.Mag.* (1917) No.8694 and (1923) No.8999; *FOC* 9:89; GRIN

67. Dunlop, G. (2006) *The History of Rodgersia pinnata* 'Superba', *The Plantsman* (NS) 5 (2) pp.93-96

CHAPTER FIVE pages 56-63

1. Sources: The primary source for Père David's third journey is his (1875) *Journal de Mon Troisième Voyage d' Exploration dans l' Empire Chinois* Paris. See also: Boutan, E. (1993) *Le Nuage et le Vitrine Une Vie de Monsieur David* Editions Raymond Chabaud Bayonne. *Revue de Deux Mondes* (1871) 93 p.632

2. David, A. (1875) II p.210

3. *Pl.Dav.*II:136, 214. For an account of these earlier discoveries see Kilpatrick, J. (2007) *Gifts from the Gardens of China*

4. David, A. (1875) I pp.87-91

5. Op.cit. pp.188-189

6. Op.cit. p.192

7. *Bot.Mag.* (1910) No.8347; *New Trees* (2009) pp.588-590

8. *Bull.Soc.Bot.France* (1891) p.413; Bean (1970) p.149; *Pl.Wils.* II pp.44-46; *New Trees* (2009) pp.47-48

9. David, A. (1875) I pp.201,208

10. David, A. (1875) I p.203

11. Bretschneider, E. (1898) II p.546; *Pl.Dav.*I:69; *Bull. Mensuel.Soc.Linn.Paris* (July 1886) pp.449-451; *Icon. Plant.* (1886) No.1539

12. Syn. *S. viciifolia Bot.Mag.* (1903) No.7883; Barham, J. and Staniforth, M. (2000) *Sophora davidii* 'Hans Fliegner' *Curtis'sBot.Mag.* 17 No.397

13. David, A. (1875) II p.131

14. *Pl.Dav.*I:211: not to be confused with *G.* 'Davidii', a garden hybrid; Bretschneider, E. (1898) II p.854

15. Scott, B. (Sept. 2004) *Père Jean Pierre Armand David CM, Oceania Vincentian* p.102

16. David, A. and Oustalet, M. (1877) *Les Oiseaux de la Chine* Paris

17. Bailes, C. (2006) *Hollies for Gardeners* Timber Press pp.241-245

18. Wilson, E.H. (1913) (facsimile edition 1986) II p.30

19. *Rev.Hort.* (1883) pp.53-56 and (1885) pp.55-57; *Bull.Soc. Bot.France* (1912) pp.197-199 *Pl.Wils.*I:104

CHAPTER SIX pages 64-85

1. Sources: Pere Delavay's letters to Père David and to Adrien Franchet at the Muséum d'Histoire Naturelle (MHN) in Paris are now kept in the Franchet Archive at the Muséum's Manuscript Department and are numbered

103-173. Letters from Père Delavay to his superiors at the headquarters of the Missions Étrangères de Paris (MEP) are kept in their Archives in the rue de Bac.
See also: Lennon, J. (2004) *Le père Jean-Marie Delavay (1834-1895) un Grand Naturalist Français au Yunnan, Bulletin de l'Association des Parcs Botaniques de France,* 38 pp.23-27. The late Jean Lennon was a rhododendron expert and his article, which was based on Père Delavay's correspondence with Franchet at the MHN, focuses on Père Delavay's rhododendron discoveries. M. Lennon quotes extensively from Père Delavay's letters, but he died before he could check the text and almost all the quotations are inaccurate. Nevertheless, his article provides an interesting and perceptive account of Père Delavay's years in Yunnan. Also: Lennon, J. (1985/60) *J-M Delavay in Yunnan (1882-1895) and his Relationship with David, Franchet and Others, Rhododendrons with Magnolias and Camellias* RHS
Henry, A. (1902) *Midst Chinese Forests, The Garden* 31 pp.3-6

2. Letter dated 'Canton 10 August' reprinted in *Union Savoisienne* 6 Oct 1869

3. Letter dd 5.2.1881 ref. 0939 J-M Delavay Archives MEP

4. *Bull.Soc.Bot.France* (1886) pp.359

5. *Bull.MHN* (1896) 2 p.149

6. Les Gets lies south-east of Geneva, opposite Mont Blanc. The town is now part of the Portes du Soleil ski area

7. Flammary, A. (1er trimestre 1929) *Le Botaniste Delavay, Revue Savoisienne* p.98

8. *Jnl.Soc.Bot.France* (1896) p.146

9. Perrier de la Battue, Eugène (1917 and 1928) *Catalogue Rasionné des Plantes Vasculaires de Savoie* I and II. Vol. II includes several specimens collected by Père Delavay: pp.19,25,88 etc.

10. Then only known as a cultivated ornamental plant: Hance, H. *Jnl.Bot.* (1880) p.262

11. *Bull.Soc.Bot.France* (1886) p.385 (not FOC); Veitch, *Hortus Veitchiana* p.391; *Pl.Wils.*I:366-367; *FOC* 19; Ahrendt, L. (1961) *Berberis and Mahonia A Taxonomic Revision, Jnl.Linn.Soc.* 57 (No.369) p.61

12. Letter dd 9.8.1882 0939 J-M Delavay Archives MEP

13. Gill, W. (1880) *The River of Golden Sand* p.296; see also Colborne Baber, E. (1886) *Travels and Researches in Western China 1881,* Royal Geographical Society Supplementary Papers 1; Hosie, A. (1890) *Three Years in Western China*; Davies, H.R. (1909) *Yun-nan The Link between India and the Yangtse* Cambridge UP

14. Franchet, A. *Bull.Soc.Bot.France* (1886) p.67; *FOC* 15:144; *Bot.Mag.* (1892) No.7216

15. (CLD486) Richards, J. (1993) *Primula* pp.201-202; also collected by Forrest, G. *Primulaceae from Yunnan and Tibet Notes RBGE* IV April 1908 1905-1909 pp.213-239, p.231

16. Farrer, R. *The English Rock-Garden* (1919) II p.169

17. *Bull.Soc.Bot.France* (1886) p.367 and (1903-04) p.623 where date is 8th July

18. *Bot.Mag.* (1890) No.7152; Mottet, S. (1914-1915) *Les Thalictrum Asiatiques, Rev.Hort.* pp.567-569; *FOC* 6:288

19. Letter dd 9.8.1882 0939 J-M Delavay Archives MEP

20. Letter 108 dd 13.1.1886 to Franchet, Franchet Archive MHN

21. Little, A.J. (1888) *Through the Yangtse Gorges* pp.232,236

22. Bird, I. (1899) *The Yangtse Valley and Beyond* Virago edition 1985 p.97

23. Pim, S. (1966) *The Wood and the Trees* p.104

24. Letters 129 dd 14.3.1888 and 125 14.12.87 Franchet Archive MHN

25. Gill, W. (1880) p.262

26. *Bull.MHN* (1896) 2 p.150

27. *Pl.Del.* p.222; *Bot.Mag.* (1920) No.8854 (as *C. serotinus*) and (1927) No.9171; *FOC* 9:91; *Pl.Del.* p.222; *Rev. Hort.* (1907) pp.256-258; *Bot.Mag.* (1915) No.8594; *FOC* 9:90; Fryer, J. and Hylmo, B. (2009) *Cotoneasters: A Comprehensive Guide* Timber Press

28. The Bai people are often called Minchia in older books

29. The Christian community created by the priests of the Missions Étrangères at Menhuoying was an enduring one. The last French missionaries only left the village in 1948 and the church and other buildings survived until 1958. The path that Père Delavay would have taken from the Eryuan road to the church is still there, running beside the fields that now cover the site where the church stood.

30. Letter 103 dd 17.51884 to David, Franchet Archive MHN

31. *Bull.Soc.Bot.France* (1886) p.387

32. in 1894: Bean p.397; Père Delavay mentions seed collecting in several letters.

33. *Jnl.Soc.Bot.France* (1894) p.285; Bretschneider, E. (1898) II p.877

34. *Bull.Soc.Bot.* (1886) p.233; *Bot.Mag.* (1898) No.7614

35. Rapport No.257 and Obituary of Père Terrasse Archives MEP

36. Bishops' Reports for 1886 and 1887 Archives MEP

37. Letter 174 dd 31.5.1883 with PS dd 4.6.1883 Franchet Archive MHN

38. *Bull.Soc.Bot.France* (1886) p.234 (syn. *R . capitatum*)

39. *Jnl.Soc.Bot.France* (1889) pp.113-114: Franchet gives the location as 'Hualapo' but Delavay's field note for 257 clearly states 'Cangshan': see Sealy, J.R. (1983)

A Revision of the Genus Nomocharis Fr, *Jnl.Linn.Soc.*87 pp.285-323, p.292; Matthews, V. (1994) No. 239 *Kew Mag/Bot.Mag* 2 1 pp.18-20; Balfour, I.B. (1929) *The Genus Nomocharis, Trans.Bot.Soc.Edin.* 27 pp.273-300

40. Hartley, A. (1931) *Nomocharis, RHSJnl* pp.15-17; Synge, P. (1980) *The Genus Nomocharis in Lilies: A Revision of Elwes' Monograph of the Genus Lilium and its Supplements* pp.204-207; Matthews, V. (1995) *The Genus Nomocharis, New Plantsman* 2 pp.195-208

41. *Jnl.Soc.Bot.France* (1898) p.220; Wilson, E.H. (1925) *Lilies of East Asia* p.13; *Bot.Mag.* (1938-39) No.9533

42. *Rev.Hort.* (1891) p.270; FOC 15:313; *Bot.Mag.* (1938-1939) No.9545; *Gard.Chron.* (Jan.1939) p.12

43. Letter 174 dd 31.5.1883 – edited copy of letter to Père David, Franchet Archive MHN

44. *Gard.Chron.* (1908) pp.396-397; *Rev.Hort.* (1912) NS.12I pp.156-157; Richards, J. (1993) pp.114-116

45. Letter 131 dd 27.7.1888 Franchet Archive MHN

46. Richards, J. (1993) p.276

47. *Bull.Soc.Bot.France* (1903) p.525; *Pl.Wils.*I:334-335; FOC 6:344 treats *C. montana* var. *sericea* and *C. spooneri* as synonyms of *C. chrysocoma*: Forrest, G. (1915-1916) *RHSJnl.* p.66; Grey-Wilson, C. (2000) *Clematis The Genus* pp.76-78,88-89

48. *Bull.Soc.Bot.France* (1885) pp.3-14 and (1886) pp.223-336

49. *Bull.Soc.Bot.France* (1885) p.266; Wright Smith, W. and Forrest, G. (1920-22-Nov.1923) *New Primulacea Notes RBGE* 13; *Bot.Mag.* (1939) No. 9527; Richards, J. (1985) *Petiolarid Primulas in Cultivation, The Plantsman* 7 6 pp.217-332; Rix, M. (2005) *Primula moupinensis, Curtis'sBot.Mag.* 22 (2)

50. *Bull.Soc.Philomathique.Paris* (1888) p.141; *Bull.Soc.Bot. France* (1894) pp.225-233; Cribb, P. (1997) *The Genus Cypripedium* RBGKew and Timber Press pp.258-260, 222-223, 198-201, *C. corrugatum* now referred to *C. tibeticum*: pp.204-209 and (2005) *Wild Orchids in Sichuan, Curtis'sBot.Mag.*22 (1) *C. sichuanense* discovered 2002 pp.71-80

51. Rolfe, R. (1896) *KewBull.* p.195 as *P. delavayi* and (1903) *Orchid Review* 11; Lancaster, R. (1982) *Five Orchids of Yunnan, RHSTheGarden* pp.425-430; Cribb. P. and Butterfield, I. (1999) *The Genus Pleione* second ed. RBG Kew. *P. yunnanensis* Rolfe was discovered by Henry in Mengzi, Yunnan - Rolfe, R. (1903)

52. Scottish Rock Garden Club (1935) *George Forrest VMH Explorer and Botanist who by his Discoveries and Plants successfully introduced has greatly enriched our Gardens 1873-1932* Edinburgh p.70

53. *Bull.Soc.Bot.France* (1887) pp.280-285

54. Cullen, J.A. (2005) *Hardy Rhododendron Species A Guide to Identification* Timber Press p.14

55. Letter 110 dd 15.6.1886 Franchet Archive MHN

56. *Bot.Mag.* (1891) No.7159, (1893) No.7301, (1894) No.7361

57. *Gard.Chron.* (1910) p.202

58. *Jnl.Soc.Bot.France* (1899) p.257; Wilson, E.H. (1926) *The Taxads and Conifers of Yunnan, Jnl.ArnoldArb.* p.46; Orr, M.V. (1933-1935) *Plantae Chinenses Forrestianae: Coniferae, Notes.RBGE* 18 p.132; Horsman, J. (1983-1984) *Spruces in Cultivation, The Plantsman* p.114; FOC 4:29; see *New Trees* (2009) pp.571-573 for a taxonomic review.

59. Franchet originally named it *P. glacialis*, but that name already belonged to another *Primula* species

60. *Bull.Soc.Bot.France* (1885) p.270 as *P. glacialis*; Richards, J. (1993) pp.170-171

61. *Bull.Soc.Bot.France* (1886) p.389

62. Grey-Wilson, C. (2000) *Poppies A Guide to the Poppy Family in the Wild and in Cultivation*, revised and updated edition pp.169-170; *Bull.Soc.Bot.France* (1886) p.390 as *Cathcartia delavayi*

63. Grey-Wilson, C. (June-July 1987) *Journey to the Jade Dragon Snow Mountains, AGS Bulletin* p.221

64. *Jnl.Soc.Bot.France* (1898) p.222; Noltie, H. (1992) *Kew Mag./Curtis'sBot.Mag.* 9 No.193 pp.51-54

65. *Bull.Soc.Bot.France* (1886) p.379: the date is given as July 1883 but this must be an error, as Père Delavay did not visit Lijiang until 1884. *Bot.Mag.* NS (1949) 68; FOC 6:249

66. Forrest, G. (1920) *Gard.Chron.* 68 p.97

67. *Bull Soc.Bot.France* (1886) pp.382-385; *Pl.Del.* p.32; Finet, A. and Gagnepain, F. (1905) *Contributions à la Flore de l' Asie Orientale*

68. Letter 139 dd 13.10.1889 to Franchet, Franchet Archive MHN

69. *Bull.Soc.Linn.Paris* (July 1886) pp.612-614; *Fruticetum Vilmorianum* (1904) pp.184-185; *Bot.Mag.* (1912) No.8459; Green, P. (1955-1958) *Revision of Osmanthus in Asia and America, NotesRBGE* 22 pp.532-553

70. *Jnl.Soc.Bot.France* (1896) p.305 as *Panax delavayi*; *Brittonia* (2001) 53:117; Hinkley, D. (2009) p.249; *New Trees* (2009) p.778

71. Letter 104 dd. 25.2.1885 to David, Franchet Archive MHN

72. Letter 129 dd 14.2.1888 to Franchet, Franchet Archive MHN

73. Hinkley, D. (2006) *Thalictrum: An Overview, The Plantsman* (NS) 5 (3) pp.78-184

74. Page, M. (2005) *The Gardener's Peony* Timber Press p.226; Stern, F.C. (1946) *A Study of the Genus Paeonia*

CHAPTER SEVEN pages 86-95

1. Quoted in Lancaster, R. (2008) *Plantsman's Paradise* p.158
2. Letter 104 dd 25.2.1885 to David, Franchet Archive MHN; Lennon, J. (2004) quoted on p.27
3. Letter 104 dd 25.2.1885 to David, Franchet Archive MHN
4. *Gentiana du Yunnan* (1884) *Bull.Soc.Bot.France* pp.373-378; Lancaster, R. (2008) p.140
5. *Bull.Soc.Bot.France* (1896) pp.483-495; *Bot.Mag.* (1922) No.8974
6. *Bull.Soc.Bot.France* (1885) p.364
7. Op.cit. p.272
8. *Bull.Soc.Bot.France* (1886) pp.61-62
9. *Rhododendrons du Thibet Oriental et du Yunnan, Bull. Soc.Bot.France* (1886) pp.223-236; op.cit. (1887) pp.280-285
10. *Jnl.Soc.Bot.France* (1895) pp.367-368
11. First described by Franchet as *S. saligna* var. *chinensis Pl.Del.* p.135; *Pl.Wils.* 2:163-164; *Bot.Mag.* (1924) No.9045; FOC does not recognise var. *chinensis* 11:330; Sealy, J. *Species of Sarcococca in Cultivation RHSJnl* (1949) 74 pp.301-304
12. Letter 106 dd 16.3.1885 to Bureau, Franchet Archive MHN
13. *Pl.Del.* p.203 as *R. polytrichus*; *Pl.Wil.I:49*; *Gard.Chron.* (1912) 151 p.167; *Bot.Mag.* (1938-9) No.9534; Bean, W. (1914-1915) *Chinese Trees and Shrubs, RHSJnl* 40 p.218
14. Franchet originally named the plant *Delavaya toxocarpa* and another collection *D. yunnanensis*, but botanists now treat *D. yunnanensis* as a synonym of *D. toxocarpa*: *Bull. Soc.Bot.France* (1886) p.462; *Pl.Del.* p.142 pl.27-28
15. Letter 107 dd 24.9.1885 Franchet Archive MHN All Père Delavay's letters are now addressed to Franchet.
16. Letter 108 dd 13.1.1886 Franchet Archive MHN
17. Letter 109 dd 4.4.1886 Franchet Archive MHN
18. *Pl.Del.* p.228; *Bot.Mag.* (1915) No.8629; *Gard.Chron.* (1912) 52 p.288: all as *Pyrus yunnanensis*
19. Franchet *Pl.Del.* p.33; Wilson, E.H. (1906) *The Chinese Magnolias, Gard.Chron.* 39 p.234, also (1920) *Romance of our Trees* New York p.151; *Bot.Mag.* (1909) No.8282; Callaway, D. (1994) *Magnolias* p.120; Gardiner, J. (2000) *Magnolias A Gardener's Guide* Timber Press pp.134-136, also (2002) *Global Magnolias, The Plantsman NS* 1 p.75
20. Letter 111 dd 19.5.1886 Franchet Archive MHN
21. *Jnl.Soc.Bot.France* (1896) p.314; Pettit, S. (2006) *Curtis'Bot.Mag.* 23 1 pp.51-55; *Rev.Hort.* (1891) pp.246-247; FOC 19:646 as *Linnaeaceae*
22. *Bull.Soc.Bot.France* (1887) p.281
23. *Bull.Soc.Bot.France* (1886) p.390 as *Cathcartia lancifolia*; Grey-Wilson, C. (2000) pp.186-188
24. *Pl.Del.* p.42; Grey-Wilson, C. (2000) pp.161-63
25. Letter 119 dd 4.7.1887 Franchet Archive MHN
26. Letter 112 dd 6.9.1886 Franchet Archive MHN
27. In 1913-1914: Scottish Rock Garden Club (1935) p.72
28. Letter 113 dd 11.11.1886 Franchet Archive MHN
29. According to Père Delavay's obituary, plague only left the district after he anointed his congregation and his church with water from Lourdes, Archives MEP
30. Hinkley, D. (2009) *The Explorer's Garden* Timber Press pp.285-286; *Pl.Wils.II:17*
31. Lancaster, R. (2008) p.115
32. Grey-Wilson, C. (June 2009) *Bailey's Blue Poppy Restored, Bull.AGS* 77 pp.215-225
33. Loc.cit pp.221-222

CHAPTER EIGHT pages 96-119

1. Henry, A. *Botanical Exploration in Yunnan, Kew Bulletin* (1897) pp.99-101
2. Letter 114 dd 12.12.1896 Franchet Archive MHN
3. Letter 112 dd 6.9.1886 Franchet Archive MHN
4. *Bull.Soc.Bot.France* (1887) p.282
5. *Bull.Soc.Bot.France* (1885) p.4 as *Saxifraga delavayi*; FOC 8:278; see RHSDatabase
6. *Pl.Del.* p.231; Letter 123 dd 6.11.1887 Franchet Archive MHN
7. Letter 124 dd 30.11.1887 Franchet Archive MHN
8. Sealy, J.R. (1958) *A Revision of the Genus Camellia* p.63; Trehane, J. (2008) *The Plantsman* 1 7 p.47; *New Trees* (2009) pp.197-198
9. *Pl.Del.* p.1: (Sept.1882) No.2 4; *Bull.Soc.Bot.France* (1903) p.537; *Pl.Wils.I:325*; Grey-Wilson, C. (2006) *Curtis'sBot.Mag.* 23 (3). No.560. pp.214-217
10. *Jnl.Soc.Bot.France* 1888 pp.67-71; Shanshan, L. *High Genetic Diversity vs. Low Genetic Differentiation in Nouelia insignis, Annals of Botany* (2006) 98 pp.583-589 www.aob.oxfordjournals.org
11. Letter 119 dd 4.7.1887 Franchet Archive MHN
12. Letter 120 dd 13.8.1887 Franchet Archive MHN
13. Davies, H. (1909) *Yun-nan The Link between India and the Yangtse* Cambridge UP
14. *Jnl.Soc.Bot.France* (1899) pp.255,261; Masters, M. (1906) *Chinese Conifers, Gard.Chron.*39 pp.212-213; *Pl.Wil.II:41-42*; Orr, M. (1933-1935) *Plantae Chinenses Forrestianae, Notes.RBGE* XVIII pp.119-158; Horsman, J. (1984-1985) *Silver firs in Cultivation The Plantsman* 6 pp.64-100
15. Craib, W. (1918-1919) *Abies delavayi in Cultivation, Notes.RBGE* XI pp.277-280; Masters, M. (1903) *Gard. Chron.* 33 p.194; Hunt, D. (1967) *Abies delavayi Group Jnl.RHS* p.263; FOC 4:48 lists only 2 varieties

16. Wilson, E.H. (1926) *The Taxads and Conifers of Yunnan, Jnl.ArnoldArb.* pp.48-49, *Pl.Wils.*II:22 et seq. for *Picea* and other conifers; Rushforth, K. (1983) *Abies delavayi and A. fabri, Intl. Dendrology Society Yearbook* pp.118-120; *New Trees* (2009) pp.48-49 for summary of current taxonomy

17. Letter 121 dd 6.10.1887 Franchet Archive MHN

18. Letter 123 dd. 6.11.1887 Franchet Archive MHN

19. Wright, D. (1980) *Philadelphus, The Plantsman* 2 (2) p.106

20. Henry, L. *Rev.Hort.* (1903) pp.12-14; *Bot.Mag.* (1923) No.9022; *P. purpurascens* is closely related but for taxonomy see *FOC* 8:398,403 and RHSDatabase

21. *Pl.Del.* p.156; Lancaster, R. (2008) pp.265-266; Hinkley, D. (2009) p.326; Letter 128 dd 26.1.1888 Franchet Archive MHN

22. *Rev.Hort.* (1895) pp.64-65 Maxim Cornu corrects erroneous information in earlier articles by M. Micheli: *Rev.Hort.* (1893) p.235 and (1894) p.244; *Garden and Forest* (1894) p.284

23. *Pl.Wil.*I:149; Lancaster, R. (2004) *Desirable Deutzias, RHSTheGarden* 129 pp.634-635; *FOC* 8:392 and RHS Database

24. Letter 493 dd 20.6.1892 from Vilmorin, Franchet Archive MHN

25. Farrer, R. (1919) *The English Rock Garden* 1 p.434; *Jnl. Soc.Bot.France* 1893 pp.137-140; *Rev.Hort.* (1893) pp.544-545; Grey-Wilson, C. (1994) *A Survey of Incarvillea in Cultivation New Plantsman* 1 pp.36-52 and (1998) *A New Look at Incarvillea subgenus Pterosceleris* 5 pp.76-98

26. Chittenden, F. (ed.) (1936) *Rock Gardens and Rock Plants, Report of RHS and AGS Conference* p.77

27. Micheli, M. (1895) *Iris delavayi, espèce nouvelle de Yunnan. Rev.Hort.* pp.398-399

28. Letter 129 dd 14.3.1888 Franchet Archive MHN

29. Letter 130 dd 26.5.1888 Franchet Archive MHN

30. Letter 126 dd 30.12.1887 Franchet Archive MHN

31. Letter 105 dd 27.10.85 to Père David, Franchet Archive MHN

32. *Jnl.Soc.Bot.France* (1899) pp.353-354; Wilson, E.H. (1926) p.43; *New Trees* (2009) pp.628-629

33. Bean (1970) p.668; *Jnl.Soc.Bot.France* (1895) p.370; *Bot. Mag.* (1922) No.8970; Hoffman, M. (2010) *An Assessment of Clethra, The Plantsman* NS 9 (3) pp.190-191

34. *Bull.Soc.Bot.France* (1886) pp.61-62; *Gard.Chron.* (1887) I p.574; *Bull.Soc.Bot.France* (1898) p.180

35. *FOC* 15:185

36. *Notes.RBGE* (1909-1913) V p.214

37. Letter 132 dd 12.9.1888 Franchet Archive MHN

38. *Jnl.Soc.Bot.France* (1892) p.317; Synge, P. (1980) *A Revision of Elwes' Monograph of the Genus Lilium and its Supplements* pp.90-91

39. *Jnl.Soc.Bot.France* (1892) p.319; Synge, P. (1980) p.133

40. Loc.sit. p.314, p.58; Wilson, E.H (1925) *The Lilies of Eastern Asia* pp.42-43; *FOC* 24:140

41. Letter 132 dd 12.9.1888 MHN; *Bull.Soc.Philomathique. Paris* (1890-1891) 8e III p.148; Père Vial wrote an account of the journey: (1887) *Voyage au Tonkin*, Notice Biographique Archives MEP

42. Colquhoun, A.R. (1883) *Across Chrysé … a Journey of Exploration through the South China Borderlands* II p.268

43. The Yi people are often referred to as Lolo in older works.

44. Farrer, R. (1919) II p.150

45. *Jnl.Soc.Bot.France* (1896) p.310; *Bot.Mag.* (1919) No.8800; Wright, D. *Climbing Honeysuckles, The Plantsman* 4 pp.236-252; Bradshaw, D. (1996) *Lonicera* pp.10-11

46. syn. *C. lacteus* W.W. Smith. *Pl.Del.* p.222

47. Fryer, J. and Hylmö, B. (2009) *Cotoneasters A Comprehensive Guide to Shrubs for Flowers, Fruit, and Foliage* Timber Press pp.47-48

48. Letters 133 dd 15.10.1888 and 134 dd 7.1.1889 Franchet Archive MHN

49. Letter 145 dd 25.4.1890 Franchet Archive MHN

50. Letters 137 dd 13.7 and 137bis 17.9, both 1889 Franchet Archive MHN

51. *Bull.Soc.Bot.France* (1886) p.109; *Pl.Del.* p.196; *Pl.Wils.*I:213, 253

52. *Jnl.Soc.Bot.France* (1896) p.307 as *Heptepleurum delavayi*; *Bot.Jahrb.* (1900) 9:486; Hinkley, D. (2009) p.249; *New Trees* (2009) p.778

53. Letter 138bis dd 3.9.1889 Franchet Archive MHN

54. Letter 141 dd 11.12.1889 Franchet Archive MHN

55. Letter 142 dd 11.1.1890 Franchet Archive MHN; *Bot. Mag.* (1893) No.7301; *Gard.Chron.* (1910) 47 p.343

56. Boucher, G. *Rev.Hort.* (1901) pp.495-497; *Bot.Mag.* New Series (1949) No.60; *Pl.Wils.*II:602; Bean, (1973)

57. Bois, D. (1904) *Fruticetum Vilmorianum* pp.103-104 as *Cornus foliolosa*; *Bot.Mag.* (1909) No.8241; various forms have been collected in the wild: McAllister, H. (2005) *The Genus Sorbus Mountain Ash and other Rowans* RBGKew pp.183-184,190-191; *New Trees* (2009) p.809

58. *Bull.Soc.Bot.France* (1893) p.113; Bois, D. (1904) pp.99-100; *Bot.Mag.* (1908) No.8218; Vilmorin, M. (1902-1903) *Some Wild Asiatic Roses, RHSJnl.* p.490; *Pl.Wils.* II:332-334; Botanists differ on the status of these roses:

FOC 9:354 has four forms, does not recognise subspecies and lists *R. omeiensis* as a separate species 9:354; GRIN has *R. sericea* f. *pteracantha*; I have followed the RHS Database; *Gard.Chron* (1905) 38 pp.260-261.

59. *Jnl.Soc.Bot.France* (1899) pp.197-200; *Pl.Wils.*II:144; Taylor, N. and Kirkham, T. (2005) *Corylus* x *vilmorinii Curtis'sBot.Mag.* 4 Plate 541 pp.225-228

60. Letter 142 dd 11.1.1890 Franchet Archive MHN

61. *Bull.Acad.Imp.Sci.Saint-Pétersbourg* (1888) 32; *FOC* 18:97

62. Letter 145 dd 25.4.1890 Franchet Archive MHN

63. *Bot.Mag.* (1900) No.7708; Wright, D. (1979) *Deutzias – A Garden Evaluation, The Plantsman* 1 3 pp.154-166; Wilson, E.H. (1 Jun. 1927) *Deutzias, Bull.ArnoldArb.*

64. Jefferson Brown, M. (2003) *Lilies a Guide to Choosing and Growing Lilies* RHS p.78 et seq.; www.chrisnorthlilies.org.

CHAPTER NINE pages 120-131

1. Letter 28 dd 7.8.1891 Franchet Archive MHN

2. Letters 146 dd 24.9.1890, 118 dd 5.5.1887, 121 dd 6.11.1887, 130 dd 26.5.1888 and 136 dd 16.6.1889, Franchet Archive MHN

3. Letter 149 dd 24.2.1891 Franchet Archive MHN; *Jnl.Soc. Bot.France* (1895) pp.399-400

4. Letter 149 dd 24.2.1891 Franchet Archive MHN

5. Letters 152 dd 26.9, 'nos chers travaux': 153 dd 17.10, 154 dd 9.11, 155 dd 22.11, all 1891, 156 dd 23.10.1892 Franchet Archive MHN

6. Letter 157 dd 4.2.1893 Franchet Archive MHN

7. Letters 158 dd 20.2, 159 dd 1.3, 160 dd 11.4 all 1893 Franchet Archive MHN

8. *Bull.Soc.Bot.France* (1892) and *Jnl.Soc.Bot.France* (1896) as Senecio, though now Cremanthodium, Ligularia etc.

9. Letter 164 dd 10.10.1893 Franchet Archive MHN

10. Letter 162 dd 15.10.1893 Franchet Archive MHN

11. Letters 34 dd 18.2 and 35 dd 19.3 both 1894 Père Bodinier to Franchet, Franchet Archive MHN

12. Letters 165 dd 29.5 and 166 29.9, both 1894 Franchet Archive MHN

13. Letter 169 dd 11.6.1895 Franchet Archive MHN

14. *Jnl.Soc.Bot.France* (1898) p.254; *Pl.Wils.*I:287; Hsu, E. and Houtman, R. (2007) *Stachyurus in Cultivation, RHS The Plantsman (NS)* 6 1 March pp.24-32; Hinkley, D. (2009) pp.302-303

15. *Bull.MHN* (1895) 1 pp.65-66

16. Ibid. p.63; Section on Podophyllum by J.H. Shaw in Stearn, W. (2002) *The Genus Epimedium* RBGKew pp.292-295; *FOC* 19 as *Dysosma delavayi*

17. *Repert.Spec.Nov.RegniVeg.* (1909) 7:339

18. *Novon* (1997) 7:263; *New Trees* (2009) pp.843-844; *Bull. Acad. Int. Géogr. Bot.* 25:42 1915

19. *Bot.Jahrb.* (19124) 7 pp.657-658; *Pl.Wils.*III:237; *New Trees* (2009) pp.751-752: as *Cyclobalanopsis glaucoides FOC* 4:400

20. *Bull.Soc.Dendrologique.France.* (1912) 23-24:58; Bean; *New Trees* (2009) pp.673-674

21. Levéillé, H. (1915-16) *Catalogue des Plantes du Yun-nan Le Mans* p.20: specimen collected May 1911; Grierson, A. (1959-1961) *A Revision of the Genus Incarvillea, Notes. RBGE* 23 pp.303-354

22. *New Trees* (2009) pp.494-495

23. *Finet et Gagnepain* (1906) *Mémoires.Soc.Bot.France* 4 pp.43-44; *Blumea* (2007) 52:562; Figlar D. (2009) *The Sinking of Michelia and Manglietia into Magnolia, RHS The Plantsman* 2 pp.118-123:cf. *Flowers of Western China* 2011

24. Doc.173 Report of Dr Deblenne, Franchet Archive MHN

25. Letter 46 17.3.1897 from Père Bodinier, Franchet Archive MHN; The Yi people used *G. delavayi* to make a waterproof fabric and Père Delavay had written an account of the process to Franchet, who published a paper on the subject: *Jnl.Soc.Bot.France* (1889) p.16

26. *Bot.Mag.* (1943-48) No.9658

27. Letters 46 dd 17.3 from Père Bodinier and 183 dd 13.5 from Père Ducloux: both 1897 Franchet Archive MHN

28. *Bull.Soc.Bot.France* (1907) pp.164-165

29. http://www.kew.org/news/paraisometrum-mileense. htm; Weitzman, A.L. et al (1998) Chinese Gesneriaceae, *Novon* 7 p.434

30. *Bull.Soc.Bot.France* (1908) p.87; syns. *M. mairei, Berberis duclouxiana*; *Pl.Wils.*I:384; Hinkley, D. (2009) p.222

31. Mazzetti-Handel, H. *A Botanical Pioneer in South West China* (1927) Vienna Translated and published David Winstanley (1996) p.88

32. Gagnepain, F. (1901) *Sur la Nouvelle Collection Ducloux du Yunnan, Bull.MHN* 7 pp.80-83; *Jnl.Soc.Bot.France* (1899) p.263; *Bot.Mag.* (1924) No.9049; Orr, M.V. (1933-35) *Plantae Chinenses Forrestianae*: Coniferae, Notes. *RBGE* XVIII pp.150-151; Farjon, A. (2005) *A Monograph of Cupressaceae and Sciadopitys* RBGKew pp.193-197

33. Letter 169 dd 11.6.1895 Franchet Archive MHN

CHAPTER TEN pages 132-145

1. *Jnl.Bot.* (1878) p.6

2. Mgr. Guichard *Rapport annuel* 1884 Archives MEP

3. Bourne, F. (1888) *Parliamentary Paper, China* (1) No.813 p.77

4. *Rapport annuel Kouy Tchéou* 1886 Archives MEP; de Launay, A. (1902) *Histoire des Missions de Chine Kouytcheou* Paris

5. Letter 45 dd 19.71896 Franchet Archive MHN

6. *Jnl.Soc.Bot.France* (1890) pp.301-306; Letter 66 dd 4.1892 from Bretschneider, Franchet Archive MHN

7. Letter 28 dd 7.8.1891 to Père Delavay, Franchet Archive MHN

8. Letter 28 dd 7.8.1891 to Père Delavay, Franchet Archive MHN

9. *Bull.MHN* (1896) 2 p.278

10. Letter 42 dd 26.11.1895 Franchet Archive MHN

11. Letter 47 dd 12.12.1897 Franchet Archive MHN

12. Letter 45 dd 19.71896 both Franchet Archive MHN

13. His title was Provicaire

14. *Bull.Acad.Internat.Géog.Bot.* (1903) p.121

15. *C. bodinieri* was actually discovered by Augustine Henry in Hubei in 1886: O'Brien, S. (2011) p.45. Moldenke, H.N. (14 Nov. 1966) *Additional Materials toward a Monograph of the Genus Callicarpa, Phytologia* pp.49-54; the specimen collected in 1905 by Père Esquirol was named *C. feddei* by H. Levéillé

16. Stearn, W.T. (2002) *The Genus Epimedium and other Herbaceous Berberidaceae* RBGKew p.107

17. Letters 48 dd 18.5 and 49 dd 16.12 both 1899 Franchet Archive MHN; *Bull.Acad.Intern.Géog.Bot.* (1901-1902) p.191

18. *Repert.Spec.Nov.Regni.Veg.* (1911) 10 p.148; *New Trees* (2009) p.191

19. *New Trees* (2009) pp.528-529

20. *Rep.Spec.Nov.Regni.Veg.* (1912) 10 p.371 *as Lindera cavaleriei*; Lancaster, R. (2010) *RHS The Garden* 135 10 p.681.

21. FOC 24:87; *New Trees* (2009) p.483; *Repert.Spec.Nov. RegniVeg.* (1912) 11 p.296 as *Celtis polycarpa*; FOC 11:218; *New Trees* (2009) pp.173-174; Archives MEP

22. Forbes, F.B and Hemsley, W.B. *Index Florae Chinensis An Enumeration of all the plants known from China Proper... Jnl.Linn.Soc* published in 3 parts: 23 (1886-1888), 26 (1889) and 36 (1902)

23. *Bull.Soc.Bot.France* (1886) Reviews

24. Letter dd 16.12.1885 DC Correspondence 151:583 Kew Archives 1966

25. Henry, A. (1902) *Midst Chinese Forests, The Garden* 31 pp.3-6; for further details concerning Henry see: Pim, S. (1966) *The Wood and the Trees A Biography of Augustine Henry* and revised edition (1984) Boethius Press Kilkenny; Nelson, E.C. (1983) *Augustine Henry and the Exploration of the Chinese Flora, Arnoldia* 43:1 pp.21-38; Morley, B. (1979) *Augustine Henry:*

His Botanical Activities in China 1882-1890, Glasra 3 pp.21-81, 52; O'Brien, S. (2011) *In the Footsteps of Augustine Henry and his Chinese Plant Collectors* Garden Art Press

26. *Jnl.Bot.* (1884) p.76

27. *Gard.Chron.* (1905) 38 p.323; Letter dd 17.9.1887 DC Correspondence 151:601 Kew Archives; O'Brien, S. (2011) p.58

28. *Jnl.Linn.Soc.* (1902) p.55; *Notes.RBGE* (1919) 11 p.278; *Pl.Wils.II:41*

29. Hemsley, W.B. (1892) *Observations on a Botanical Collection made in Western China, with Descriptions of Some New Chinese Plants from Various Collections, Jnl.Linn.Soc.Bot.* pp.298-321; *Bot.Mag.* (1912) No.8471

30. *Jnl.Linn.Soc.Bot.* (1889-1902) 26 p.22; Cox, P. and Cox, K.E. (1997) *Encyclopaedia of Rhododendron Species* Glendoick Publishing pp.166-167; Cullen, J. (2005) *Hardy Rhododendron Species* Timber Press p.156

31. *Icon.Plant.* (1890) XX No.1934

32. Engler, A. *Pflanzenr.* (1907) IV 241 (Heft 30) p.33; *New Trees* (2009) pp.829-830

33. *The Chinese Recorder* (1873) 4 p.107, (1875) 6 p.236, (1886) 17 p.242, (1899) 30 pp.581-583; Bretschneider, E. (1898) 1:954; *History of Protestant Missions* (1901); *Neue Deutsche Biographie*

34. Pratt, A.E. (1892) *To the Snows of Tibet through China*; Letter dd 14.2.1889 DC Correspondence 151:631 Kew Archives; O'Brien, S. (2011) p.68

35. As recounted by Alice Henry in Pim, S. (1966) p.45; O'Brien, S. (2011) p.80

36. *Icon.Plant* (1891) No.1961; *Jnl.Linn.Soc.Bot.* (1901-1904) 35 pp.556-559

37. Morley, B. (1980) *Augustine Henry, The Garden* pp.285-289; Andrews, S. and Nelson, E.C. (1986) *Augustine Henry's Plants in Kew Gardens, Kew Magazine* 33 (3) pp.136-140

38. O'Brien, S. (2011) p.190

39. Letter dd September 1897 in Nelson, E.C. (1982-1983) p.33

40. O'Brien, S. (2011) p.237

41. Henry, A. (1897) *A Budget from Yunnan, Kew Bulletin* pp.407-414

42. Spongberg, S. (1990) *A Reunion of Trees* Harvard UP p.172; Cox, E. (1946) *Plant-hunting in China* p.138

43. Shephard, S. (2003) *Seeds of Fortune A Gardening Dynasty* pp.237-240

44. Nelson, E.C. (1983) p.34

45. Op.cit. p.29

46. Correvon H. (ed. Barron L.) (1930) *Rock Gardens and Alpine Plants* p.7; Rockley, Lady (1936) *Rock Gardening*

History, RHS.Jnl. pp.11-19; Thomas G. S. (1989) *The Rock Garden and its Plants from Grotto to Alpine House*

47. Maclean, B. (2004) *George Forrest Plant Hunter* Antique Collectors' Club pp.25-31
48. Nelson, E.C. (1983) p.35
49. Barker, D. (2007) *Epimediums and other Herbaceous Berberidaceae* Hardy Plant Society, revised edition p.42; Rice, G. (March 2008) *Raising Rising Stars, RHSTheGarden* pp.160-165

CHAPTER ELEVEN pages 146-163

1. 'As this part of China is not very well known to botanists …. interesting specimens might be obtained.' Augustine Henry in a letter to Thistleton-Dyer, 20th March 1885 DC Correspondence 151:578 Kew Archives
2. *De Incarnatione Verbi Dei* 3 Quoted in Mcgrath, A. (2001) *A Scientific Theology* Edinburgh and New York *Nature* p.253
3. Letter 196 dd 2.6.1891 Franchet Archive MHN: Père Farges' letters to Franchet are numbered 195-204.
4. Little, Mrs Archibald (1902) *The Land of the Blue Gown* pp.125-126
5. La Tourette, K. (1929) *A History of Missions in China* SPCK p.344
6. Obituary, Archives MEP
7. Letter 197 dd 5.2.1892 Franchet Archive MHN
8. *Bull.Linn.Soc.Paris* (1893) p.1067; *Kew.Bull.* (1898) pp.313-317; Whittaker, P. (2005) *Hardy Bamboos* Timber Press pp.105-112; *Pl.Wils.*II:64
9. O'Brien, S. (2011) *In the Footsteps of Augustine Henry* Garden Art Press p.50
10. *Jnl.Soc.Bot.France* (1892) pp.233-235; *Rev.Hort.* (1893) p.447 and (1900) pp.270-274; *Bot.Mag.* (1902) No.7848; *Pl.Wils.*I:344; *Gard.Chron.* (1923) 73 p.161; *FOC* 6:440; Christenhusz, M. (2012) *An Overview of Lardizabalaceae, Curtis'sBotMag.* 9:3 p.246
11. Letter 493 dd 20.6.1892 from Vilmorin, Franchet Archive MHN
12. Mottet, S. (1899) *Rev.Hort.* pp.475-476; *Bot.Mag.* (1900) No.7715; Rafill, C. (1938) *L. davidii, duchartre (sic) and its varieties, Gard.Chron.*104 p.230.
13. Wilson, E.H. (1925) *The Lilies of Eastern Asia* pp.85, 67-69, 85
14. *L. fargesii* was discovered by A. Henry: O'Brien, S. (2011) p.77
15. *Les Lis de la Chine et du Thibet dans l'herbier du Muséum de Paris, Jnl.Soc.Bot.France* (1892) pp.305-321;

Synge, P. (1980) *Lilies: A Revision of Elwes' Monograph of the Genus Lilium and its supplements* p.71-73; Jefferson-Brown, M. (2003) *Lilies A Guide to Choosing and Growing Lilies* RHS pp.74-78

16. Letter 164 dd 10.10.1893 from Père Delavay, Franchet Archive MHN
17. Letter 195 dd 18.3.1893 Franchet Archive MHN; Fournier 1932 p.57
18. Letter 198 dd 10.1.1894 Franchet Archive MHN; Franchet, A. (1894) *Plantes Nouvelles de la Chine Occidentale, Jnl.Soc.Bot.France*
19. Op.cit. p.293 as *Acer nikoense* var. *griseum*
20. Op.cit. p.294; *Repert.Spec.Nov.Regni.Veg.* (1905) 1 p.6; *New Trees* (2009) pp.88-90; *A. fargesii* Veitch ex Rehder is a synonym for *A. fabri*
21. O'Brien, S. (2011) pp.50, 86
22. *Jnl.Soc.Bot.France* (1899) p.256; *Gard.Chron.* (1905) 38 p.355; Master, M. (1906) *Chinese Conifers, Gard. Chron.* 39 p.212; *Pl.Wils.*II:42-43, 48; Orr, M. (1933-1935) *Plantae Chinenses Forrestianae: Coniferae, Notes. RBGE* 18 pp.119-158; Horsman, J. (1984-1985) *Silver Firs in Cultivation, The Plantsman* 6 pp.78-79; *FOC* 4:47 lists three varieties, but other botanists recognise four: see GRIN/The Plant List
23. O'Brien, S. (2011) p.86; *Jnl.Soc.Bot.France* (1899) p.258 as *Abies brachytyla*
24. *Pl.Wils.*II:33-36; *Bot.Mag.* (1922) No.8969; Horsman, J. (1983-84) *Spruces in Cultivation, The Plantsman* 5 108; *FOC* 4:31-32; Wilson, E.H. (1926) *The Taxads and Conifers of Yunnan, Jnl.ArnoldArb.* pp.37-68
25. *Pl.Wils.*II:57-60; Wilson, E.H. (March 1924) *New and Rare Conifers, The Garden* pp.141-142; *FOC* 4:73
26. *Pl.Wils.*II:37-38; Wilson, E.H. (1926) p.50
27. *Jnl.Soc.Bot.France* (1899) p.264; O,Brien, S. (2011) p.143
28. *Bot.Mag.* (1910) No. 8347
29. *Rev.Hort.* (1896) p.497; Bean, W.J. *Garden Notes of New Trees and Shrubs, Kew.Bull.* (1909) pp.353-354; *Bot.Mag.* (1949) No.53; Wharton, P. and Lancaster, R. (2007) *The Reintroduction of the Goat Horn tree Carrierea calycina, The Plantsman* NS 6(4) pp.250-255
30. *Nouv.Arch.MHN* (1894) Ser.3 VI p.195; *Bot.Mag.* (1936) No.9458; O'Brien, S. (2011) p.50
31. Burkill, I.H. *Jnl.Linn.Soc.Bot.* (1899) p.497; *Jnl.Soc.Bot.France* (1899) pp.205-208; Ashburner, K. (1980-1981) *Betula A Survey, The Plantsman* 2 pp.31-62; Ashburner, K. and McAllister, H. (2013) *The Genus Betula* pp.172-176, 261-266, 381-382; O'Brien, S. (2011) pp.85-86
32. *Jnl.Soc.Bot.France* (1899) p.206; *New Trees* (2009) pp.169-170; Lancaster, R. (2010) *RHSTheGarden* 135 (10)

p.681; Ashburner, K. and McAllister, H. (2013) pp.176-181

33. *FOC* 18:10; *Bull.MHN* (1896) 2 pp.279-280; Flanagan, M. and Kirkham, T. (2009) *Wilson's China A Century On* RBGKew p.216; O'Brien, S. (2011) pp.77,155

34. *New Trees* (2009) pp.204-205, 274

35. *con.Plant.* (1905) No.2787

36. O'Brien, S. (2011) p.60

37. O'Brien, S. (2011) p.87; *Jnl.Soc.Bot.France* (1895) p.390 as *R. fargesii*; *Pl.Wils.*II:540 and I:542 as *R. discolor*, which is still maintained as a separate species by *FOC* 14:338

38. *Pl.Wils.*I:524

39. *Jnl.Linn.Soc.Bot* (1899) XXVI p.528-529; *Pl.Wils.*III:47; Newsome, C. (1992) *Willows The Genus Salix* pp.74-75; O'Brien, S. (2011) pp.80-1,11

40. *Jnl.Soc.Bot.France* (1898) p.255; *Pl.Wils.* I:77; *Bot.Mag.* (1946) No.9670; Andrews, S. (1986) *The Ilex fargesii Complex, Kew Magazine* 3 (3) pp.122-135; Bailes, C. (2006) *Hollies for Gardeners* Timber Press pp.196-200; O'Brien, S. (2011) p.87

41. *Jnl.Soc.Bot.France* (1895) p.371; *Bot.Mag.* (1935) No. 9413

42. *Jnl.Soc.Bot.France* (1895) p.369; *Pl.Wils.*I:502; Hoffman, M. (2010) *An Assessment of Clethra, The Plantsman* NS 9 (3) p.191; the taxonomic status of *C. bodinieri, C. cavaleriei* and *C. fabri* is still uncertain: cf. *FOC* 14:238, GRIN and The Plant List

43. Now a synonym of *C. trichotomum* var. *trichotomum*; *Pl.Wils.*III:376-377

44. *Jnl.Soc.Bot.France* (1896) p.283 and *Pl.Dav.*I:124

45. *Jnl.Soc.Bot.France* (1896) pp.281-282 (syns *D. fargesii* and *D. bodinieri*); *Rev.Hort.* (1897) p.486 as *D. corymbiflora* and 1895 p.65; *Gard.Chron.* (1898) 23 pp.265-267; Lemoine, É. (April 1902) *Monographie Horticole du Genre Deutzia* Paris pp.8-16; *Bot.Mag.* (1909) No.8255

46. *Ornamental Flowering Trees and Shrubs Report of the RHS Conference April 26-29 1938 RHS.Jnl.* (1939) pp.121-123; See GRIN *Species Records for Deutzia* for summary of Lemoine's hybrids and current synonymy

47. Wright, D. (1979) *Deutzias – A Garden Evaluation, The Plantsman* 1 3 pp.154-166; *Pl.Wils.*I:12

48. Compare RHS Horticultural Database with FOC and GRIN

49. Thomas, G.S. (1992) *Ornamental Shrubs and Climbers* p.310; Hu, S.Y. (1955) *Monograph of the Genus Philadelphus, Jnl.ArnoldArb.* p.355; *Bot.Mag.* (1922) No.8941; Taylor, J. (1989-1990) *Four Lemoine Genera, The Plantsman* II pp.229-240

50. *Pl.Wils.*I:235-236; Grey-Wilson, C. (2000) *Clematis The Genus* p.45

51. syns. *C. x fargesoides* 'SUMMER SNOW'; Savill, R. (2005) *Clematis Breeding in the Crimea, The Plantsman* (NS) 4 (2) pp.84-91

52. *Jnl.Soc.Bot.France* (1894) p.278; (1984-1985) *Some notes on Actinidias and their Propagation, The Plantsman* 6 pp.167-180

53. Loc.sit

54. Vilmorin, M. (1898) *Incarvillea grandiflora, Rev.Hort.* pp.330-331; Grierson, A. (1959-1961) *A Revision of the Genus Incarvillea, Notes.RBGE.* 23 p.341; Grey-Wilson, C. (1994) *A Survey of Incarvillea in Cultivation, New Plantsman* 1 p.47

55. *Jnl.Soc.Bot.France* (1898) pp.301 et seq. Franchet only knew of 6 American species; *FOC* 5:250,252 and RHS Database recognise *A. chinense* and *A. debile*, although GRIN does not

56. *Jnl.Soc.Bot.France* (1898) pp.281-283; Stearn, W.T. (2002) *The Genus Epimedium and other Herbaceous Berberidaceae* RBG Kew p.72 et seq.

57. *Bull.Soc.Bot.France* (1886) p.109; Hinkley D. (1999) *The Explorer's Garden* Timber Press p.99 et seq.; Stearn, W.T. (2002) p.91; Lancaster, R. (2008) pp.92-93

58. White, R. (April 1996) *Epimedium: Dawning of a New Era, RHSThe Garden* pp.208-214; Barker, D. (March 2007) *Epimediums and other Herbaceous Berberidaceae* Hardy Plant Society revised ed. p.42; Rix, M. (April 2007) *The Second Coming, Gardens Illustrated* 124 pp.64-71; Rice, G. (March 2008) *Raising Rising Stars, RHSTheGarden* pp.160-165

59. subsp. *pseudointegrifolia*: Grey-Wilson, C. (2000) *Poppies A Guide to the Poppy Family in the Wild and in Cultivation* revised and updated edition p.196

60. *Jnl.Soc.Bot.France* (1894) p.267; Cribb, P. (1997) *The Genus Cypripedium* RBG Kew/Timber Press pp.258-260 and (2005) *Wild Orchids in Sichuan, Curtis'sBot.Mag.* 1 (22) p.74

61. *Rev.Hort.* (1911) pp.197-199; *Bot.Mag.* (1920) No. 8861; Gusman, G. and L. (2006) *The Genus Arisaema* revised edition ARG Gantner Verlag

62. Père David might have sent seeds as Decaisne mentioned in a letter to Hooker in 1871 that the *Davidia* was in cultivation at A. Leroy's nursery at Angers, but nothing more is heard of these plants: *Garden and Forest* (1889) II p.123

63. Letters 198 dd 10.1.1894, 199 dd 30.1.1895, 200 dd 28.6.1895 and 201 dd 8.1.1896 Franchet Archive MHN

64. *Rev.Hort.* (1902) pp.377-379; Bean (1973); Letter 202 dd 16.10.1896 Franchet Archive MHN

65. *Icon.Plant.* (1891) No.1961; Wilson, E.H. (1926) *Aristocrats of the Garden* Boston pp.286-295; Spongberg,

S. (1990) p.185-191; Briggs, R. (1993) *Chinese Wilson*

66. *Gard.Chron.* (1903) 33 p.236
67. *Bot.Mag.* (1912) No.8432; *Pl.Wils.*II:255-257; Wilson, E.H. (1986) *A Naturalist in Western China* Cadogan Books Edition p.43
68. Letter 204 dd 12.10.1900 Franchet Archive MHN; Obituary Archives MEP
69. del Tredici, P. (1998) *The First and Final Flowering of Muriel's Bamboo, Arnoldia* 58 (3) pp.11-17
70. See Lancaster, R. (2008) p.16
71. *New Trees* (2009) pp.869-871

CHAPTER TWELVE pages 164-175

1. Norman, H. (1895) *Peoples and Politics of the Far East* pp.304-305
2. Obit. *Acta Ordinis Fratrum Minorum* (1901) 20 p.164; Ghilardi, F. (1921) *Il P. Guiseppe Giraldi Missionario in Cina (1848-1901)* Tipografia del Collegio di S.Bonaventura Quaracchi and (1924) *Un Missionario Scienziato Le Missioni Francescane* 9:271-273; Sabatelli, G. (1973) *Un missionario studioso di botanica, Osservatore Romano* 3/8 p.5; Clauser, M. et al. (2002/2003) *Sulle Collezioni Storiche dell'Orto Botanica di Firenze: Antonio Biondi e le Introduzioni di Piante fra la Fine dell'Ottocento e l'Inizio del Novecento, Museologica Scientifica* 19 (1) 121-139. Padre Giraldi's Herbarium is kept at the Museo Botanica, Florence.
3. Sabatelli, G. (1973)
4. Information from Paolo Cuccuini, Botany Department, Museo di Storia Naturale, Florence
5. Engler, A. *Pflanzenr.* (1902) IV 163 8:80; *FOC* 11:539, cf. *Flowers of Western China* (2011)
6. Bean p.266; *New Trees* (2009) p.125; cf. *FOC*, GRIN
7. *Pl.Wils.*2:577-578; *Bot.Mag.* (1920) No.8833; Osborn, A. (1922) *Gard.Chron.* 72 pp.310-311
8. *Bot.Jahrb.* 29 p.593; *Gard.Chron.* (1903) 33 p.81; *Bot. Mag.* (1914) No.8563; Pampanini, R. (1910) *Le Piante Vascolari raccolte dal. R.P. C. Silvestri nell'Hupeh durante gli anni 1904-1907 Nuove Giornale Botanico Italiano* NS 17 p.721; *Icon.Plant.* (1913) No.2937; Wilson, E.H. (1916) *Aristocrats of the Garden* Boston p.193. *FOC* 19:646:Linneaceae. Graebner based the genus on the specimen collected in 1895.
9. Specimens had been received from Potanin in Gansu in 1885
10. *Bot.Jahrb.* 29 p.531; McKelvey, S.D. (1928) pp.151-157; Fiala/Vrugtman (2008) pp.96-97
11. O'Brien, S. (2011) p.45; *Bot.Mag.* (1916) No.8682 as *C. bodinieri* var. *giraldii*; Moldenke, H.N. (Nov. 1966)

Additional Materials toward a Monograph of the Genus Callicarpa, Phytologia 14 pp.54-60; *FOC* 17:11 and 17:19, The Plant List

12. Hinkley, D. (2009) p.119
13. *Bot.Jahrb.* 29 p.444; Bean (1973)
14. *Bot.Mag.* (1917) No.8732; *RHS.Jnl.* (1961) 86 p.485; Brickell, C. and Matthew, B. (1976) *Daphne Genus in the Wild and in Cultivation* AGS p.104
15. Nelson, E.C. (1988) *Of Rosa hugonis and Father Hugh, Kew Magazine* 5 (1) pp.39-43: p.42 for the full Latin text of Padre Giraldi's letter dd 1.2.1896 to the British Museum
16. *Bot.Mag.* (1905) No.8004; *The Garden* (1908) p.313 and (1916) pp.514-516; *Pl.Wils.*2:330-331; Lauener, L.A. (1996) p.211; Haw, S.G. (1996) *Notes on some Chinese and Himalayan Rose Species of Section Pimpinellifoliae, New Plantsman* 3 pp.143-146; *FOC* 9:252, The Plant List
17. NHM Acquisition registers 1890-1908; Nelson, E.C. (1988)
18. *Pl.Dav.*II:30; *Bot.Mag.* (1911) No.8359; *FOC* 12:34, The Plant List.
19. Engler, A. *Bot.Jahrb.* (1900-1901) 29 pp.169-659: Diels' 'central' region included Sichuan and stretched as far west as Kanding so was not strictly confined to the provinces usually indicated by the term 'central'.
20. See Pampanini, R. (1910) for Padre Silvestri's plants, *Cotoneaster silvestrii* p.288 had been collected by Henry: *Pl.Wils.*I:169; Clauser, M. et al (2002(2003); Obit. *AOFM* (1955) 74 p.21
21. Fiala, J. revised Vrugtman, F. (2008) *Lilacs A Gardener's Encyclopaedia* Timber Press pp.270-280; McKelvey, S.D. (1928) *The Lilacs A Monograph* New York pp.182-186, 194-195
22. Pettit, S. and Upson, T. (2006) *Rosa* x *pteragonis* 'Cantabrigiensis', *Curtis'sBot.Mag.* 23 (4) pp.30-36

CHAPTER THIRTEEN pages 176-189

1. Pratt, A.E. (1892) *To the Snows of Tibet through China* p.136
2. For accounts of the Tibetan missions, see de Launay, A. n.d (c.1901) *Histoire de la Mission du Thibet* Paris; d'Orlèans, H. (1891) *Les Missionaires Français au Thibet Extrait du Corréspondent* Paris; Fournier, P. (1932) pp.111-120
3. Cooper, T.T. *Travels in Western China and Eastern Tibet, Proceedings of the Royal Geographical Society of London* (1869-1870) 14 (5) p.338, (1867-1868) 12 (5) p.339, and (1871) *Travels of a Pioneer of Commerce in Pigtail and Petticoats* pp.277, 312
4. *Notice nécrologique* de Père Soulié written by Mgr. Giraudeau, Archives MEP; various papers Box 1649

Soulié:Archives Martyrs, Archives MEP

5. The eldest was murdered by pirates and another was murdered in Burma; his stepmother became a nun Archives MEP

6. Letter 35 dd 19.3.1894 from Père Bodinier, Franchet Archive MHN

7. Wilson, E.H. (1913) *A Naturalist in Western China* Cadogan Books 1986 edition pp.206-207

8. Gill, W. (1880) *The River of Golden Sand* II p.79

9. *Notice nécrologique*; de Launay, A. (c.1901) pp.221-235

10. Rockhill, W.W. (1891) *Land of the Lamas* New York and London

11. Mesny had collected some plants in Guizhou in 1880 for Henry Hance, including *Jasminum mesnyi*, which was named after him (*Jnl.Bot.* (1882) 20 p.37)

12. Huc, R. E. (1850) *Souvenirs d'un Voyage à travers la Tartarie et le Thibet et la Chine pendant les années 1844, 1845 et 1846 Paris*; Colborne Baber, E. (1886) *Travels and Researches in Western China, Royal Geographic Society Supplementary Papers* I Pt.1 pp.1-201; Gill, W. (1880); Desgodins, C.H. (1885) *Le Thibet d'après Correspondance des Missionaires* Paris

13. Pratt, A.E. (1892) p.136. Antwerp Edgar Pratt was born in the Isle of White in 1852 and initially started out as a draper before devoting himself to natural history. He married in 1882 and his wife Alice and their two children accompanied him to China in 1888. They stayed at Yichang, where he met Augustine Henry. Alice was pregnant and after giving birth to a son in 1889, she returned to England with the children. He died in 1924. His parents seem to have had a penchant for unusual names: he had a brother called Florence Charles and a sister called Vienna.

14. *Bot.Mag.* (1903) No.7883

15. Bonvalot, G. (1892) *De Paris au Tonkin à travers le Tibet Inconnu … cent huit illustrations gravées d'àpres photographies prises par le Prince Henri d'Orléans* Paris

16. Directors' Correspondence 151:641 dd 15.12.1890 Kew Archive; NHM Acquisitions Register 1891-92; O'Brien, S. (2011) *In the Footsteps of Augustine Henry* p.192

17. Hemsley, W. B. (1892) *Observations on a Botanical Collection made by Mr A.E. Pratt in Western China, with Descriptions of some new Chinese Plants from Various Collections, Jnl.Linn.Soc.Bot.* 29 pp.298-321; *Jnl. Soc.Bot.France* (1891) several articles

18. *Jnl.Linn.Soc.Bot.* (1892) 29 p.305; *Bot.Jahrb.* (1901) 29 p.401; *Pl.Wils.*1:55

19. *Jnl.Soc.Bot.France* (1895) p.389; *Bot.Mag.* (1935) No.9414; Cox, P. and K. (1997) *Encyclopaedia of Rhododendron Species* Glendoick Publishing pp.166-167, 170-71; *FOC* 14:384

20. *Jnl.Linn.Soc.Bot.* (1892) 29 p.320 syn. *C. corrugatum* Fr

21. *Jnl.Soc.Bot.France* (1891) p.91; Grey-Wilson, C. (2000) pp.168-169

22. Tebbit, M., Liden, M. and Zetterlund, H. (2008) *Bleeding Hearts, Corydalis and their Relatives* Timber Press

23. *Jnl.Soc.Bot.France* (1891) p.153 and (1898) p.221; *Icon. Plant.* (1893) No.2219 as *Fritallaria lophophorum*; Wilson, E.H. (1915) *The Lilies of Eastern Asia* p.104; Synge, P. (1980) *Lilies: A Revision of Elwes' Monograph of the Genus Lilium and its supplements* p.151

24. Grierson, A. (1959-1961) *A Revision of the Genus Incarvillea, Notes.RBGE.23* p.339; Grey-Wilson, C. (1994) *A Survey of Incarvillea in Cultivation, New Plantsman* 1 p.40

25. *Jnl.Soc.Bot.France* (1903) pp.524-526; *Pl.Wils* I:334-335; *FOC* 6:344 as *C. chrysocoma*

26. Letter 726 dd 18.10.1893 Franchet Archive MHN

27. Franchet, A. (1890-1891) *Diagnoses d' Espèces Nouvelles provenant d' une Collection de Plantes du Thibet Chinois envoyée au Muséum par M. l' Abbé Soulié, Bull. Soc.Philomatique.Paris* 3 pp.140-150

CHAPTER FOURTEEN pages 190-201

1. E.A. Finet's description of Père Soulié *Jnl.Soc.Bot.France* (1907-1908) 2ème série 1 pp.14-21

2. Little, A.J. (1901) *Mount Omi and Beyond* p.189; Gill, W. (1880) II p.322

3. Gill, W. (1880) *The River of Golden Sand* I p.49

4. d'Orléans, H. (fevrier 1891) *De Paris au Tonkin par Terre, Revue des Deux Mondes* offprint p.48

5. Rockhill, W.W. (1894) *Diary of Journey through Mongolia and Tibet in 1891 and 1892* Smithsonian Institution Washington p.364

6. *Gard.Chron.* (1906) 39 pp.139, 166

7. *Jnl.ArnoldArb.* (1968) 49 p.80

8. Letters 718 n.d. and 719 dd 25.10.1891 Franchet Archive MHN

9. Letter 724 dd 31.7.1893 Franchet Archive MHN

10. Letters 722 dd.17.12.1892 and 723 dd 17.7.1893 Franchet Archive MHN

11. Letter 721 dd 29.11.1891 Franchet Archive MHN

12. Letter 723 dd 17.7.1893 Franchet Archive MHN

13. ibid.

14. Letter 722 dd 17.12.1892 Franchet Archive MHN

15. Letter 725 dd 28.08.1893 Franchet Archive MHN

16. Letter 724 dd 31.7.1893 Franchet Archive MHN

17. *Jnl.Soc.Bot.France* (1895) p.395

18. op.cit. p.390 as *R. lucidum*

19. op.cit. p.393

20. *Bull.Soc.Bot.Belg.* (1896) 35 (2) p.21; *Fruticetum Vilmorianum* (1904) p.85; *Bot.Mag.* (1907) No.8158; Wilson, E.H. (1913-1986 edition) p.155 and *Pl.Wils.*2:314

21. *Jnl.Soc.Bot.France* (1895) pp.448,450

22. *Bull.Soc.Bot.Genève* (1909) Sèr.2 (1) pp.191,195,375; *Jnl. Soc.Bot.France* (1896) pp.373, 376

23. *Jnl.Linn.Soc.Bot.* (1889) 26 p.121; *Rev.Hort.* (1897) p.394 and (1898) pp.132-134; *Fruticetum Vilmorianum* (1904) p.189; *Pl.Wils.*1:567-568

24. *Pl.Dav.*II:103; *Rev.Hort.* (1898) pp.132-134; *Bull.Soc.Bot. France* (1904) p.46; Stuart, D. (2006) *Buddlejas* Timber Press/RHS; RHS Buddleja Trial Report (2009-2010) at rhs.org.uk/planttrials

25. Letter 726 dd 18.10.1893

26. ibid. and 728 dd. 9.4.1894 Franchet Archive MHN

27. Letters 727 dd 3.2.1894 and 729 dd 30.4.1894 Franchet Archive MHN

28. Letter 730 dd 1.7.1894 Franchet Archive MHN

29. Letter 731 dd 1.9.1894 Franchet Archive MHN

30. Soulié, R.P.J (1897) *De Ta-tsien-lou à Tse-kou (rive droite du Mékong) 11 octobre −7 decembre 1894, Bull. Soc.Géographie* 7ème serie 38 pp.30-80. The date of Père Soulié's journey is sometimes erroneously shown as 1891 (cf. Fournier, P. (1932) p.119) which I think results from a mistake Mgr Giraudeau made in Père Soulié's obituary (Archives MEP) where he gives the date of Père Courroux's death as 1891,

CHAPTER FIFTEEN pages 202-213

1. McLean, B. (2004) *George Forrest Plant Hunter* Antique Collectors Club: letter to Clementina Traill, Oct. 1905 quoted p.71

2. Launay, A. (n.d. c.1903) *Histoire de la Mission du Thibet* Lille-Paris pp.68-71; d' Orléans, H. (1898) *From Tonkin to India by the Sources of the Irawadi January '95-January'96* Trans. H. Bent p.220

3. Launay, A. (c.1903) pp.264-265, 297

4. d' Orléans, H. (1898) p.220

5. Letter 732 dd 17.5.1895 Franchet Archive MHN

6. Kingdon-Ward, F. (1913) *The Land of the Blue Poppy* Cambridge UP pp.70-73

7. Ibid.

8. *Bull.Soc.Bot.France* (1887) p.283; *Jnl.Soc.Bot.France* (1898) p.263; FOC 14:283 as var. *heliolepis*

9. *Jnl.Soc.Bot.France* (1898) p.263

10. Letter 732 dd 17.5.1895 Franchet Archive MHN

11. *Jnl.Soc.Bot.France* (1898) pp.68-70; Compton, J. et al. (1998) *Taxon* 47:613

12. FOC 6:143

13. Kingdon Ward, F. (1913) p.61 *A. vaginata* also has pink flowers.

14. *Bull.Soc.Bot.France* (1885) p.268; Kingdon Ward, F. (1913) p.72

15. *Jnl.Soc.Bot.France* (1898) pp.149-150; Evans, W. E. (1925) *Revision of Genus Nomocharis, NotesEBG* 15 June pp.19-33; Balfour, I.B. (1929) *The Genus Nomocharis, Trans.Bot.Soc.Edin* 27 pp.273-300; *Bot.Mag.* (1933) No.9296; Matthews, V. (1995) *The Genus Nomocharis, New Plantsman* 2 pp.195-208; *Flowers of Western China* (2011) p.547

16. W.E. Evans in Woodcock, H. and Coutts, J. (1935) *Lilies* p.225

17. Synge, P. (1980) *The Genus Nomocharis in Lilies: A Revision of Elwes' Monograph of the Genus Lilium and its Supplements* pp.204-207; Sealy, J.R. (1983) *A Revision of the Genus Nomocharis Fr, Jnl.Linn.Soc.* 87 pp.285-323

18. Sealy. J.R. (1950) *Kew Bull.* 5:296; Kingdon-Ward, F. (1913) p.102

19. *Bull.Soc.Bot.France* (1898) p.180: the location is given as 'prope Sela, haud procul ab Yerkalo' but this must be a mistake as Père Soulié was not at Caka'lho on 15th July; Handel-Mazzetti, H. (1927, 1996) *A Botanical Pioneer in South West China* Vienna translated and published David Winstanley Essex p.107

20. Received between 1896-1903 Herbier MHN

21. Letter 732 dd 17.5.1895 Franchet Archive MHN

22. Launay, A. (c.1903) p.353; Fournier, P. (1932) p.120: *Spelaeonis troglodytoides souliei* and *Actinodura souliei*

23. d' Orléans, H. (1898) p.217

24. Op.cit. pp.221-225

25. Soulié, J. (1897) *De Ta-tsien-lou à Tse-kou (rive droite du Mékong) 11 octobre −7 decembre 1894, Bull.Soc. Géographie* 7ème series 38 pp.30-80.

26. Letter 171 dd 27.10.1895 Franchet Archive MHN

27. d'Orléans, H. (1898) pp.166, 428

28. Franchet, A. (1897) *Les Rodgersia, Rev.Hort.* pp.174-176; Cullen, J. (1975) *Taxonomic Notes on the Genus Rodgersia, Notes.RBGE* 34 pp.113-123; Booker, H. (2008) *Prince Henri's Rodgersia, The Plantsman* NS 7 (4) pp.252-254

29. *Jnl.Soc.Bot.France* (1898) p.220; Synge, P. *Lilies* (1980) pp.84-86; McRae, E. (1998) *Lilies A Guide for Growers and Collectors* Timber Press pp.137-139

30. Le Lièvre, A. (1994) *Prince Henri of Orléans Explorer and Planthunter* (1867-1901), *New Plantsman* 1 (4) pp.238-247; Howgego, R. (2008) Prince Henri died of a liver abscess in Saigon in 1901.

CHAPTER SIXTEEN pages 214-223

1. Handel-Mazzetti, H. (1927) *A Botanical Pioneer in South West China* Translated and published David Winstanley 1996 p.87

2. Goré, P. *La Mission Catholique du Thibet [dans le Territoire de Batang], Une Mission Thibetain, La Mission Catholique au Thibet dans les Marches Yunnaneses*: undated typescripts; all Box 1649 Soulié Archives Martyrs and Bishop's Report 1904: Archives MEP

3. Letter dd 26.11.1904 published in No.19 *Revue Religieuse du Diocese de Rodez* 12 May, Martyrs' Box 1649 Archives MEP

4. Soulié, J.A. (1904) *Géographie de la Principauté de Batang, Bull.Soc.Géog.France* 1 pp.87-104

5. *Bull.Herb.Boissier* (1908) Ser.2 8 pp.363-370

6. *Jnl.Soc.Bot.France* (1900) pp.60-63; *Bull.Soc.Bot.France* (1900) pp.158-172

7. *Rev.Hort.* (1902) pp.379-381 and (1907) pp.256-258; *Bot. Mag.* (1914) No.8571; Leveillé, H. (1915) *Bull.Acad.Intl. Géographie Botanique* 25 p.45: as *C. franchetii* FOC 9:96; Fryer, J. and Hylmö, B. (2009) *Cotoneasters A Comprehensive Guide to Shrubs for Flowers, Fruit, and Foliage* Timber Press pp.242-243

8. *Gard.Chron.* (1912) 52 p.288; Bean (1970)

9. *Fruticetum Vilmorianum* (1904) p.119; *Bull.Soc.Bot. France* (1904) p.cliii; Bois, D. (1906) *Descriptions de Plantes Nouvelles* Paris; *Bot.Mag.* (1909) No.8284; Fryer and Hylmö (2009) pp.222-223

10. *Bull.Herb.Boissier* (1905) sér. 2 5 p.449: *Pl.Wils.*I:360, III:437; Ahrendts, L. (1961) *Berberis and Mahonia A Taxonomic Revision, Jnl.Linn.Soc.* 57 pp.57, 77-78; *Pl.Dav.*2:13

11. Maclean, B. (2004) *George Forrest Plant Hunter* Antique Collectors Club p.56

12. Launay, A. (c.1903) p.332: referred to as Lou-tse-kiang by the missionaries

13. Handel-Mazzetti, H. (1927-1996) p.94; Kingdon-Ward, F. (1913) p.4 p.69-70

14. McLean, B. (1997) *A Pioneering Plantsman A.K. Bulley and the Great Planthunters* pp.69-70

15. McLean, B. (2004) p.61

16. Launay, A. (c.1903) pp.339-340. Père Mussot (1854-1905) arrived in Tibet in 1883. He served at Chapa, Kangding

and Lentsy. He did some botanising and sent specimens back to Paris.

17. Obituaries: Père Mussot and Père Soulié; Goré, P. *La Mission Catholique du Thibet dans le territoire de Batang* and *Une Siècle d'Apostat au Thibet*, both undated typescripts, Box 1649 Soulié Archives Martyrs, Archives MEP

18. McLean, B. (2004) p.62

19. Kingdon-Ward, F. (1913) *The Land of the Blue Poppy* Cambridge UP p.46

20. Forrest, G. (1910) *The Perils of Plant Collecting, Gard. Chron.* 47 pp.325-326, 344; *Jnl.Hort.* (1912) 64 pp.34-36; McLean, B. (2004) pp.61-73

21. Scottish Rock Garden Club (1935) *George Forrest VMH Explorer and Botanist* p.33; see Geiger, J. (2009) *The Third Man Factor*

22. *Notes.RBGE* (1908) 4 p.221; *Bull.Soc.Bot.France* (1885) 32 p.266

23. *Gard.Chron.* (1906) 39 p.138; *Kew Bulletin* (1906) p.159; *The Garden* (1916) pp.514, 516; *Bot.Mag.* (1910) No.8338 and (1931) No.9248; Stuart Thomas, G. (1998) *The Rose Book* pp.88-89; Harkness, P. (2003) *The Rose A Colourful Inheritance* p.57

24. CIM Register Vol.88 and CIM China Council Minutes No.74: China Inland Mission Records, SOAS; *Chinese Recorder* (1899) 30 pp.100-102; Rijnhart, S.C. (1901) *With the Tibetans in Tent and Temple* New York; Shelton, F.B. (1912) *Sunshine and Showers on the Tibetan Border* Cincinnati

CHAPTER SEVENTEEN pages 224-231

1. Vincent, H. (1892) *Newfoundland to Cochin China* p.367

2. Handel-Mazzetti, H. (1927) *A Botanical Pioneer in South West China* Vienna Translated and published David Winstanley (1996) p.87

3. Handel-Mazzetti, H. (1927-1996) p.121: now a synonym for *P. glabra* subsp. *genestieriana*

4. *Bull.MHN* (1914) 20 p.303

5. Forrest, G. (1910) *The Perils of Plant Collecting, Gard. Chron.* 47 p.344; *New Trees* (2009) pp.206-207. Yellow-fruiting S. 'Joseph Rock' appears to be a chance natural hybrid between *S. commixta* and *S. monbeigii*.

6. *Notes.RBGE* (1918-1919) 11 p.205; *Ornamental Flowering Trees and Shrubs, Report of the Conference held by the Royal Horticultural Society April 26-29 1938* pp.118-120; Wright, D. (1979) *Deutzias – An Evaluation, The Plantsman* 1 3 p.161

7. Handel-Mazzetti, H. (1996) p.33

Index

Page numbers for illustrations are in **bold**.

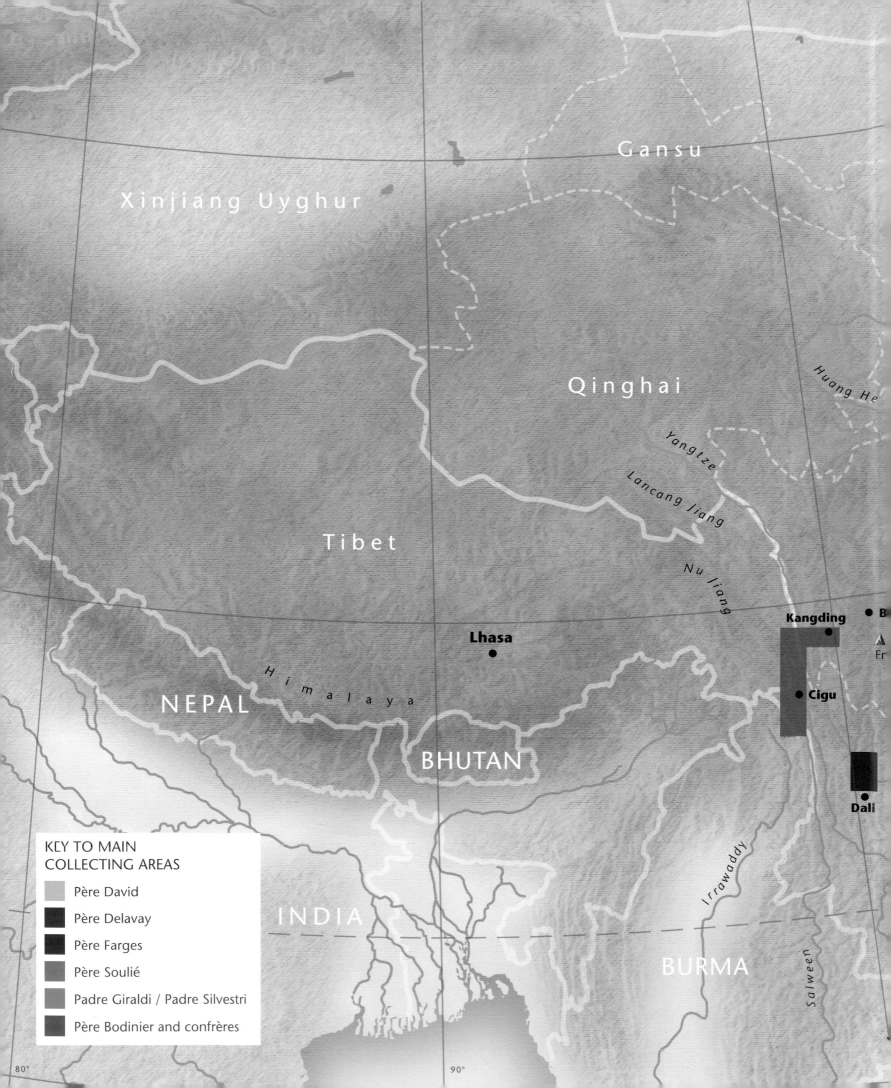

KEY TO MAIN
COLLECTING AREAS

Père David

Père Delavay

Père Farges

Père Soulié

Padre Giraldi / Padre Silvestri

Père Bodinier and confrères